Structure Property Correlations for Nanoporous Materials

Structure Property Correlations for Nanoporous Materials

Abhijit Chatterjee

CRC Press
Taylor & Francis Group
Boca Raton London New York

CRC Press is an imprint of the
Taylor & Francis Group, an **informa** business

CRC Press
Taylor & Francis Group
6000 Broken Sound Parkway NW, Suite 300
Boca Raton, FL 33487-2742

First issued in paperback 2017

© 2010 by Taylor and Francis Group, LLC
CRC Press is an imprint of Taylor & Francis Group, an Informa business

No claim to original U.S. Government works

ISBN-13: 978-1-4200-8274-6 (hbk)
ISBN-13: 978-1-138-11402-9 (pbk)

This book contains information obtained from authentic and highly regarded sources. Reasonable efforts have been made to publish reliable data and information, but the author and publisher cannot assume responsibility for the validity of all materials or the consequences of their use. The authors and publishers have attempted to trace the copyright holders of all material reproduced in this publication and apologize to copyright holders if permission to publish in this form has not been obtained. If any copyright material has not been acknowledged please write and let us know so we may rectify in any future reprint.

Library of Congress Cataloging-in-Publication Data

Chatterjee, Abhijit.
   Structure property correlations for nanoporous materials / Abhijit Chatterjee.
      p. cm.
   Includes bibliographical references and index.
   ISBN 978-1-4200-8274-6 (hardcover : alk. paper)
   1. Porous materials. 2. Nanostructured materials. I. Title.

TA418.9.P6C456 2010
620.1'16--dc22   2009038085

Visit the Taylor & Francis Web site at
http://www.taylorandfrancis.com

and the CRC Press Web site at
http://www.crcpress.com

Contents

Foreword

Nanomaterials are a major focus of nanoscience and technology. It is a growing field of study attracting tremendous interest, insight, and effort in research and development around the world for its multidomain applications. *Nanoporous* materials are nanostructured materials, which possess unique surface, structural, and bulk properties with applications in fields such as ion exchange, separation, catalysis, sensor applications, and molecular isolation and purification in biology. Nanoporous materials are of scientific and technological importance because of their immense capability to adsorb and interact with atoms, ions, and molecules on the large interior surfaces of the nanometer-sized pores. They also offer new opportunities in inclusion chemistry, guest–host syntheses, and molecular manipulations and reactions at a nanoscale for making nanoparticles, nanowires, and other quantum nanostructures.

Research on nanoporous materials can benefit from computer modeling/simulation studies not only to explain the experimental observations but to design new materials of interest. Materials with targeted properties can even be designed computationally. Modeling techniques can help in formulating the synthesis strategies and characterizing the materials from laboratory to pilot plant and further to plant production.

There is a need, in this context, for a book that introduces, reviews the literature, and discusses the correlation between the structure and property relationships of nanoporous materials and that can serve as a "bridge" between experiment and theory, simulation/modeling while addressing key issues in the area of nanoporous materials. This book is novel in its kind and its timely publication will be of interest to experimentalists in this area. Commercially available software is used throughout the book, which enhances its utility to the nonspecialist. The author has many years of research and experience in exploring, learning, and understanding the field of molecular modeling of nanoporous materials, especially those areas of interest in catalysis.

Dr. Paul Ratnasamy
INSA-Ramanujan Research Professor
National Chemical Laboratory
Dr. Homi Bhabha Road
Pune, 411 008 India

Preface

Nanoporous materials consist of a regular organic or inorganic framework supporting a regular, porous structure. Due to the unique structure–property relation, nanoporous materials have huge potential for a wide range of applications.

Molecular modeling techniques are nonexperimental reasonings to explain or predict chemical phenomena, much of which has in fact fallen under the heading of computational chemistry; i.e., the application of computation to the solution of problems in chemistry. Chemical theorists have also used the power of statistical mechanics to provide a bridge between the microscopic phenomena of the quantum world and the macroscopic bulk properties.

To compare theoretical calculations with experimental, synthesizing a new material is always a challenge, starting from the choice of raw material, followed by several characterization techniques to determine the bulk and physical properties of the material. Moreover, the application of the material — for instance, as catalyst — demands molecular-level understanding to describe the mechanism, which is unknown most of the time and hence it is difficult to optimize the respective parameters. The optimization process is time consuming, requires expensive instruments, and still lacks molecular visualization for precise prediction. To solve this problem, computer simulation is an effective measure and is cost effective, convenient to study anytime, and has the potential to include complex, hazardous, real-world processes. Hence, I believe that computer simulation is an integral part of applied science, especially for materials.

This book aims to direct the experimentalist toward the capability of simulation as well as the level of accuracy on which one can depend on the technology. That belief then only will turn one's interest to really design the novel material of interest or explain the chemical phenomenon underneath by comparing the experiments with computer simulation.

To provide a comprehensive overview of the area of nanoporous materials, Chapter 1 begins with a definition of nanoporous materials followed by classifications, the importance of nanomaterial, current applications, challenges, and future prospects. The definitions in relation to porous materials and nanoporous materials are necessary to understand the concept of nanoporous materials in context to their applications. Following this introduction, the classification and scope of nanoporous materials are presented. The properties and their characterization and measurement methods are briefly described before major applications in various fields are reviewed. Finally, in this chapter, key scientific and engineering issues and future directions are identified as challenges and opportunities for researchers in this field.

In Chapter 2 the focus is on defining molecular modeling and showcasing different methodologies to solve the nanomaterial problem. The main focuses are on atomistic simulation techniques with Monte Carlo and molecular mechanics. Finally, the challenges and the future of these techniques within this domain are also discussed.

In Chapter 3 the focus is on introducing a density functional theory (DFT) technique to the readers. The methodology is introduced with a context of the basic wave mechanics, followed by an explanation of density matrix. The chemical potential is then explained to address the issue with the modelling of chemical bonds. This is the main motivation of modeling to reproduce and explain the experimental scenarios. The challenges and the future of the DFT technology are described at the end of this chapter.

Chapter 4 aims to define and explain localized reactivity descriptors. The local softness and hardness paradigm along with philicity concepts has been explained to provide a basic idea of the concept, as well as to provide a background to discuss and rationale for the applications of these methodologies to understand the experimental postulates.

Structural complexity evolves during synthesis, and therefore one way of capturing such complexity within atomistic models is to "simulate synthesis," as explained in the Chapter 5. The discussion starts with the nucleation and growth of nanoporous material, which is where the synthesis process starts in an experimental workroom. This involves selection of raw materials, which includes changing the chemical composition and optimizing the bulk structure of the varied composition including the role of metal incorporation. Further demand is to look into the effect of pressure and temperature in the reaction process. This is performed to mimic the synthesis condition and to rationalize the synthesis process of nanoporous materials.

Experiments have their own limitations; thus, the combination of experiment and computer simulation is necessary to obtain a desired material with specific functionalities for proper applications. Chapter 6 covers the area of the structural characterization of the material mainly in terms of chemical composition, spectroscopic analysis, mechanical stability, and porosity to show the capability of simulation to justify and validate experimental observations. This also can provide a lot of space for experimentation where analysis is ambiguous.

The design of new nanoporous materials in terms of their specific applications like surface adsorption phenomena, sensors, or catalysts for various organic syntheses is the main theme of Chapter 7. Nanoporous materials are widely used as catalysts because of their large internal surface area and the consequent presence of controllable large voids. Microporous materials like zeolites are mainly used as heterogeneous redox catalysts in the petroleum industry, in various shape selective reactions, and in separation. In discussing the catalytic activity of the microporous material, it must be mentioned that the transition metal–substituted microporous materials (TMSM) with aluminosilicate or aluminophosphate framework are covering a large part of the catalysis. TMSM mainly take part in the various oxidative transformations in the presence of a mild oxidizing agent like hydrogen peroxide or oxygen. A number of applications in waste treatment processes, including removal of heavy metals and radioactive species, as well as ammonia, different phosphates, and toxic gases from water, soil, and air, are due to the unique structural and surface physico-chemical properties of microporous materials, such as excellent absorption and ion-exchange capacities. The details of these applications are described in Chapter 8, "Application of Nanoporous Materials." So, each of the functions demands a specific surface property such as pore architecture, pore size, surface area, and acidity or basicity of the matrix. Considering the surface properties, adsorption is generally used to characterize the surface structure. However, it is a complicated process and it is difficult to obtain a molecular-level scenario on an experimental basis. Thus, simulation can play a significant role in simulating the actual situation occurring at the molecular level by comparing the binding energy or the attachment energy and by comparing the surface chemistry and producing a result to show the probability of physisorption and chemisorption.

Nanoporous materials combine with the advantages of porous materials. These tiny materials provide a huge surface area, controllable pore sizes, morphology, and capable of any surface-related applications. Due to their considerably small-size porous structure, material properties have increased compared to their bulk counterpart. Therefore, it is obvious that the revolutionary properties of nanoporous materials make them a strong contender for its wide range of applications. The main interest of Chapter 8 is in providing some promising applications of nanoporous materials, such as (a) photonic crystals, (b) bio-implants, (c) sensors, and (d) separation.

Now, the designed material, which is synthetically viable and technically applicable in terms of pore architecture, metal loading and other related parameters as established through experimental technique and computer simulation technology needs to be tested further to confirm the functionality. In Chapter 9 we focus on the catalytic activity of the nanoporous material. The discussion covers related topics of catalysis starting from shape-selective reactions within nanopores, chemical adsorption reactions, cracking, and the mechanistic aspect of a reaction using transition state theory, especially the activation barrier, the intrinsic reaction coordinate, and the effect of solvent on catalytic reactions. This chapter provides a detailed explanation of the methodology of simulation, so that one can follow these cases and approach their own problems for a solution.

I hope this book will convey the significance of the combination of traditional experimental work and molecular simulation to the researcher. I think this book will stimulate the interest of bench chemists in molecular modeling and remove the hurdle of understanding what simulation can provide and how one can successfully design a matrix for a specific application.

Acknowledgments

I would like to take this opportunity to thank all my mentors and collaborators, Prof. Sukalayan Basu, Dr. Rajappan Vetrivel, Dr. Paul Ratnasamay, Prof. Akira Miyamoto, Prof. Kazuo Tori, Dr. Takashi Iwasaki, Dr. Toshihige Suzuki, Prof. Fujio Mizukami, Prof. Paul Geerlings, Prof. John Newsam, Prof. Alfred Mortier, Prof. Kimiko Hirao, Dr. Sourav Pal, Dr. Asit K. Chandra, and Dr. Debasis Bhattacharya.

My special thanks go to Dr. Rajappan Vetrivel for introducing me to the amazing world of computer simulation application in the field of nanoporous catalysis and material science.

I am grateful from the bottom of my heart to my father, Prof. K. N. Chatterjee, for making me fascinated with chemistry from my school days and my mother for her continuous encouragement and contemplation.

I wish to specially thank my wife, Dr. Maya Chatterjee. The writing of this book would not have been possible without her involvement and continuous inspiration. She has contributed as a collaborator by participating in a very fruitful scientific discussion to share her experimental experience with nanoporous material synthesis and reaction, which enriched the book immensely. I am thankful to my daughter for her patience and help with proofreading. I wish to thank all of my extended family members.

I also wish to thank all my colleagues in Accelrys for their support and consideration.

Author

Abhijit Chatterjee was born in Chandananagore, a little town near Kolkata, India. He was a student of Kanailal Vidyamandir and received his Master's and Ph.D. from Burdwan University, West Bengal, India. After graduating from University with his Ph.D. in physico-analytical chemistry in 1992, he moved to the National Chemical Laboratory, Pune, India, and worked there until 1995 as a Research Associate in the Catalysis Division. He exposed himself to the simulation of catalytic material. He has traveled around the world and collaborated with many groups in catalysis before settling down in Japan as a researcher in the field of computational chemistry and established himself in this field with more than 100 papers in international journals of repute; he has presented invited talks at many international and national symposiums, and written chapters in books published by Elsevier, Taylor & Francis, and Wiley. He served as a member of the editorial board of the *International Journal of Molecular Science*. He is a member of many leading scientific societies (ACS, IOP, IZA, etc.). His research interest is focused on density functional theory and its application on different materials, especially related to catalysis (zeolite, clay, oxides). He also explored other materials of interest like composites, semiconductors, and metal clusters to rationalize the structure–property correlation and finally was involved in deriving reactivity index, Fukui function, and simpler algorithms to help experimentalists to design a novel material.

1 Basic Aspects of Nanoporous Materials

In recent years, nanomaterials have been a major focus of nanoscience and nanotechnology. It is a self-growing field of study attracting tremendous interest, insight, and effort in research and development around the world for its multidomain applications. Nanoporous materials are a type of nanostructured materials that possess unique surface, structural, and bulk properties. This underlines their important usage in various fields of research such as ion exchange, separation, catalysis, sensor applications, biological molecular isolation, and purification. Nanoporous materials are also of scientific and technological importance because of their great ability to adsorb and interact with atoms, ions, and molecules on their large interior surfaces and in the nanometer-sized pore space. They offer new opportunities in areas of inclusion chemistry, guest–host synthesis, molecular manipulations, and reactions in the nanoscale for making nanoparticles, nanowires, and other quantum nanostructures.

To provide a comprehensive overview of the area of nanoporous materials, this chapter will begin with a definition of nanoporous materials followed by their classifications, the importance of nanomaterials, current applications, and, lastly, challenges and future prospects. The basic concepts and definitions in relation to porous materials and nanoporous materials are necessary for understanding the concept of nanoporous materials as well as for the better understanding in context to their applications. Following this introduction, the classification and scope of nanoporous materials will be presented. The properties and their characterization and measurement methods will be briefly described before major applications in various fields are reviewed. Finally, key scientific and engineering issues and future directions are identified as challenges and opportunities for researchers in this field.

Porous inorganic solids have found great utility as catalysts and sorption media because of their large internal surface area and the presence of voids of controllable dimensions at the atomic, molecular, and nanometer scales. With increasing environmental concerns worldwide, nanoporous materials have become more important and useful for the separation of polluting substances as well as the recovery of useful ones. The prospective applications include their use as templates for the production of electrically conducting nanowires and for highly selective biosensors and biomembrane materials. Inorganic–organic or hybrid nanoporous crystalline materials have recently attracted much attention and increasing interest due to their potential in gas separation and hydrogen storage. This chapter will cover recent developments in the synthesis, characterization, and property evaluation of new nanoporous inorganic materials and some hybrid solids.

The idea of the book is to walk together with experimentalists and follow their way of approaching the nanoporous domain. The book therefore will proceed to show the techniques available in the realm of simulation to approach this world of science. Chapters highlighting synthesis of nanoporous materials, their characterization, porosity measurement, and adsorption as a phenomenon will follow this, as well as some projected applications with a futuristic approach and finally the reaction mechanism. All of this will be done in comparison with experiments and a simulation recipe will be prescribed. There will be some emphasis on the silica and phosphate-based frameworks by using hydrothermal and microwave procedures with X-ray diffraction (XRD), spectroscopy (infrared [IR], nuclear magnetic resonance [NMR]), and electron microscopy characterization techniques in brief. Moreover, the functionalization of nanoporous materials by physical and/or chemical treatments; studies of their fundamental properties, such as catalytic effects or adsorption; and their applications will be described based on both theoretical studies and experimental studies in the following chapters.

1.1   DEFINITION

The journey of nanoscience starts with the famous lecture of Professor Feynman, "There's plenty of room at the bottom," in 1959 [1]. However, the real burst of nanoscience came after the 1990s, and now it has opened a wide door that does not leave any major research area outside. Nanomaterials are an integral part of nanoscience. The broad definition of nanomaterials is those materials, which possess a size that is smaller than one micron (μm) in at least one dimension. Actually, the definition does not restrict the material in question to any fixed minimum or maximum size but logically it must be between the micro- and atomic/molecular scale. According to the definition, it's a small world, and the prefix *nano* is a Greek word meaning dwarf or small. It should be mentioned that nanomaterials are not a simple miniaturization but are somewhere between the bulk and quantum scale. Now, the question is, why is the world running for the smallest material, the nanomaterials? The answer lies in their amazing and exclusive properties, which are completely different from bulk materials of the same composition. The uniqueness of their characteristic properties originates due to their smaller size and consequently the large surface area–to-volume ratio, high surface energy, and spatial confinement. Nanoporous materials are an important part of the nanomaterials, which contain unique porous surface structure, large porosity, and a pore size generally between 1 and 100 nm. These types of materials find wide applications in various fields ranging from sorption, ion exchange, catalysis, host–guest interaction, etc.

What makes nanoscale building blocks interesting is that by controlling the size in the range of 1–100 nm and the assembly of such constituents, one could alter and prescribe the properties of the assembled nanostructures. As Professor Roald Hoffmann, the chemistry Nobel Laureate put it, "Nanotechnology is the way of ingeniously controlling the building of small and large structures, with intricate properties; it is the way of the future, with incidentally, environmental benignness built in by design." [2] Nanostructured materials may possess nanoscale crystallites, long-range ordered or disordered structures, or pore space. Nanomaterials can be designed and tailor-made at the molecular level to have the desired functionalities and properties. Manipulating matter at such a small scale with precise control of its properties is one of the hallmarks of nanotechnology.

Porous materials are like music: the gaps are as important as the filled-in bits. The presence of pores (holes) in a material can render all sorts of useful properties that the corresponding bulk material would not have. Generally, porous materials have porosity (volume ratio of pore space to the total volume of the material) between 0.2 and 0.95. Pores are classified into two types: open pores, which connect to the surface of the material, and closed pores, which are isolated from the outside. In functional applications such as adsorption, catalysis, and sensing, closed pores are not of any use. In separation, catalysis, filtration, or membranes, penetrating open pores is often required. Materials with closed pores are useful in sonic and thermal insulation or lightweight structural applications. Pores have various shapes and morphology such as cylindrical, spherical, and slit types. Pores can also take more complex shapes, such as hexagonal. Pores can be straight, curved, or have many turns and twists, thus having a high porosity. According to the International Union of Pure and Applied Chemistry (IUPAC), micropores are smaller than 2 nm in diameter, mesopores are in the range of 2 to 50 nm, and macropores have a pore diameter larger than 50 nm. However, this definition is somewhat in conflict with the definition of nanoscale objects. Nanoporous materials are a subset of porous materials, typically having large porosities (greater than 0.4) and pore diameters between 1 and 100 nm. In the field of chemical functional porous materials, it is better to use the term *nanoporous* consistently to refer to this class of porous materials having diameters between 1 and 100 nm. For most functional applications, pore sizes do not normally exceed 100 nm. Nanoporous materials actually encompass some microporous materials and all mesoporous materials. So, what are the unique properties of those materials? Nanoporous materials have a specifically high surface-to-volume ratio, with a high surface area and large porosity, of course, and very ordered, uniform pore structure. They have a very versatile and rich surface composition and surface properties that can be used for functional applications such as catalysis, chromatography, separation, and sensing. Many inorganic nanoporous materials are made of oxides. They are often nontoxic, inert, and chemically and thermally stable; in certain applications the thermal stability requirement is very stringent, so a highly thermally stable catalyst is necessary.

1.2 CLASSIFICATION

Porous materials consist of a network of interconnected pores of controllable dimension in the atomic, molecular, and nanometric scale. Porous materials are creating quite a stir among the material scientists because the size of the channel or the voids can be designed according to the potential applications. As mentioned before, porous materials can be grouped into three different categories depending on their pore diameter, such as (1) microporous where the pore size is less than 2 nm, (2) mesoporous with pore diameters in the range of 2–50 nm, and (3) macroporous, in which the pore size is larger than 50 nm. Nanoporous materials of pore sizes between 1 and 100 nm, generally categorized as bulk materials and membranes, can have open pores and closed pores. Bulk materials can be carbon, silica, metal oxide, organic metal composite, and organic silica composite where the membranes are mainly developed with zeolite. In addition, mesoporous and inorganic/organic hybrid materials are secured in their place in the field of nanoporous materials.

To describe the fundamental properties of nanoporous materials it is necessary to have a microscopic understanding of the material. Molecular simulation along with the experimental tool has played a significant role, which will be discussed further. Material classification is based on the experimental numbers with the pore architecture, but it is an experimentalist's dream to walk through the channels to see where the active site is and how one can tune the active site, whether the structure contains multiple channels and other interesting features of the nanopore through visualization using simulation. Then, if one wants to probe more of the structure property-based correlations, one needs the help of methods to calculate them, which are based on what you want and to what level of accuracy.

1.3 KEY MATERIALS OF INTEREST

The term *nanoporous material* covers a vast area that deals with materials of pore size greater than 1 nm. This includes the conventional microporous crystalline aluminosilicates like zeolites, clay, periodic mesoporous materials, metal organic frameworks, porous polymers, etc. The main interest of this book will be zeolites and mesoporous materials, aluminum phosphate (AlPO), and metal organic frameworks.

Zeolites are a naturally occurring mineral. More than 150 types of zeolites have been synthesized and 40 types of zeolites are found in nature. Although natural zeolites occur in large quantities, they offer only a limited range of atomic structures and properties. The most common zeolites are silicalite-1, ZSM-5, and zeolite X, Y, A, and zeolite beta. Zeolites are generally crystalline aluminosilicate with a three-dimensional rigid and crystalline structure consisting of interconnected tunnels and cages. The general formula of zeolite is $M_{m/z} \times mAlO_2 \times nSiO_2 \times qH_2O$. In the crystalline structure of zeolite, the metal atoms (mainly Si and Al) are surrounded by four oxygen atoms to form an approximate tetrahedral structure. The metal atom as a cation sits at the center of the tetrahedron. This tetrahedral metal is called a *T-atom*. The framework structure of zeolite is typically anionic and the presence of the charge-compensating cation makes it neutral. When the cations are replaced with the proton zeolite can act as strong solid acid catalysts. Zeolites contain microscopically small channels, so sometimes they are referred to as a *molecular sieve*. Why are these zeolites of key interest to theoretical chemists in the last few decades? The answer is that zeolites are amazing materials because the physical and chemical properties and the crystal structures of zeolites can be varied easily by changing the chemical composition. Zeolites are extremely active catalysts but their small channel size restricts them to the smaller molecules. An example with the most used zeolite, ZSM-5, which is an MFI-type of zeolite, is shown in Figure 1.1. This zeolite, developed by Mobil Oil, is an aluminosilicate zeolite with high silica and low aluminum contents. Its structure is based on channels with intersecting tunnels and a regular pore structure. The maximum size of molecular or ionic species that can enter into the channel is dictated by the dimension of the channel. These channels are conventionally referred to according to their ring size; for example, an eight-membered ring represents a closed loop built with tetrahedrally coordinated 8 silicon or aluminum with oxygen. It is not necessary for the

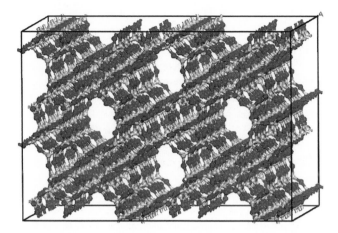

FIGURE 1.1 The model of ZSM-5, an MFI-type of zeolite nanoporous material. The straight channel and zigzag channel can be seen. (See CD for color image.)

zeolite to maintain a symmetrical ring structure; therefore, the pores in zeolites are not always cylindrical. Due to the porous structure, zeolites can accommodate different types of cations, such as Na^+, K^+, and Ca^{+2}, etc., and are loosely bound so that they can easily be ion-exchanged. The uniform microcrystalline structure of zeolites has always driven researchers toward their catalytic applications. Several changes of the surface and the framework were attempted by incorporating transition metal ion or acidity of the matrix. It is well known that catalysis chemistry of the ZSM-5 is mainly based on the acidic property of the matrix. The substitution of Al^{3+} in place of the tetrahedral Si^{4+} silica requires the presence of an added positive charge. When this is H^+, the acidity of the zeolite is very high. However, it is difficult to reach the uniform acidity of a zeolite matrix. The main reason for not achieving uniformity of acid sites rests with the chemical reactivity of tetrahedral framework aluminum ions to water, especially those adjacent to acidic O---H groups, and the instability of the primary, defect crystal structure formed upon aluminum hydrolysis in the presence of steam [3]. Though we have yet to synthesize the zeolite with uniform acidity, it is a widely accepted catalyst for shape-selective reactions in the petroleum industry. Due to the shape-selective property of zeolites, they can provide an effective solution for environmental pollution by minimizing the formation of pollutants. Variation of the silica-alumina ratio, incorporation of transition metal ions, surface area, pore sizes, pore architecture, and cations produces a wide spectrum of zeolites to act as catalysts or adsorbents. The basic advantage of zeolites for use as heterogeneous catalysts is easy recovery and recycling, which is cost effective and reduces waste and by-product formation. The use of zeolite catalysts provides better activity and selectivity for different types of reactions. The uniform porous microcrystalline structure drives zeolites as effective catalysts. The interaction between the reactant molecule and the catalyst surface is an important phenomenon to achieve the catalytic activity for heterogeneous catalyst, as it involves catalyst and reactant in different phases. Thus it is necessary to have a molecular-level understanding of the process, for example, isomerization of meta-xylene to *para*-xylene over zeolite. Xylene isomerization is an industrially important reaction that allows the conversion of less useful *meta*-xylene into useful *para*-xylene. It is difficult to picture the reaction pathway, and computer simulation provides an advanced technique to predict the mechanism of the reaction. A model can be generated from the database of structures in the Material Studio® of Accelrys. Materials Studio is a flexible client-server software environment that brings some of the world's most advanced materials simulation and modeling technology to any PC. Materials Visualizer is the core module in Materials Studio, providing modeling, analysis, and visualization tools. Materials Visualizer provides fast interactive tools that enable the client to construct graphical models of molecules, crystalline materials, surfaces, interfaces, layers, and polymers. The user can then manipulate, view, and analyze these models.

The acidic zeolite promotes carbocation isomerizations. The shape of the reactant plays an important role in product selectivity. Perhaps *para*-xylene has a favorable shape that allows it to diffuse rapidly through the zeolite structure, whereas *meta*-xylene takes longer to pass through the zeolite and thus has more opportunity to be converted into *para*-xylene. Secondly, the orientation of reactive intermediates within the zeolite channels specifically favors *para*-xylene. The list of computer simulations of zeolite covers a wide range, especially for catalytic activity. Details will be provided in Chapter 9.

Another example of microporous materials is aluminophosphate (AlPO), in which the vertex is shared by the AlO_4 and PO_4 tetrahedra. Before the discovery of mesoporous materials, aluminum phosphate zeolite was the spectacular material has pores with 18 rings and thus causing a pore diameter greater than 12 Å. Aluminum phosphate zeolite is an artificial material in which tetrahedral AlO_4 and PO_4 are linked alternately in three dimensions to make a regular pore structure with no defects. This structure makes $AlPO_4$ electronically neutral with neither countercations nor Brønsted acid sites like standard natural zeolites. Consequently, $AlPO_4$ is expected to have adsorption, catalysis, and separation properties different from standard zeolites. Isomorphous substitution of different transition metal like Fe, Co, Ti, Mn, Sn, and Zn in this matrix makes it a highly selective catalyst for industrial application. An exemplary structure model of AlPO-8 generated by the same modular environment is shown in Figure 1.2. All the structures are based on the International Zeolite Association database [4].

FIGURE 1.2   The model of AlPO-8, another phosphate-based nanoporous material where the pore architecture is visible. (See CD for color image.)

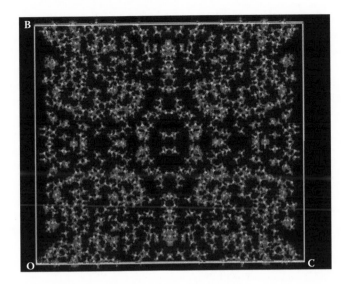

FIGURE 1.3 A model structure of mesoporous material. (For color image see CD)

In 1992, Kresege et al., researchers from Mobil Oil Company, first synthesized mesoporous material (2–50 nm pore diameter) using a self-assembly process involving electrostatic interaction between positively charged quaternary ammonium ions and inorganic anions as a framework precursor [5]. Following their discovery, several extended routes of synthesis such as neutral templating route [6] and electrostatic assembly approach [7] have been used. The simple mechanism behind the formation of porous structures is the hydrolysis of the silica precursor followed by condensation on the walls of the template and subsequent removal of the template, leaving behind the porous structure. Figure 1.3 represents a model structure of mesoporous material. Changing the composition of the synthesis mixture can change the type of self-assembled process and leads to the formation of hexagonal (MCM-41), cubic (MCM-48), or lamellar (MCM-50) [8]. Among mesoporous materials, MCM-41, with a honeycomb structure, is the most studied because of its easy synthesis, and its pore sizes, which can be modified by changing the chain length of the surfactant or by hydrothermal restructuring [9]. In zeolitic material, the template molecule is a single molecule or ion, but self-assembled templates required for the synthesis of mesoporous material allow the formation of a variety of structures depending on the synthesis condition, such as rod, sheet, or three-dimensional structures, mentioned before. The increase in the surfactant ($C_nH_{2n+1}NMe_3$) chain length increases the pore-to-pore spacing; for example, when $n = 8$ changes to $n = 16$, the pore diameter changes from 18 to 37 Å [10]. In addition, using a block copolymer it is possible to increase the pore size and an ordered nanostructured material can be obtained [11]. They have large surface areas and are a great contender for catalysis of larger molecules. Postsynthetic modification of the surface or the pore by silylation can make MCM-41 an ideal material for adsorption. Siliceous mesoporous materials are relatively inert with low hydrothermal stability due to the amorphous nature of the walls, which severely hinders their practical applications. To provide additional functions to the siliceous material, Si can be easily substituted by other heteroatoms to improve catalytic activity and hydrothermal stability. Furthermore, mesoporous materials seem to be ideal hosts for metal nanoparticles due to their high surface areas to immobilize catalytically active species on or provide nanosize confinement inside the pore system. Unlike zeolites, molecular simulation in the area of mesoporous materials is still in its infancy, possibly because of the amorphous structure, though it shows X-ray diffraction peak in the lower angle region. Molecular simulations are often used to understand and predict properties of materials based on their intermolecular interactions. Ravikovitch and coworkers [9] studied in detail the adsorption and phase behavior of the fluid confined in MCM-41 and other nanoporous materials by using nonlocal density functional theory (NLDFT) and molecular simulation to rationalize, because there are some novel phase behaviors of the fluid confined in slit and cylindrical pores at low temperature. Their research indicates that NLDFT generally predicts the reversible and hysteretic adsorption isotherms and full-phase diagrams of capillary condensation including the points of phase equilibrium and spinodal transitions, in good agreement with the Monte Carlo (MC) simulation and reference experiments [12]. Using these interaction models, it is possible to simulate the adsorption property of the mesoporous material and hence it is possible to obtain a fundamental understanding of the relation between interactions and material properties, which is often impossible to obtain from experiments or theory alone.

After the discovery of mesoporous materials, Huo et al. [13] extended the surfactant (both positive and negatively charged) templating method to the synthesis of nonsiliceous porous materials, in particular metal oxides. It has been observed that the nature of the porous structure depends on the surfactant charge and the inorganic ion involved in synthesis. However, it still remains a challenge to discover the new reaction path to stabilize the structure of these materials.

Apart from traditional nanoporous materials, metal organic frameworks (MOFs) represent one of the major breakthroughs in the area of nanoporous materials. They have vast potential to explore several domains. MOFs are crystalline porous materials with microscopic holes and a well-defined low-density structure (Figure 1.4). Depending on the metal and the clusters, MOFs can be one-, two-, or three-dimensional structures. This type of material can be tailored simply by changing the metal, ligand, and linker that constitute the framework. Based on the nature of the secondary building unit or the linker, formation of new materials with distinctive properties takes place. MOFs have been produced with a wide range in terms of surface areas, from 500 to 4500 m^2 g^{-1} (equivalent to a football field in one gram of sample) and the flexibility of the system results in better sorption properties compared to the rigid porous system. Adsorption in zeolite is a surface phenomenon. The Knudsen or gas phase diffusion is dominated by the size of the pores present in the matrix. However, in the case of MOFs, as the pore geometries are more extensive, there may be an increased affinity for adsorption on alternative sites. Thus, they are already established as storage materials for explosive gases like hydrogen and methane and are also used in catalysis [14] and small molecule sensing [15].

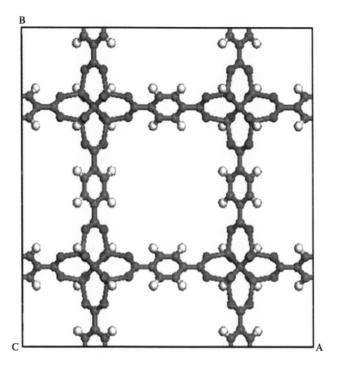

FIGURE 1.4 Framework of MOF-5 displaying free access to the nanosized voids. (See CD for color image.)

Characterization of MOFs is very difficult and expensive, so a computational design technique is required to design new MOFs. For example, DFT calculations have been performed on the aluminum-containing MIL-53, an MOF, to know the charge distribution in the structure, which is an efficient material for the adsorption of carbon dioxide [16].

1.4 MAJOR APPLICATIONS

Nanoporous materials have different pore architectures and the pores can be tunable with metal incorporation, which will again result in variation in the acidity of the materials. This results in their widespread applicability in the field of shape-selective catalysis and separation technology. The porous structures make these materials available for membrane materials and suitable for application in composite materials. There are multiple uses for larger pore structures like mesopores, where the zeolitic materials are used to grow in the walls of the mesopore to improve the surface area and activity of catalysis. Nanoporous materials work for gas storage materials with selective adsorption and desorption capabilities. These materials also have very high mechanical stability, which will allow them to be used in the packaging industry. The porous architecture makes these materials suitable for photonic materials, which we will discuss more in the futuristic application models in Chapter 8 and 9. We will now elaborate on each of the applications in brief.

1.4.1 CATALYTIC APPLICATIONS

From a practical point of view there are many applications for nanoporous materials, but in industry they are mainly used as catalysts, for gas separation, and as ion exchange materials. The different catalytic activities and ion exchange properties mainly depend on the different synthesis conditions despite the fact that their chemical compositions are the same, but framework structure varies along with the chemical composition.

The global market for nanocatalysts is projected to approach $3.7 billion in 2007 and $5.6 billion in 2010, at an average annual growth rate (AAGR) of 6.3%. Commercially well-established nanocatalysts such as industrial enzymes, zeolites, and transition metal nanocatalysts accounted for about 98% of global sales in 2003. Newer types, such as transition metal oxides, metallocene, carbon nanotubes, and others, are expected to more than triple their combined market share by 2010, to 6.8%. The refining/petrochemical sector was the largest user in 2003, with more than 38% of the market, followed by chemicals/pharmaceuticals, food processing, and environmental remediation. Here again, it is the smaller end-user segments whose consumption is growing the fastest.

More efficient catalytic processes require improvement in catalytic activity and selectivity. Both aspects will rely on a tailored design for catalytic materials with desired microstructures and active site dispersion. Nanoporous materials offer such possibilities in this regard with the controlled large and accessible surface area of a catalyst but avoiding the standalone fine particles. The traditional methods of impregnation of metal ions in nanoporous supports are not as effective in achieving high dispersion of active centers, whereas incorporation in template synthesis or intercalation are more advanced techniques rendering high activity, due to the high surface area of the active components and selectivity due to the narrow pore size distribution.

Zeolites are a major part of the group of nanoporous materials, extremely useful as catalysts for several important reactions such as cracking, isomerisation, and hydrocarbon synthesis. Zeolites can promote a diverse range of catalytic reactions, including acid-base and metal-induced reactions and can be used as the active support for active metals. It plays an important role as the shape-selective catalyst either by transition-state selectivity or by exclusion of competing reactants on the basis of molecular diameter. Actually, the reaction takes place inside the pores, which allows a strong control over the product. Synthetic zeolites are considered one of the most important catalysts in petroleum refining, synthetic fuels, and petrochemical production.

The experiment finds its own limitation to explore the activity selectivity yield of the product deactivation of the catalysts and all. Molecular simulation becomes a powerful tool to investigate the process at the microscopic level where the experiment may not reach or is unable to access and visualize the problem to ratify. For instance, the active site of zeolite catalysts is mainly the bridging hydroxyl proton, whose activity can be tunable with metal substitution. Molecular simulation makes it feasible to get an insight on the catalytic properties of zeolite resulting from that by comparing the deprotonation energy and knowledge of the exact location of this active site [17]. Whenever there is a change in the structure, one needs to visualize the structure and understand the change in property, which is where simulation comes into play. Simulation also helps to determine the change in porosity after structural modification, which will help one decide which type of molecules to use to perform a reaction. Another important and appealing criterion for the usage of computer simulation is the ability to determine the reaction mechanism by transition state calculation, in particular the new reaction route to reach the particular products, where experiment still remains very much predictive in terms of the reaction. Moreover, computational methods are helpful to investigate the product selectivity of a reaction. In Chapter 9 we will discuss the chemical adsorption, catalytic property covering the shape selectivity, cracking, and activation barrier in combination with experiment along with examples.

Compared to the resin, the application of nanoporous materials as ion exchangers is limited. They are mainly used in the softening of water, removal of the strong radionuclides because some of the porous matrices have considerable resistance to radiation, and also in wastewater treatment to remove heavy metal ions. The ion exchange capacity of zeolite-type materials is directly related to the quantity of Al present in the framework as well as the nature of the cationic species in the solution, solvent, and temperature. The charge-balancing cations are, in general, exchangeable. The molecular sieving, ion exchange, and catalytic properties of zeolites are largely influenced by the nature of these cations. Ion exchange studies of zeolites are usually carried out in aqueous solution and in solid state. The exchange behavior can be shown through isotherms. These isotherms may be obtained graphically and present concentrations of a given ion, both in the zeolitic phase and in solution, in an equilibrium system. Molecular simulation can be used to study the possibility of ion exchange in different types of zeolite. It is also possible to check the effect of Al content, which influences the ring structure and consequently the exchangeable cations, and the obtained results are comparable with the experimental data.

1.4.2 Application as Separator, Sensor, and Storage

As the regulatory limits on environmental emissions become increasingly stringent, industries have become more active in developing separation technologies that can remove contaminants and pollutants from waste gas and water streams. Adsorption processes and membrane separation are two dominating technologies that have attracted continuous investment in R&D. Adsorbent materials and membranes (typically nanoporous materials) are increasingly being applied and new adsorbents and membranes are being invented and modified for various environmental applications such as the removal of SO_2, NO, and VOC emissions. New adsorbent materials with well-defined pore sizes and high surface areas are being developed and tested for potential use in energy storage and environmental separation technologies. Zeolites are used to adsorb a variety of materials. Considering the characteristic three-dimensional channels available for guest molecules, the large internal surface and high thermal stability make them appropriate candidates for molecular sieving.

This property can be fine-tuned by varying the structure and changing the size and number of cations around the pores. The applications including the molecular sieving property are drying, purification, and separation. Nanoporous materials can remove water to very low partial pressures and act as an effective desiccant. Zeolites can remove volatile organic chemicals from air streams and separate isomers and mixtures of gases. Historically, molecular simulation offers a smart means to investigate the sitting and segregation effects of a molecular sieve [18]. Computer simulation using grand canonical Monte Carlo (GCMC) calculation has been successfully used to elucidate the mechanism of adsorption. Successful prediction is possible for the adsorption isotherm of different gases like nitrogen, oxygen, argon, etc. In addition, density functional theory (DFT) allows the calculation of the equilibrium density profile for all locations inside the pores. The combination of DFT and MC is useful for analysis of pore size, which can be validated with experimentally obtained pore size by the BJH method, which will be discussed in detail later.

In the case of phosphate materials, Na^+-SAPO-34 sorbents are ion-exchanged with several individual metal cations to study their effect on the adsorption of similarly sized light gases. Measurements of pure component adsorption equilibria, with emphasis on CO_2, are performed at different temperatures (273–348 K) and pressures (<1 atm). Adsorption isotherms for CO_2 in M^{n+}-SAPO-34 materials displayed a nonlinear behavior and did not follow the typical pore-filling mechanism. In general, the overall adsorption performance of the exchanged materials increased as follows: $Ce^{3+} < Ti^{3+} < Mg^{2+} < Ca^{2+} < Ag^+ < Na^+ < Sr^{2+}$. The strontium exchanged materials excelled at low-pressure ranges, exhibiting very sharp isotherm slopes at all temperatures. For example, the Sr^{2+}-SAPO-34 were capable of removing as much as 2.8 wt% at a CO_2 partial pressure of 10^{-3} atm at room temperature. Isosteric heat of adsorption data confirmed that the Sr^{2+} species were responsible for the strong surface interaction, and therefore it is plausible to state that cations occupy exposed sites (SII′) in the material. In addition, all divalent cations were found to interact more with the sorbate when compared to the other charged species. Adsorption isotherms for trivalent exchanged cations (Ce^{3+}, Ti^{3+}), on the other hand, showed what appears to be partial blockage of pore windows (occupancy of SIII), is actually resulting in very low adsorption capacities. The surface interactions are analyzed according to electrostatic and nonspecific contributions. Due to strong ion–quadrupole interactions, all the sorbent materials exhibited higher affinity for CO_2 over the other gases tested (i.e., CH_4, H_2, N_2, and O_2). Mathematical modeling to estimate binary component adsorption performance during vacuum pressure sorbate adsorption (VPSA) corroborated that Sr^{2+}-SAPO-34 sorbents are by far the best option for CO_2 removal from CH_4 mixtures, especially at low concentrations [19].

As mentioned before, although microporous zeolites are widely used as commercial catalysts, their application is intrinsically limited due to the smaller channel size. Introduction of mesoporous materials with larger pores and large surface area made their use in the field of catalysis of larger molecules possible. However, the catalytic activity and hydrothermal stability of mesoporous materials are low compared to microporous zeolites because of the amorphous nature of the mesoporous wall. Modification of the mesoporous wall structure, increasing thermal stability, and the incorporation of the acid sites enables them to be used as catalysts (to be discussed in Chapter 9) for different types of reaction such as alkylation, acylation, cyclization, Aldol condensation, Knoevenagel condensation, oxidation, reduction, isomerization, disproportionation, polymerization, esterification, protection, deprotection, acetalization, etc.

Nanoporous materials have the ability to act as sensors because they have a large surface area, which is very sensitive to the environment; e.g., temperature, pressure, humidity. So these types of materials can be used as gas or humidity sensors, sensors for biological applications, and bio-mimicking materials.

The future of the fuel industry is a hydrogen economy. Hydrogen will be the dominant fuel and will be converted into electricity in fuel cells, leaving only water as an end product. Hydrogen has roughly three times the energy content of gasoline on a per weight basis. On a volumetric basis, however, hydrogen's energy content is only 8 megajoules (MJ) per liter for the cryogenic liquid compared with gasoline's 32 MJ/L. DOE's 2010 targets call for developing a hydrogen storage system with an energy density of 7.2 MJ/kg and 5.4 MJ/L. Energy density refers to the amount of usable energy that can be derived from the fuel system. The figures include the weight and size of the container and other fuel delivery components — not just the fuel (C&EN, August 22, 2005, 42–47). Fuel cell development has been very rapid in recent years. However, there are many technological challenges before fuel cells become commercially viable and widely adopted. Most of the problems are associated with materials notably related to electrocatalysts, ion-conducting membranes, and porous supports for the catalyst. Both the high surface area and the opportunity for nanomaterial consolidation are key attributes of this new class of materials for hydrogen storage devices. A recent review has discussed in length the application domain of hydrogen storage with nanomaterials [20]. Nanostructured systems including carbon nanotubes, nano-magnesium-based hydrides, complex hydride/carbon nanocomposites, boron nitride nanotubes, TiS_2/MoS_2 nanotubes, polymer nanocomposites, and metal organic frameworks are considered to be potential candidates for storing large quantities of hydrogen. Recent investigations have shown that nanoscale materials may offer advantages if certain physical and chemical effects related to the nanoscale can be used efficiently. Every government is looking for storage material for atomic or molecular hydrogen and nanoporous materials are a good candidate. The synergistic effects of nanocrystallinity and nanocatalyst doping on the metal or complex hydrides for improving the thermodynamics and hydrogen reaction kinetics are key.

Certain nanoporous materials such as carbon nanotubes and zirconium phosphates have already shown promise for application in fuel cells. Currently there are no optimal systems for hydrogen storage. Hydrogen can be stored in gaseous, liquid, or, more recently, solid forms. Nanostructured materials such as carbon nanotubes again show promise as adsorbents. Despite many controversial reports in the literature, hydrogen storage in carbon nanotubes may become competitive and useful. Another type of nanostructured carbon is templated by using three-dimensional ordered mesoporous silicates. However, the modification of the wall structure, improved hydrothermal and chemical stability, expansion of functionality, and incorporation of active site have been crucial to their emerging applications as efficient catalysts, in biosensors, for drug delivery, in the semiconductor industry, for gas separation, for energy storage, etc.

The unique and tunable properties of carbon-based nanomaterials enable new technologies for identifying and addressing environmental challenges. A recent review critically assesses the contributions of carbon-based nanomaterials to a broad range of environmental applications: sorbents, high-flux membranes, depth filters, antimicrobial agents, environmental sensors, renewable energy technologies, and pollution prevention strategies [21].

Although various carbon-based adsorbents are accepted as storage material for hydrogen, it is difficult to engineer the material systematically and to identify the specific hydrogen-binding site. The MOF-5 structure motif and related compounds provide ideal platforms on which to adsorb gases, because the linkers are isolated from each other and accessible from all sides to sorbate molecules. The scaffolding-like nature of MOF-5 and its derivatives leads to extraordinarily high apparent surface areas (2,500 to 3,000 $m^2 \, g^{-1}$) for these structures, which can adsorb hydrogen up to 4.5 wt% (17.2 hydrogen molecules per formula unit) at 78 K and 1.0 wt% at room temperature and pressure of 20 bar. On a practical level, the preparation of MOFs is simple, inexpensive, and of high yield [22].

Very highly porous materials, such as zeolites, carbon materials, polymers, and metal-organic frameworks, offer a wide variety of chemical composition and structural architectures that are suitable for the adsorption and storage of many different gases, including hydrogen, methane, nitric oxide, and carbon dioxide. However, the challenges associated with designing materials to have sufficient adsorption capacity, controllable delivery rates, suitable lifetimes, and recharging characteristics are not trivial in many instances. The different chemistry associated with the various gases of interest makes it necessary to carefully match the properties of the porous material to the required application [23]. In addition, other nanomaterials and novel sorbent systems like carbon nanotubes, fullerenes, nanofibers, polyaniline nanospheres, metal organic frameworks, etc., need to be explored, but this is beyond the scope of this book.

This endeavor of adsorption behavior demands simulation to be an active ingredient. We will talk about the adsorption capability of nanoporous materials with a special emphasis on zeolitic and mesoporous materials and how simulation can help the cause in Chapter 7.

1.4.3　Biological Applications

Nanomaterials that are assembled and structured on the nanometer scale are attractive for biotechnology applications because of the potential to use material topography and the spatial distribution of functional groups to control proteins, cells, and tissue interactions and also for bioseparations. Bionanotechnology is all about creating nanomaterials or biomaterials for biological applications. Many studies are underway in fundamental understanding and exploiting the nature of nanoscale systems and processes to develop improved chemical separations and isolation media using nanoporous materials. Moreover, integrating engineered and self-assembled materials into useful devices ranging from biosensors to drug delivery systems and development of new products and biomedical devices by manipulating biomolecule enzymes, other proteins, and biochemical processes at the nanoscale has also been exploited.

Nanoporous materials are currently being developed for use in implantable drug delivery systems, bio-artificial organs, and other novel medical devices. Advances in nanofabrication have made it possible to precisely control the pore size, pore distribution, porosity, and chemical properties of pores in nanoporous materials. As a result, these materials are attractive for regulating and sensing transport at the molecular level. In this work, the use of nanoporous membranes for biomedical applications is reviewed. The basic concepts of underlying membrane transport are presented in the context of design considerations for efficient size sorting. Desirable properties of nanoporous membranes used in implantable devices, including biocompatibility and anti-befouling behavior, and to remove the unwanted accumulation of microorganisms are also discussed. In addition, the use of surface modification techniques to improve the function of nanoporous membranes is reviewed. An intriguing possibility involves functionalizing nanoporous materials with smart polymers in order to modulate biomolecular transport in response to pH, temperature, ionic concentration, or other stimuli. These efforts open up avenues to develop smart medical devices that respond to specific physiological conditions as mentioned in a recent review [24].

Proteins have been used by nature for billions of years to create the incredibly complex nanoscale structures within a living cell. Molecular-scale scaffolding, cables, motors, ion pores, pumps, coatings, and chemically powered levers composed primarily of proteins are all found in nature. Proteins provide superior catalytic abilities over traditional inorganic catalysts and the simplified reaction conditions of enzymes require less complex engineering than catalytic reactors. Because nanoporous materials are porous and some are biocompatible, they afford the capability to build enzymatic nanomaterials that mimic natural biological reactions. Immobilizing recombinant enzymes into nanoporous materials can be used for long-lifetime biological reactors for a variety of applications.

Mesoporous silicates (MPS) have an ordered pore structure with dimensions comparable to many biological molecules. They have been extensively explored as supports for proteins and enzymes in biocatalytic applications. Since their initial discovery, novel synthesis methods have led to precise control over pore size and structure, particle size, chemical composition, and stability, thus allowing the adsorption of a wide variety of biological macromolecules, such as heme proteins, lipases, antibody fragments, and proteases, into their structures. A recent work discusses the application of ordered, large-pore, functionalized mesoporous silicates to immobilize proteins for biocatalysis [25]. This still remains a challenge; we have given some more examples in the futuristic application model in a later chapter in the book. Mesoporous structure determination still remains a challenge, and simulation can explore things when one can define the atomic structure as molecular modeling can describe property based on the molecular structure. We will provide some examples to deal with mesoporous matrixes in Chapter 6.

1.4.4 POLYMER APPLICATIONS

In the midst of this era of nanotechnology and shrinking device size, block copolymers have evolved from their use in traditional application areas (e.g., adhesives, additives, and elastomers) to enable the development of materials for emerging and more advanced technologies. The block copolymers were used to generate nanoporous polymers, which can be used either directly, in applications such as membrane filtration, or subsequently as a template for the formation of other nanostructured materials.

The application of nanoporous materials covers the areas of (1) nanolithography and new alignment technologies, (2) monoliths, (3) new fabrication techniques, and (4) membranes. A summary and perspective on the current direction of the stimulating area of block copolymer-derived nanoporous materials was highlighted in a recent review [26].

The applications of nanoporous materials are not only restricted to the abovementioned areas and can be extended to the porous electrodes for fuel cells, highefficiency thermal insulators, electrode materials for batteries, and porous electronic substrates for high-speed electronics.

1.5   THE CHALLENGES

In the science of nanoporous materials, there are many challenges and opportunities ahead of us. For example, in catalysis, one of the key goals is to promote reactions to have a high selectivity with a high yield. To meet this goal, tailoring a catalyst particle via nanoparticle synthesis and self-assembly, so that it catalyzes only a specific chemical conversion with higher yield and greater energy efficiency is imperative. For application of nanoporous materials in the field of adsorption and catalysis, it is necessary to have a narrow pore size distribution of desired pore size. Traditional amorphous nanoporous materials such as silica gels, alumina, and activated carbons are limited in shape selectivity because of their broad pore size distribution and fixed pore geometries. Microporous zeolites and pillared clays are the only class of nanocrystalline materials with uniform pores. However, their pore sizes are limited to below 1.0–1.2 nm. The mesostructured molecular sieves and oxides that have been developed since the invention of MCM-41 by Mobil scientists have shown great promise for separation, catalysis, and biological applications where large molecules are involved. Supramolecular templating techniques and processing have revolutionized the synthesis and application opportunities of nanoporous materials. There are many templating pathways in making mesostructured materials. New synthesis strategies are constantly being revealed and trialled for improving the pore size range, chemical composition, and thermal and hydrothermal stabilities.

The major challenges in the nanoporous materials are designing of smart materials with predetermined structure, composition, and properties for specific applications such as the extension of the supramolecular templating method to the formation of the pores larger than 10 nm for their application in enzyme catalysis, bioseparation, and biosensing, where larger substrates are required. A combination of crystalline wall structure like micropores and the pore size of mesopores would be useful in overcoming the current limitations with existing zeolitic pore structures for selective petrochemical and fine chemical processing. The main targets for the synthesis of the nanoporous material would be (1) the tailoring of pore size to attain compositional flexibility, structural stability, and the degree and homogeneity of dispersion; (2) better understanding of the structure property relationship; (3) synthetic and theoretical characterization: (a) nanoporous materials require proper characterization for scientific and industrial purpose and experimental techniques are still limited due to size; (b) it is necessary to have homogeneous distribution of the metal active site for mesoporous materials for optimized performance as catalyst; [27]; and a combination of computational chemistry and experiment is essential because it is difficult to handle nanoporous materials only experimentally. Molecular simulation enables scientists to design the material without any real chemicals, which can save resources and less exposure to toxic chemicals leads to greener science.

1.6 THE FUTURE

Nanomaterials will have a profound impact on many industries including microelectronics, manufacturing, medicine, clean energy, and the environment. In these industries, there are many examples of applications of microporous zeolites and molecular sieves as nanoscale catalysts and gas separation membranes. Expanding the pore dimensions to the mesopore range will increase the scope of their applications in these fields. In particular, mesoporous materials will have wider applications in biological separation, biosensors, and nanoreactors for conducting multiple and controlled biological reactions on microchips. In the fields of clean energy production and storage, nanoporous materials as catalysts, storage media, and electrode materials will have tremendous potential in enabling process innovations in areas such as gas-to-liquid conversion, hydrogen production, alternative solar cells, fuel cells, and advanced batteries. In the environmental field, nanoporous and nanocrystalline semiconductors are the key to cost-effective photocatalytic purification of water and air, economic removal and recovery of organic vapors, greenhouse gas reduction, and utilization. In health care, biomaterials for orthopedic and cardiovascular applications, tissue repair, biosensors, and controlled drug delivery are likely to be developed and applied in the near future. All of which will depend on the development of new nanoporous substrates or coatings. The future of nanoporous materials is promising, with many great opportunities. The area of nanoporous materials is interdisciplinary, including every branch of science. The electronic industry, especially in semiconductors such as nitrides as LEDs, laser diodes, and quantum dots, and wires for these materials, has already begun such applications. Moreover, the adsorbent property offers potential use in environmental remediation. In the future, it would be expected that nanoporous materials cover all possible areas, including chemical, medical, electronics, etc. Material chemists have a major role in the proper design of materials considering every facet of nanoporous materials in combination with theory, and computer simulation.

Simulation has its own challenges, too. With the development of powerful hardware, the future of simulation is very bright. Ten years ago, simulation was only used for validating experiments or for visualization, but within this ten tears people now know the structure of porous material they can change tune and design a material by only simulation and then go to bench to make their material, but there is still a sizable gap between experiment and theory. The simulation we are talking about here is atomistic simulation, which can deal with smaller systems more accurately using quantum mechanical calculations and with a maximum size of a million atoms with a very qualitative understanding through mesoscale-type of simulation. The demand is now to run accurate calculations for a large system. This is not that challenging if one can see a small scenario and have the capability of logic and vision to extrapolate for a larger system, which is how many modelers have survived even with a small calculation. But the challenge is, if one knows the properties of the material and the behavior required, can one design or propose a structure with the desired material. Genetic algorithm is one of the choices where the modelers, physicists, and chemists are working together to design future materials.

REFERENCES

1. Lecture of Richard Feynmann, December 29th 1959 at the annual meeting of the American Physical Society at the California Institute of Technology (Caltech) was first published in the February 1960 issue of Caltech's *Engineering and Science*.
2. Roald Hoffmann, Foresight Update 20, Foresight Institute, Palo Alto, CA.
3. J. A. Rabo, *Appl. Catal.*, 229 (2002) 7.
4. J. N. Richardson, E. T. C. Vogt, *Zeolites*, 12 (1992) 13.
5. C. T. Kresege, M. E. Leonowicz, W. J. Roth, J. C. Vartuli, J. S. Beck, *Nature*, 359 (1992) 710.
6 P. T. Tanev, T. J. Pinnavaia, *Science*, 267 (1995) 865.
7. Q. S. Huo, D. I. Margolese, U. Ciesla, P. Y. Feng, T. E. Gier, P. Sieger, R. Leon, P. M. Petroff, F. Schuth, G. D. Stucky, *Nature*, 368 (1994) 317.
8. A. Firouzi, D. Kumar, L. M. Bull, T Besier, P. Sieger, Q. Huo, S. A. Walker, J. A. Zasadzinski, C. Glinka, J. Nicol, *Science*, 267 (1995) 1138.
9. (a) Q. S. Huo, D. I. Margolese, G. D. Stucky, *Chem. Mater.*, 8 (1996) 1147; (b) A. Sayari, P. Liu, M. Kruk, M. Jaroniec, *Chem. Mater.*, 9 (1997) 2499.
10. J. S. Beck, J. C. Vartuli, W. J. Roth, M. E. Leonowicz, C. T. Kresge, K. D. Schmitt, C. T. W. Chu, D. H. Olson, E. W. Sheppard, *J. Am. Chem. Soc.*, 114 (1992) 10834.
11. D. Y. Zhao, Q. S. Huo, J. L. Feng, B. F. Chmelka, G. D. Stucky, *J. Am. Chem. Soc.*, 120 (1998) 6024.
12. (a) P. I. Ravikovitch, D. Wei, W. T. Chueh, G. L. Haller, A. V. Neimark, *J. Phys. Chem. B*, 101 (1997), 3671; (b) P. I. Ravikovitch, S. C. O. Domhnaill, A. V. Neimark, F. Schuth, K. K. Unger, *Langmuir*, 11 (1995) 4765; (c) P. I. Ravikovitch, A. Vishnyakov, A. V. Neimark, *Phys. Rev. E*, 64 (2001) 011602.
13. (a) Q. Huo, D. I. Margolese, U. Ciesla, P. Feng, T. E. Gier, P. Sieger, R. Leon, P. M. Petroff, F. Schuth, G. D. Stucky, *Nature*, 368 (1994) 317; (b) Q. Huo, D. I. Margolese, U. Ciesla, D. G. Demuth, P. Feng, T. E. Gier, P. Sieger, A. Firouzi, B. F. Chemlka, F. Schuth, G. D. Stucky, *Chem. Mater.*, 6 (1994) 1176.
14. (a) J. S. Seo, D. Whang, H. Lee, S. I. Jun, J. Oh, Y. J. Jeon, K. Kim, *Nature*, 404 (2000) 982; (b) S. H. Cho, B. Q. Ma, S. T. Nguyen, J. T. Hupp, T. E. Albrecht-Schmitt, *Chem. Comm.*, (2006) 2563.
15. B. Zhao, X. Y. Chen, P. Cheng, D. Z. Liao, S. P. Yan, Z. H. Jiang, *J. Am. Chem. Soc.*, 126 (2004) 15394.
16. N. A. Ramsahye, G. Maurin, S. Bourrelly, P. Llewellyn, T. Loiseau, G. Ferey, *Phys. Chem. Chem. Phys.*, 9 (2007) 1059.
17. (a) E. G. Deroune, J. G. Feripat, Zeolites, 1985, 5, 165; A. Redondo, J. P. Hay, *J. Phys. Chem.*, 97 (1993) 11754; (b) K. P. Schroder, J. Sauer, M. Lesslie, C. R. A. Catlow, *Zeolites*, 12 (1992) 20; (c) C. R. A. Catlow, W. C. Mackrodt (eds.), *Computer Simulation of Solids*, Springer, Berlin (1982); (d) C. R. A. Catlow (ed.), *Modelling of Structure and Reactivity in Zeolites*, Academic Press, London (1992).

18. (a) R. L. June, A. T. Bell, D. N. Theodorou, *J. Phys. Chem.*, 96 (1992) 1051; (b) R. Q. Snurr, A. T. Bell, D. N. Theodorou, *J. Phys. Chem.*, 97 (1993) 13742.
19. M. E. Rivera-Ramos, G. J. Ruiz-Mercado, A. J. Hernández-Maldonado, *Ind. Eng. Chem. Res.*, 47 (2008) 5602.
20. M. U. Niemann, S. S. Srinivasan, A. R. Phani, A. Kumar, D. Y. Goswami, E. K. Stefanakos, *J. Nanomater.*, (2008) Article ID 950967.
21. M. S. Mauter, M. Elimelech, *Environ. Sci. Techn.*, 42 (2008) 5843.
22. N. L. Rosi, J. Eckert, M. Eddaoudi, D. T. Vodak, J. Kim, M. O'Keeffe, O. M. Yaghi, *Science*, 300 (2003) 1127.
23. R. E. Morris, P. S. Wheatley, *Angew. Chem. Int. Ed.*, 47 (2008) 4966.
24. S. P. Adige, L. A. Curtiss, J. W. Elam, M. J. Pellin, C.-C. Shih, C. Shih, S. J. Lin, Y. Y. Su, S. D. Gittard, J. Zhan, R. J. Aryan, *J. Min. Met. Mater. Soc.*, 60 (2008) 1047–4838 (Print), 1543–1851 (Online).
25. S. Hudson, J. Cooney, E. Manger, *Angew. Chem. Int. Ed.*, 47 (2008) 8582.
26. D. A. Olson, L. Chen, M. A. Hillmayer, *Chem. Mater.*, 20 (2008) 869.
27. D. R. Rolison, *Science*, 299 (2003) 1698.

2 Molecular Modeling

Molecular modeling is a method that combines computational chemistry techniques with graphics visualization for simulating and predicting three-dimensional structures, chemical processes, and physicochemical properties of molecules and solids. Molecular modeling has been used as a tool to support experiments, mainly the experimentally unknown, or to explain a certain behavior that cannot be explained otherwise. One of the strengths of molecular modeling is to be able to predict properties of materials before experiments are performed. This banks valuable time and resources and expedites the development process. The other most important aspect of molecular modeling is the ability to generate data from which one may gain insight and thereby rationalize the behavior of a large class of molecules. Although it sounds simple, there is still a gap between experiment and theory. The intention of this book is to bring these two together, to show how these two are compatible and to enhance the confidence of the experimentalist to understand the technique and application to gain insight and enhance the development process. In this chapter the focus is to define molecular modeling and to showcase different methodologies to solve the nanomaterial problem with the main focus on atomistic simulation techniques with Monte Carlo and the methodology of molecular mechanics. Finally, the challenges and the future of these techniques within this domain are discussed.

2.1 GENERAL INTRODUCTION

Computational chemistry may be defined as the application of mathematical and theoretical principles to the solution of chemical problems. Molecular modeling, a subset of computational chemistry, concentrates on predicting the behavior of individual molecules within a chemical system. The most accurate molecular models use *ab initio* or first principle electronic structure methods, based upon the principles of quantum mechanics, and are generally very computer intensive. However, due to advances in computer storage capacity and processor performance, molecular modeling has been a rapidly evolving and expanding field, to the point that it is now possible to solve relevant problems in an acceptable amount of time.

The types of predictions possible for molecules and reactions include:

1. Heats of formation
2. Bond and reaction energies
3. Molecular energies and structures (thermochemical stability)
4. Energies and structures of transition states (activation energies)
5. Reaction pathways, kinetics, and mechanisms
6. Charge distribution in molecules (reactive sites)
7. Substituent effects

8. Electron affinities and ionization potentials
9. Vibration frequencies (IR and Raman spectra)
10. Electronic transitions (UV/visible spectra)
11. Magnetic shielding effects (NMR spectra)

Prediction of these properties has many applications in nanoporous materials research, including studies of synthesis pathways, reaction products, initiation mechanisms, etc.

Molecular modeling is the general term used to describe the use of computers to construct molecules and perform a variety of calculations on these molecules in order to predict their chemical characteristics and behaviors. The term molecular modeling is often used synonymously with the term *computational chemistry*. Computational chemistry is a broader term, referring to any use of computers to study chemical systems. Some chemists use the term *computational quantum chemistry* to refer to the use of computers to perform electronic structure calculations, where the electrons in a chemical system are calculated.

Molecular modeling is employed to calculate a wide range of properties of individual atoms and molecules. Chemists typically want to know three things about an individual molecule: its chemical structure (number and type of atoms, bonds, bond lengths, angles, and dihedral angles), its properties (basic characteristics of the molecule, such as molecular energy, enthalpy, and vibration frequencies), and its activity (those characteristics that describe how the molecule behaves in the presence of other molecules, such as its nucleophilicity, electrophilicity, and electrostatic potentials).

The computational chemist must learn and apply a variety of methods to specific modeling situations. The purpose of this book is to focus on four general methods: molecular mechanics (MM), *ab initio* quantum chemical methods, semiempirical quantum chemical methods, and density functional theory (DFT) — the newest method. Depending on how broadly one defines molecular modeling or computational chemistry, there are a number of other methods that can be considered.

A computer simulation is valuable because it is applied to a precisely defined model for the material of interest. The model is actually a combination of two interactive components: one for the interactions among the molecules making up the system and another for interactions between the molecules and their environment:

Simulated model = model of molecular interactions
+ model for system–environmental interactions

The model for molecular interactions is contained in an intermolecular force law or, equivalently, an intermolecular potential energy function. This potential function implicitly describes the geometric shapes of individual molecules or, more precisely, their electrons. Thus, when one specifies the potential function, and establishes the symmetry of the molecules. A detailed characterization of intermolecular potential functions may be given analytically or numerically; in any case, a quantitative form for the potential function defines a molecular model and hence the form must be chosen before a simulation can be performed. In most simulations the intermolecular potential energy is taken to be a sum of isolated pair interactions; this assumption is called a pair-wise additivity and hence can be written as:

$$V = \sum \sum_{i<j} u(r_{ij})$$

where $u(r_{ij})$ is a pair potential energy function whose form is known and r_{ij} is the scalar distance between molecules i and j. Because no dissipative forces act among molecules, intermolecular forces are conservative, and therefore the force on molecule I is related to the potential by:

$$F = -\frac{\partial V(r^N)}{\partial r_i}$$

where $\delta/\delta r$ represents the gradient operator.

The second part of the model is with boundary conditions, which describes how the molecules interact with their surroundings. Characteristics of boundary conditions are largely dictated by the physical situation to be simulated; however, some freedom usually exists in the way in which boundary conditions are realized.

2.2   METHODS

Accurate simulation of atomic and molecular systems generally involves the application of quantum mechanical theory. However, quantum mechanical techniques are computationally expensive and are usually only applied to small systems between 10 and 1000 atoms or small molecules. It is difficult to model large systems such as a condensed polymer containing thousands of monomers in this way. The objective behind the simulation of larger systems is often to extract bulk (statistical) properties, such as diffusion coefficients or Young's moduli, which depend on the location of the atomic nuclei or, more often, an average over a set of atomic nuclei configurations. Under these circumstances the details of electronic motion are lost in the averaging processes. Thus, bulk properties can be extracted if a good approximation of the potential in which atomic nuclei movement is available. In addition, if there are any methods available that can generate a set of system configurations, which may not follow the exact dynamics of the nuclei, but are statistically consistent with a full quantum mechanical description.

There are a number of potentials (or force fields) and distribution generating techniques available and they are collectively referred to as *classical simulation methods*. The term *classical* is used because the approach is based on some of the earliest simulation generated configurations by integrating the Newtonian (classical) equations of motion. All molecular modeling techniques can be classified under three general categories: (1) quantum mechanical calculations, (2) molecular mechanics, and (3) Monte Carlo simulation.

2.2.1 Quantum Mechanical Calculations

In quantum mechanics there are three major divisions in the calculation method in terms of theory and implication. The first one is *ab initio* electronic structure methods, based upon quantum mechanics, which therefore provide the most accurate and consistent predictions for chemical systems. However, *ab initio* methods are extremely computer intensive. This method is capable of high-accuracy predictions over a wide range of systems. Rapid advances in computer technology are making *ab initio* methods increasingly practical for use with realistic chemical systems. The limitation is the convergence of the size of the system with time. The second category is density functional theory (DFT) methods, which are much faster than the ab initio code due the simplistic localized atomic orbital summation method with charge density-based calculation methods. Finally, the simplest one is semiemprical methods, in which a portion of the calculation comes from experimental data, and the rest comes from mathematics. The major advantage of semiempirical methods is that they are faster and able to perform calculations on larger molecules. Table 2.1 shows a comparison between different methods of molecular simulations.

In this chapter we will cover two methods: the Monte Carlo method of simulation and molecular mechanics along with molecular dynamics. This book will only deal with one major theory of quantum mechanical calculation, which is density functional theory; the *ab initio* and other higher accuracy theory remain outside the scope of this book. The book is to guide experiments and help experimentalists rationalize the interaction and design new matrices, which demands handling of a larger number of atoms and a similar and accurate theory is best a theory, which is capable of balancing size of system and accuracy will be chosen. That is why density functional theory or its combination with molecular mechanics like quantum mechanics and molecular mechanics, called QM/MM methods, is best. Density functional methodology will be described in detail in the following chapter for better understanding.

2.3 MONTE CARLO SIMULATION

Monte Carlo molecular modeling is the application of Monte Carlo methods to molecular problems. These problems can also be modeled by the molecular dynamics method. The difference between these two approaches relies on statistical mechanics rather than molecular dynamics. Instead of trying to reproduce the dynamics of a system, it generates states according to appropriate Boltzmann probabilities. It is therefore also a particular subset of the more general Monte Carlo method in statistical physics.

Monte Carlo molecular modeling employs a Markov chain procedure in order to determine a new state for a system from the previous one. According to its stochastic nature, this new state is accepted at random. Each trial usually counts as a move. The avoidance of dynamics restricts the method to studies of static quantities only, but the freedom to choose moves makes the method very flexible. These moves must only satisfy a basic condition of balance in order to describe the equilibrium properly, but detailed balance and a stronger condition are usually imposed when designing new algorithms. An additional advantage is that for some systems, such as the Ising model, which lacks a dynamical description and is only defined

TABLE 2.1
Comparing the Capability of Different Methods

Method	Advantages	Disadvantages	Best for
Ab initio			
Uses quantum physics Mathematically rigorous No empirical parameters	Useful for a broad range of systems Does not depend on experimental data Calculates transition states and excited states	Computationally expensive	Small systems (tens of atoms); electronic transitions; systems without experimental data; systems requiring high accuracy
DFT			
States that all ground-state properties are functional of the charge density ρ.	Useful for a broad range of systems Does not depend on experimental data Calculates transition states and excited states	Computationally less expensive than *ab initio* methods with a little compromise with accuracy	Reasonable systems (thousand of atoms); structure property correlation, activity, reaction etc.
Semiempirical			
Uses quantum physics Uses experimental parameters Uses extensive approximations	Less demanding computationally than *ab initio* methods Calculates transition states and excited states	Requires *ab initio* or experimental data for parameters Less rigorous than *ab initio* methods	Medium-sized systems (hundreds of atoms); electronic transitions
Molecular mechanics			
Uses classical physics Relies on force field with embedded empirical parameters	Computationally cheap, fast, and useful with limited computer resources Can be used for large molecules like enzymes	Does not calculate electronic properties Requires *ab initio* or experimental data for parameters	Large systems (thousands of atoms); systems or processes that do not involve bond breaking
Monte Carlo			
Relies on statistical mechanics rather than molecular dynamics Instead of trying to reproduce the dynamics of a system, it generates states according to appropriate Boltzmann probabilities	Very cheap, fast, and useful with limited computer resource, as well with large experimental data and a wish to generate a logistic guess	Does not need parameters to fit, does not produce electronic properties	Large systems (thousands of atoms); systems or processes that do not involve bond breaking

by an energy prescription, the Monte Carlo approach is the only feasible approach. The Ising model is defined as a model that consists of a lattice of "spin" variables with two characteristic properties: (1) each of the spin variables independently takes on either the value +1 or the value −1; and (2) only pairs of nearest-neighboring spins can interact. The study of this model in two dimensions forms the basis of the modern theory of phase transitions. The great success of Monte Carlo method in statistical mechanics has led to various generalizations such as the method of simulated annealing for optimization, in which a fictitious temperature is introduced and, which is then gradually lowered.

Experimentally, a molecular system is described by a small number of parameters, such as volume and temperature. The collection of molecular configurations that satisfy this partial knowledge is called an *ensemble* of configurations. An ensemble is described by a distribution function, ρ_m, which represents the probability of each configuration, m, in the ensemble. The probability of a configuration, m, in the canonical ensemble is given by

$$\rho m = C \exp[-\beta E_m] \tag{2.1}$$

where C is an arbitrary normalization constant, β is the reciprocal temperature, and E_m is the total energy of configuration m [1].

The reciprocal temperature is given by:

$$\beta = \frac{1}{k_B T} \tag{2.2}$$

where k_B is the Boltzmann constant and T is the absolute temperature.

The total energy of configuration m is calculated according to the following sum:

$$E_m = E_m{}^{AA} + E_m{}^{AS} + U_m{}^{A} \tag{2.3}$$

in which $E_m{}^{AA}$ is the intermolecular energy between the adsorbate molecules, $E_m{}^{AS}$ is the interaction energy between the adsorbate molecules and the substrate, and $U_m{}^A$ is the total intramolecular energy of the sorbate molecules. The intramolecular energy of the substrate is not included because its structure is fixed throughout the simulation; thus, energy contributions are fixed and vanish, becausee only energy differences play a role in these type of calculations.

The total intramolecular energy, U^A, is the sum of the intramolecular energy of all adsorbates of all components:

$$U^A = \sum_{\{N\}_m} u_{intra} \tag{2.4}$$

where $\{N\}_m$ denotes the set of adsorbate loadings of all components in configuration m.

When one starts a MC simulation, the starting configuration will take several steps to adjust to the current temperature. A simulation is, therefore, separated into an equilibration and a production stage. The properties returned at the end of the run are based on the production stage only. In the equilibration and production stages of a simulation, each step starts with the selection of a step type using the weights set at the start of the run. The step type can be either a translation or a rotation. After a step type is selected, a random component is chosen and the step type is applied to a random species of that component.

The Metropolis Monte Carlo method used in simulation samples the configurations in an ensemble by generating a chain of configurations, m, n,..., where the probability of transition from m to n is π_{mn}. Thus, if configuration m is sampled with a frequency ρ_m, then, on average, $\rho_m\pi_{mn}$ of them are transformed to n. Likewise, $\rho_n\pi_{nm}$ of configurations n is transformed to m. Clearly, these fluxes must be the same to preserve the density, ρ; otherwise, there would be a net flow from m to n (or vice versa) and this would increase ρ_n (or ρ_m) and a different ensemble would be sampled. Thus, the following detailed balance condition for equilibrium is obtained:

$$\rho_m\pi_{mn} = \rho_n\pi_{nm} \qquad (2.5)$$

In the Metropolis Monte Carlo method, the step that transforms configuration m to n is a two-stage process. First, a trial configuration is generated with probability α_{mn}. Then, either the proposed configuration, n, is accepted with a probability P_{mn} or the original configuration, m, is retained with a probability $1 - P_{mn}$. The overall transition probability, π_{mn}, is thus obtained from:

$$\pi_{mn} = \alpha_{mn}P_{mn} \qquad (2.6)$$

It is easy to verify by substitution that the following choice for the acceptance probability satisfies the equation

$$\pi_{mn} = \alpha_{mn}P_{mn} \qquad (2.7)$$

For example, the probability of displacing a molecule from 5 to 5.5 Å in the direction x is the same as the probability of displacing a molecule from 5.5 to 5 Å. Clearly, the probability of displacing a molecule from 5 to 5.5 Å in the direction x could equally be halved, provided that twice as many of these attempts are accepted, precisely as dictated by Equation (2.7). This expression can be simplified further. In the traditional Metropolis Monte Carlo method [2], trial configurations are generated without bias; i.e., $\alpha_{mn} = \alpha_{nm}$. As such, the corresponding acceptance probability (Equation (2.7)) becomes:

$$P_{mn} = \min\left[1, \frac{\rho_n}{\rho_m}\right] \qquad (2.8)$$

Thus, transitions of a configuration m to a more likely one ($\rho_n > \rho_m$) are always accepted, but transitions to configurations with a lower probability ($\rho_n < \rho_m$) are less likely to be accepted.

In the configurational bias Monte Carlo method [3], trial configurations are generated with a bias to low energies. For each step, the attempt rates α_{mn} and α_{nm} are calculated and retained in the acceptance rule (Equation (2.8)).

A bias is introduced toward high-density ρ to avoid having to attempt configurations with low-density ρ, which are likely to be rejected by the acceptance test. By calculating the bias factor, α_{mn}, of each attempt, and applying Equation (2.7), the biased Monte Carlo method is guaranteed to sample the same ensemble as the Metropolis Monte Carlo method, albeit in some cases much more efficiently.

This expression can be simplified further, depending on the ensemble being simulated and the selected Monte Carlo method.

$$P_{mn} = \min\left[1, \frac{a_{nm}}{a_{mn}} \frac{\rho_n}{\rho_m}\right]$$ (2.9)

Transforming a configuration is a two-stage process. First, a trial configuration is generated with probability α_{mn}. Then, either the proposed configuration, n, is accepted with a probability P_{mn} or the original configuration m is retained with a probability $1 - P_{mn}$. In the configurational bias Monte Carlo method [3], trial configurations are generated with a bias to low energies. For each step, the attempt rates α_{mn} and α_{nm} are calculated and retained in the acceptance rule (Equation (2.9)).

The acceptance probability is inversely proportional to the attempt rate; hence, simply increasing the transition to a particular configuration n will just decrease its acceptance rate. To enhance the acceptance probability, it is essential to construct the bias such that the trial configurations generated have, on average, a higher density ρ_n. Let us think of a multiple configuration scenario.

Instead of generating one trial configuration, a total of K trial configurations $n^{(1)}, \ldots, n^{(K)}$ is generated. Each configuration $n^{(k)}$ is given a weight $w^{(k)}$. One configuration is then selected with a probability proportional to its weight (w_n). With w_n, the weight of the selected configuration, the attempt rate of a configuration n is given by:

$$a_{mn} = \frac{w_n}{\frac{1}{K}\sum_{k-1}^{K} w^{(k)}} = \frac{Kw_n}{w_n + \sum_{k-1}^{K-1} w^{(k)}}$$ (2.10)

The sum in the last expression runs over all trial configurations that were not selected. The attempt rate is normalized such that $\alpha_{mn} = 1$ if all configurations have the same weight. By construction, configurations with higher than average weights are more likely to be selected; hence, in general, $\alpha_{mn} > 1$. Of course, it is still possible, but less likely, to select a trial configuration with a lower than average weight, $\alpha_{mn} < 1$.

Following Equation (2.9), the attempt rate of the reverse step, α_{nm}, is also required to determine the probability of accepting a biased attempt, which is the attempt rate for transformation of a configuration, n, into the known configuration, m. It is calculated exactly as in Equation (2.10), the only difference being that the selected configuration m is given; hence, only $K - 1$ trial configurations have to be generated to determine the attempt rate.

The procedure above applies mainly to one degree of freedom and needs to be repeated for each degree of freedom if there is more than one degree of freedom. Hence, for each degree of freedom, K states are generated and their weights are determined. One state is drawn from the weighted distribution and the attempt rate, $\alpha_{mn,d}$, for this selection is calculated using Equation (2.10). After all degrees of freedom have been assigned a value, the resulting configuration is the candidate for the acceptance step. The total attempt rate of configuration is the product of the attempt rates for each degree of freedom:

$$a_{mn} = \prod_{d-1}^{D} a_{mn,d} \tag{2.11}$$

where d is the total number of degrees of freedom. The attempt rate of the reverse step, α_{nm}, is calculated in the same way. As above, an unweighted selection would result in $\alpha_{mn} = 1$.

The configurational bias Monte Carlo method not only recognizes translational and rotational degrees of freedom but also torsional degrees of freedom for each torsion monitor defined on the molecule.

The probability of a configuration, m, and the reciprocal temperature in the canonical ensemble is given by [1]:

$$\rho_m = C \exp[-\beta E_m] \tag{2.12}$$

where C is an arbitrary normalization constant, β is the reciprocal temperature (Equation (2.1)), and E_m is the total energy of configuration m (Equation (2.2)).

Substitution into Equation (2.8) leads to the acceptance probability:

$$P_{mn} = \min\left[1, \frac{a_{nm}}{a_{mn}} \exp\left[-\beta(E_n - E_m)\right]\right] \tag{2.13}$$

where E_m is the total energy of configuration m (Equation (2.2)). Multiple steps can be included in a conformational bias simulation to incorporate the desired phenomenon in the simulation like regrowth, twisting, rotation, translation, etc.

Temperature in Monte Carlo Simulations is added through simulated annealing, both for the equilibration and production steps. Simulated annealing is a metaheuristic algorithm for locating a good approximation to the global minimum of a given function in a large search space [4]. The concept is drawn from the process of annealing in metallurgy, where microcrystalline materials are heated and then slowly cooled in a controlled manner to increase crystallite size and reduce the number of defects in the crystal lattice. At high temperatures, the molten material is disordered because the kinetic energy forces atoms to explore higher energy states, such as substitution or defect sites. The system is cooled very slowly such that, at any given time, it is approximately in thermodynamic equilibrium. A slow rate of cooling increases the probabilities that the atoms will find configurations with lower energy, corresponding to more regular positions in the crystal lattice. As cooling proceeds, the system becomes more ordered and finally freezes into a ground state. If the system is not heated to a sufficiently high temperature or if it is cooled too quickly, lattice defects may become trapped and the system will quench in metastable states, corresponding to local energy minima.

Any function can be minimized in analogy with this physical process. The object function plays a role equivalent to the energy of the crystal to be annealed. The function variables correspond to the atom configurations, with the global minimum solution being the ground state. The algorithm to solve the minimum is controlled by a parameter equivalent to the temperature in physical annealing. In order to minimize the object function, it is evaluated along a sequence of states. At each step, the function is evaluated for a neighboring state and compared with the current value. If the value has improved, this replaces the current solution. If the value has worsened, it may still replace the current solution, depending on how much worse the new solution is within a tolerance set by the control parameter. Decreasing the control parameter has the effect of reducing the probability that such steps in the "wrong" direction are accepted. Hence, the above sampling method combined with a slowly decreasing control parameter provides a way to find a global minimum, starting from an arbitrary initial state.

The simulated annealing method uses a canonical Monte Carlo sampling of search space during which the temperature is gradually decreased. Each step of the Monte Carlo sampling will attempt to change the current state of the system into a randomly chosen neighboring state. If the energy decreases as a result of this step, the new state will be the starting point for the next step. On the other hand, if the energy increases, the attempt is followed by a decision about whether to accept the selected neighboring state or to use the current state as the starting point for the next step. This decision is made on the basis of a random number. The probability of accepting a state of higher energy depends on the energy difference between the states and the temperature of the system.

An essential feature of the sampling method is that it allows the system to explore states that are higher in energy. Although this momentarily worsens the object function, the system is able to escape from local minima and search for better ones. To steer the simulation in the right direction, the probability of accepting a less favorable state must decrease with the value of the function to be minimized; otherwise, the simulation would spend too much time in areas that are of no interest. The Monte Carlo method provides just such a scenario. In this method, the probability of accepting an energy increase, $E(s') - E(s) > 0$, is equal to $\exp\left\{\frac{-[E(s')-E(s)]}{kT}\right\}$, which clearly decreases with the energy difference. The temperature, T, multiplied by the Boltzmann constant, k, sets the scale of the energy barriers that can be overcome.

In the simulated annealing method, the temperature is slowly decreased during the course of the simulation. Initially, when the temperature is high, large energy increases are acceptable, allowing the system to explore a broad region of the search space, while ignoring small ripples in the energy surface. As the temperature decreases, steps that lead to an increase in energy are increasingly disfavored, thereby steering the system to step to neighboring states with a lower energy.

Eventually, when the temperature is very low, the system is forced to evolve to the local minimum in the current region of the search space.

At the end of an annealing simulation, the system will have reached a state corresponding to a local minimum of the object function in an area of the search space. This state is an approximation to the global minimum in the search space. By repeating the annealing process, it is possible to improve on the approximation. So, starting from the local minimum reached in the previous run, the system is heated up again and allowed to explore a broad region around this local minimum. Then, the temperature is decreased and the system is annealed in another state of minimal energy. Repeating the temperature cycles can give solutions that are closer to the global minimum solution.

The Monte Carlo simulation is designed for the study of individual systems, allowing us to find low-energy adsorption sites on both periodic and nonperiodic substrates or to investigate the preferential adsorption of mixtures of adsorbate components or pack molecule is different orientation or look into other related phenomena. Adsorbates are typically molecular gases or liquids and substrates are usually porous crystals or surfaces, such as zeolites, carbon nanotubes, and amorphous structures such as silica gel or activated carbon. It is possible to use MC methods to identify the possible adsorption configurations by carrying out Monte Carlo searches of the configurational space of the substrate–adsorbate system as the temperature is slowly decreased. These methods can be extended to simulate a pure adsorbate (or mixture of adsorbate components) absorbed in a sorbent framework; i.e., a three-dimensional periodic structure with pores of a size and shape suitable to accommodate the sorbate molecules. Typical examples of sorbents include microporous and mesoporous materials such as zeolites, aluminophosphates, clays, nanotubes, polymer membranes, silica gels, activated carbons, and metal–organic frameworks. Characterizing the sorption behavior of these materials is of importance in the fields of catalysis and separation technology.

Several pieces of information can be obtained from these types of simulation, which include adsorption isotherms, binding sites, binding energies, global minimum sorbate locations, density and energy fields, energy distributions, sorption selectivity, solubility, isosteric heats, and Henry's constant. Two of the main features that are exhaustively used by the experimentalist are adsorption isotherms and Henry's constant (from online manual of SORPTION module of Accelrys Inc. Material Studio version 4.4.).

- Adsorption isotherms: Allow one to perform a series of fixed-pressure simulations at a set temperature in a single step. A fixed-pressure simulation is performed for a specified number of steps between the start and end fugacity set at the start of the Monte Carlo run. This is as if running multiple fixed-pressure calculation within the framework of atoms in the bulk. The pressure of the reservoir depends on the temperature and the fugacities. Because the latter are fixed during the simulation, so is the pressure. For an ideal gas reservoir, the fugacity equals the partial pressure and the reservoir pressure can thus be obtained by summing the fugacity over all components. Fixed pressure simulation always starts with an empty framework (the three-dimensional periodic structure that was in focus at the start of the Monte Carlo run). The starting configuration will take several steps to adjust to the specified fugacity and temperature. A simulation is therefore separated into an equilibration and a production stage. You can specify how many steps should be used for both stages. The properties returned at the end of the run are based on the production stage only.

- Henry's constant: In this calculation the coefficient in Henry's law, H, is calculated. Henry's law states that when a framework is in equilibrium with a reservoir containing some component, the loading of this component in the framework, N, is proportional to the fugacity of the component in the reservoir, f:

$$N = Hf$$

Because loading is dimensionless, the Henry constant, H, carries the inverse units of the fugacity; that is, if the fugacity is specified in kPa, the Henry constant will have units of loading per cell per kPa. Henry's law does not apply to any reservoir but has been proved to hold for an ideal gas reservoir. Because every reservoir behaves ideally in the limit of zero pressure, a convenient definition of the Henry constant is

$$H = \lim_{P \to 0} \frac{N}{f}$$

where P is the pressure of the reservoir. In the zero pressure limits, the fugacity equals the partial pressure and the pressure can thus be obtained from the sum of fugacities over all components. According to the equation written above, in the limit of zero fugacity, the loading is proportional to the fugacity. The slope of the adsorption isotherm is the Henry constant. However, to determine the adsorption isotherm and to assess whether the linear regime is applicable, the average loadings at several fugacities close to zero have to be determined, which is computationally expensive. The alternative will be to calculate the average loading by grand canonical ensemble and then imposing the zero pressure limits.

$$H = \beta V \exp[\beta \mu_{intra}] < \exp[-\phi E_m] >_u$$

where V is the volume of the framework, β is the reciprocal temperature (Equation (2.1)), E_m is the total energy of configuration m (Equation (2.2)), and μ_{intra} is the intramolecular chemical potential (Equation (2.11)).

Monte Carlo methods are simpler methods though difficult to understand the probability part in one reading, but they generate useful numbers by quenching the numbers fast, one uses the methodology when the interest is to see the packing of molecule sorbents in a void or at least within the domain of nanoporous materials.

2.4 MOLECULAR MECHANICS

Molecular mechanics (MM) is often the only feasible means with which it is possible to model very large and nonsymmetric chemical systems such as proteins or polymers. Molecular mechanics is a purely empirical method that neglects explicit treatment of electrons, relying instead upon the laws of classical physics to predict the chemical properties of molecules. As a result, MM calculations cannot deal with problems such as bond breaking or formation, where electronic or quantum effects dominate. Furthermore, MM models are wholly system dependent; MM energy predictions tend to be meaningless as absolute quantities and are generally useful only for comparative studies. Despite these shortcomings, MM bridges the gap between quantum and continuum mechanics and has been used quite extensively to study mesoscopic effects in energetic materials.

2.4.1 Potential Energy Surface

The complete mathematical description of a molecule, including both quantum mechanical and relativistic effects, is a formidable challenge, due to the small scales and large velocities involved. Therefore, in the following discussion, these intricacies are ignored; instead, the focus is on general concepts. This is possible because MM and molecular dynamics are based on empirical data that implicitly incorporate all the relativistic and quantum effects. Because no complete relativistic quantum mechanical theory is suitable for the description of molecules, this discussion starts with the nonrelativistic, time-independent form of the Schrödinger description.

The Schrödinger equation is

$$H\psi(R,r) = E\psi(R,r) \tag{2.13}$$

where H is the Hamiltonian for the system, Ψ is the wavefunction, and E is the energy. In general, Ψ is a function of the coordinates of the nuclei (R) and of the electrons (r). Although the Schrödinger equation is quite general, it is too complex to be of any practical use, so approximations are made. Noting that the electrons are several thousands of times lighter than the nuclei and therefore move much faster, Born and Oppenheimer [5] proposed the following approximation: the motion of the electrons can be decoupled from that of the nuclei, giving two separate equations. The first of these equations describes the electronic motion:

- Equation for electronic motion or the potential energy surface depends only parametrically on the positions of the nuclei:

$$H\psi(r;R) = E\psi(r;R) \tag{2.14}$$

The direct solution of Equation (2.14) is in the domain of quantum mechanics and we will discuss that in detail in the following chapter.

The second equation describes the motion of the nuclei on this potential energy surface $E(R)$.

- Equation for nuclear motion on the potential energy surface

$$H\Phi(R) = E\Phi(R) \tag{2.15}$$

Solving Equation (2.15) is important for the structure or time evolution of a model. As written, Equation (2.15) is the Schrödinger equation for the motion of the nuclei on the potential energy surface. In principle, Equation (2.14) could be solved for the potential energy E, followed by the Equation (2.15). However, the effort required to solve Equation (2.14) is extremely complex, so usually an empirical fit to the potential energy surface, commonly called a *force field* (V), is used. Because the nuclei are relatively heavy objects, quantum mechanical effects are often insignificant in that case Equation (2.15) can be replaced by Newton's equation of motion as shown in Equation (2.16).

$$-\frac{dV}{dR} = m\frac{d^2R}{dt^2} \tag{2.16}$$

The solution of Equation (2.16) using an empirical fit to the potential energy surface $E(R)$ is called *molecular dynamics (MD)*. *Molecular mechanics* ignores the time evolution of the system and instead focuses on finding particular geometries and their associated energies or other static properties. This includes finding equilibrium structures, transition states, relative energies, and harmonic vibration frequencies.

The coordinates of a structure combined with a force field create an *energy expression* (or *target function*). This energy expression is the equation that describes the potential energy surface of a particular structure as a function of its atomic coordinates.

The potential energy of a system can be expressed as a sum of valence (or bond), cross term, and nonbond interactions:

$$E_{total} = E_{valence} + E_{crossterm} + E_{nonbond} \tag{2.17}$$

The energy of valence interactions is generally accounted for *diagonal* terms:

- Bond stretching (bond)
- Valence angle bending (angle)
- Dihedral angle torsion (torsion)
- Inversion, also called out-of-plane interaction terms, which are part of nearly all force fields for covalent systems

A Urey-Bradley (UB) term may be used to account for interactions between atom pairs involved in 1–3 configurations (i.e., atoms bound to a common atom):

$$E_{valence} = E_{bond} + E_{angle} + E_{torsion} + E_{oop} + E_{UB} \tag{2.18}$$

The basic assumptions of typical molecular mechanics methods are listed below.

1. Each atom (i.e., electrons and nucleus) is represented as one particle with a characteristic mass.
2. A chemical bond is represented as a spring, with a characteristic force constant determined by the potential energy of interaction between the two participating atoms. Potential energy functions can describe intramolecular bond stretching, bending and torsion, or intermolecular phenomena such as electrostatic interactions or van der Waals forces.
3. The potential energy functions rely on empirically derived parameters obtained from experiments or from other calculations.

Current molecular mechanics models are characterized by the set of potential energy functions used to describe the chemical forces. These force fields depend upon (1) atomic displacements (i.e., bond lengths); (2) atom types; that is, the characteristics of an element within a specific chemical context (e.g., a carbonyl carbon versus a methyl carbon); and (3) one or more parameter sets relating atom types and bond characteristics to empirical data. The purpose of a force field is to describe the potential energy surface of entire classes of molecules with reasonable accuracy. In a sense, the force field extrapolates from the empirical data of the small set of models used to parameterize a larger set of related models. Some force fields aim for high accuracy for a limited set of elements, thus enabling good predictions of many molecular properties. The broadest possible coverage of the periodic table with necessarily lower accuracy is the other aim.

Modern (second-generation) force fields generally achieve higher accuracy by including *cross-terms* to account for such factors as bond or angle distortions caused by nearby atoms. These terms are required to accurately reproduce experimental vibration frequencies and, therefore, the dynamic properties of molecules. Cross-terms can include the following: stretch-stretch, stretch-bend-stretch, bend-bend, torsion-stretch, torsion-bend-bend, bend-torsion-bend, and stretch-torsion-stretch. The energy of interactions between nonbonded atoms is accounted for by:

- van der Waals (vdW) force
- Electrostatic (Coulomb) force
- Hydrogen bond (hbond) terms in some older force fields

Different force fields use different functional forms to describe similar interactions. Table 2.2 lists some common functional forms, the units of their parameters, and the forms used by the standard force fields distributed with Materials Studio version 4.4 of Accelrys Inc. grouped by interaction type. The variables that are used in these functional forms are geometric measures that describe the spatial configuration of interacting particles.

For simulations that use force fields, the interactions of a system of particles are governed by an analytic expression that represents the potential energy

TABLE 2.2

List of Common Functional Forms with the Units of Their Parameters, and the Forms Used by the Standard Forcefields Distributed with Materials Studio Grouped by Interaction Type.

Potential Name Bond Stretch	Functional form	Units	Forcefield
Harmonic	$E = \dfrac{K_b}{2}(R-R_0)^2$	K_b : kcal/mol/Å² R_0 : Å	Dreiding Universal pcff COMPASS Custom
Cubic	$E = \dfrac{K_b}{2}(R-R_0)^2[1+d(R-R_0)]$	K_b : kcal/mol/Å² R_0 : Å d : Å⁻¹	
Quartic	$E = \dfrac{K_b}{2}(R-R_0)^2[1+c(R-R_0)+d(R-R_0)^2]$	K_b : kcal/mol/Å² R_0 : Å c : Å⁻¹ d : Å⁻²	pcff COMPASS
Morse	$E = D_0[\exp(-\alpha(R-R_0))-1]^2 - S$ where $\alpha = \sqrt{\dfrac{K_b}{2D_0}}$ and S is a constant shift value	D_0 : kcal/mol R_0 : Å S : kcal/mol K_b : kcal/mol/ Å² a : Å⁻¹	cvff
Angle Bend			
Harmonic	$E = \dfrac{K_0}{2}(\theta-\theta_0)^2$	K_0 : kcal/mol/ rad² θ_0 : °	Dreiding cvff pcff COMPASS Custom
Cubic	$E = \dfrac{K_0}{2}(\theta-\theta_0)^2[1+d(\theta-\theta_0)]$	K_0 : kcal/mol/ rad² θ_0 : ° d : /°	
Quartic	$E = \dfrac{K_0}{2}(\theta-\theta_0)^2\left[1+c(\theta-\theta_0)+d(\theta-\theta_0)^2\right]$	K_0 : kcal/mol/ rad² θ_0 : ° c : /° d : /°/°	pcff COMPASS

TABLE 2.2 (CONTINUED)
List of Common Functional Forms with the Units of Their Parameters, and the Forms Used by the Standard Forcefields Distributed with Materials Studio Grouped by Interaction Type.

Potential Name Bond Stretch	Functional form	Units	Forcefield
Cosine harmonic	$E = \dfrac{K_0}{2}\left(\dfrac{[\cos(\theta) - \cos(\theta_0)]^2}{[\sin(\theta_0)]^2} \right)$ $\theta_0 = 0, E = K_0(1 - \cos[\theta])$ for $\theta_0 = 180, E = K_0(1 + \cos[\theta])$	K_0 : kcal/mol θ_0 : °	Universal
Cosine periodic	$E = C(1 - B[-1]^n \cos[n\,\theta])$	C : kcal/mol B : - n : -	
Torsion			
Dihedral	$E = \displaystyle\sum_j \left\{ \dfrac{B_j(1 - d\cos[n_j\phi])}{2} \right\}$ Where ϕ = angle between the DK and JKL planes	B_j : kcal/mol d_j : - n_j : -	Dreiding cvff Universal Custom
Scaled dihedral	Same as Dihedral except that the barrier will be scaled by the number of torsions about the common bond J-K.	B_j : kcal/mol d_j : - n_j : -	
Shift dihedral	$E = \displaystyle\sum_j \left\{ \dfrac{B_j(1 + \cos[n_j\phi - \phi_0])}{2} \right\}$ Where ϕ_0 = initial phase shift	B_j : kcal/mol d_j : - n_j : - θ_0 : °	pcff COMPASS
Inversion (out-of-plane)			
Umbrelia	$E = \dfrac{C}{2}(\cos w - \cos w_0)^2$ For $w_0 \neq 0$ $E = K_w(1 - \cos w)$ For $w_0 = 0$ $C = \dfrac{K_w}{\sin^2(w_0)}$ where W = angle between the IL bond and its projection on the IJK plane	Kw : kcal/mol W_0 : °	Dreiding Universal Custom

(continued)

TABLE 2.2 (CONTINUED)
List of Common Functional Forms with the Units of Their Parameters, and the Forms Used by the Standard Forcefields Distributed with Materials Studio Grouped by Interaction Type.

Potential Name Bond Stretch	Functional form	Units	Forcefield
Mean- Wilson	$E = \dfrac{K_w}{2} w_{av}^2$ Where W_{av} = the average of the three umbrella inversions, as defined above, about atom I.	K_w : kcal/mol/ rad^2 W_{av} : °	cvff pcff COMPASS
Stretch-stretch R-R	$E = K_{ss}(R_{ij} - R_{ij0})(R_{jk} - R_{jk0})$	K_{ss} : kcal/mol/Å^2 R_{**0} : Å	cvff pcff COMPASS
Stretch-bend-stretch R-THETA	$E = (\theta - \theta_0)(K_{ij}[R_{ij} - R_{ij0}] + K_{jk}[R_{jk} - R_{jk0}])$	K_{**} : kcal/mol/ Å/rad R_{**0} : Å θ_0 : °	cvff pcff COMPASS
R-COSINE	$E = (\cos\theta - \cos\theta_0)(C_{ij}[R_{ij} - R_{ij0}] + C_{jk}[R_{jk} - R_{jk0}])$	C_{**} : kcal/mol/Å R_{**0} : Å θ_0 : °	
Bend-bend THETA- THETA	$E = K_{bb}(\theta_{jil} - \theta_{jil0})(\theta_{kil} - \theta_{kil0})$	K_{bb} : kcal/mol/ rad^2 θ_{***0} : °	cvff pcff COMPASS
COSINE- COSINE	$E = C(\cos\theta_{jil} - \cos\theta_{jil0})(\cos\theta_{kil} - \cos\theta_{kil0})$	C : kcal/mol θ_{**0} : °	
Torsion-stretch R_FOURIER	$E = (R - R_0)(V_a \cos[n_a\phi] + V_b \cos[n_b\phi] + V_c \cos[n_c\phi])$	V_* : kcal/mol/Å R_{**0} : Å n_* : -	pcff COMPASS
Torsion-bend-bend THETA2- COSPHI	$E = K_{tbb}(\theta_{ijk} - \theta_{ijk0})(\theta_{jkl} - \theta_{jkl0})\cos\phi$	K_{tbb} : kcal/mol/ rad^2 θ_{***0} : °	cvff pcff COMPASS
COSINE2- COSPHI	$E = C_{tbb}(\cos\theta_{ijk} - \cos\theta_{ijk0})(\cos\theta_{jkl} - \cos\theta_{jkl0})\cos\phi$	C_{tbb} : kcal/mol θ_{***0} : °	

TABLE 2.2 (CONTINUED)
List of Common Functional Forms with the Units of Their Parameters, and the Forms Used by the Standard Forcefields Distributed with Materials Studio Grouped by Interaction Type.

Potential Name Bond Stretch	Functional form	Units	Forcefield
Bend-torsion-bend			
THETA-FOURIER	$E = (\theta_{ijk} - \theta_{ijk0})\sum_{d} V_{jd}\cos[n_d\phi] + (\phi_{jkl} - \theta_{jkl0})\sum_{d} V_{kd}\cos[n_d\phi]$	V_{**} : kcal/mol/ rad θ_{***0} : ° n_* : -	pcff COMPASS
Stretch-torsion-stretch			
R_FOURIER	$E = (R_{ij} - R_{ij0})\sum_{d} V_{jd}\cos[n_d\phi] + (R_{kl} - R_{kl0})\sum_{d} V_{kd}\cos[n_d\phi]$	V_{**} : kcal/mol/Å R_{**0} : Å n_* : -	pcff COMPASS
Separated-stretch-stretch			
R-R	$E = K(R_{ij} - R_{ij0})(R_{kl} - R_{kj0})$	K : kcal/mol/ Å² R_{**0} : Å	pcff COMPASS
van der Waals			
LJ 9 4	$E = D_0\left[\dfrac{4}{5}\left(\dfrac{R_0}{R}\right)^9 - \dfrac{9}{5}\left(\dfrac{R_0}{R}\right)^4\right]$	D_0 : kcal/mol R_0 : Å	Custom
LJ 9 6	$E = D_0\left[2\left(\dfrac{R_0}{R}\right)^9 - 3\left(\dfrac{R_0}{R}\right)^6\right]$	D_0 : kcal/mol R_0 : Å	pcff COMPASS Custom
LJ 12 4	$E = D_0\left[\dfrac{1}{2}\left(\dfrac{R_0}{R}\right)^{12} - \dfrac{3}{2}\left(\dfrac{R_0}{R}\right)^4\right]$	D_0 : kcal/mol R_0 : Å	Custom
LJ 12 6	$E = D_0\left[\left(\dfrac{R_0}{R}\right)^{12} - 2\left(\dfrac{R_0}{R}\right)^6\right]$	D_0 : kcal/mol R_0 : Å	Dreiding Universal cvff MS Martini Custom
LJ 12 10	$E = D_0\left[5\left(\dfrac{R_0}{R}\right)^{12} - 6\left(\dfrac{R_0}{R}\right)^{10}\right]$	D_0 : kcal/mol R_0 : Å	Custom
Exponential 6	$E = D_0\left[\dfrac{6}{(y-6)}\exp y\left(1 - \dfrac{R}{R_0}\right) - \dfrac{y}{(y-6)}\left(\dfrac{R_0}{R}\right)^6\right]$	D_0 : kcal/mol R_0 : Å y : -	

(continued)

TABLE 2.2 (CONTINUED)
List of Common Functional Forms with the Units of Their Parameters, and the Forms Used by the Standard Forcefields Distributed with Materials Studio Grouped by Interaction Type.

Potential Name	Functional form	Units	Forcefield
Bond Stretch			
Morse	$E = D_0[X^2 - 2X]$ where $X = \exp\left(-\left(\dfrac{y}{2}\right)\left(\dfrac{R}{R_0} - 1\right)\right)$	D_0 : kcal/mol R_0 : Å y : -	Custom
Buckingham	$E = D_0 \exp(-\beta R) - \dfrac{C_6}{R^6}$	D_0 : kcal/mol β_0 : Å^{-1} C_6 : (kcal/mol) Å^6	Custom
Soft Harmonic	$E = \dfrac{a}{2} R_c \left(1 - \dfrac{R}{R_c}\right)^2$	a : kcal/mol/ Å R_c : Å	Custom
Hydrogen Bond			
LJ 12 10	$E = D_0\left[5\left(\dfrac{R_0}{R}\right)^{12} - 6\left(\dfrac{R_0}{R}\right)^{10}\right]\cos^4(\phi)$	D_0 : kcal/mol R_0 : Å	Dreiding Custom
Coulombic			
LIN-R-ε	$E = C\dfrac{q_i q_j}{\varepsilon R^2}$	q_i, q_j: e C: (kcal/mol) Å^2/e^2 ε : -	Dreiding Universal custom
CONST-ε	$E = C\dfrac{q_i q_j}{\varepsilon R}$	q_i, q_j: e C: (kcal/mol) Å/e^2 ε : -	cvff pcff COMPASS MS Martini
ERFC(R)	$E = C\dfrac{q_i q_j \, \mathrm{erfc}\left(\frac{R}{\beta}\right)}{\varepsilon R}$ where "erfc" is a complimentary error function	q_i, q_j: e C: (kcal/mol) Å/e^2 ε : -	
Three body			
Modified Stillinger-Weber (MSW)	$E = K_{jik} \exp(u + v)[\cos\theta_{jik} - \cos\theta_{jik,0}]^2$ where $u = \dfrac{Y_{ij}}{(r_{ij} - R_{ij,0})}$ $v = \dfrac{Y_{ij}}{(r_{ik} - R_{ik,0})}$	K: kcal/mol R$_{**}$, r$_{**}$, Y$_{**}$: Å θ: °	

Note: Custom = User Defined

surface, the energy expression. For large systems, the energy expression can consist of many terms and so commercial software provides a way of constructing the energy expression. A key concept in this construction is that of the force field type, also referred to as the *potential type*, the *force field atom type*, or simply the *atom type*.

As the name suggests, the force field type gives an indication of the nature and properties of a given particle in a simulation. If the simulation is atomistic, then the principal determinant of the force field type is the element to which the atom belongs. The force field type also gives an indication of the nature of the local microchemical environment of a given atom (or, more generally, particle). For example, a carbon atom in ethane has a different local environment to one in benzene and so each has a different force field type. A number of properties can be used to define a force field type and the definition may include a combination of the following properties:

- Element (if particle is an atom)
- Type of bonds (for example, single, double, resonant, etc.)
- Number of other particles to which the given particle is bonded
- The type of particles to which the given particle is bonded
- Hybridization
- Formal charge

Because the force field type assigned to a given particle depends only on its local environment, any symmetry that the simulated system possesses must be reflected in the assigned force field types. For example, the four hydrogen atoms of a methane molecule are related by symmetry and so they are all assigned the same force field type.

As a simple example of a complete energy expression, consider the following equation, which might be used to describe the potential energy surface of a water structure, e.g. water:

- Example energy expression for water

$$V(R) = K_{ok}\left(b - b_{ok}^0\right)^2 + K_{oh}\left(b' - b_{oj}^0\right)2 + K_{hoh}\left(\theta - \theta_{hoh}^0\right)^2 \qquad (2.20)$$

In this example, the force field defines:

- Bond lengths (b) and bond angles (θ)
- The functional form (a simple quadratic in both types of coordinates)
- The force constants (K)
- The *reference* O–H bond length (b^0) and H–O–H angle (θ) are the values for an ideal O–H bond and H–O–H angle at zero energy, which is not necessarily the same as their *equilibrium* values in a real water molecule.

Equation (2.19) is an example of an energy expression as set up for a simple molecule. Equation (2.20) is an example of the corresponding general, summed force field function:

- Example of force field function

$$V(R) = \sum_b D_b[1 - \exp(-a(b - b_0))]^2 + \sum_e H_e(\theta - \theta_0)^2 + \sum_\phi H_\phi[1 + s\cos(n\phi)]$$

$$+ \sum_x H_x X^2 + \sum_b \sum_{b'} F_{bb'}(b - b_0)(b' - b_0') + \sum_e \sum_e F_{ee}(\theta - \theta)(\theta - \theta_0')$$

$$+ \sum_e \sum_e F_{be}(b - b_0)(\theta - \theta_0) + \sum_e \sum_e F_{ee'\phi}(\theta - \theta')(\theta - \theta_0)\cos\phi$$

$$+ \sum_\chi \sum_{\chi'} F_{\chi\chi'} XX' + \sum_i \sum_{i>j} \left[\frac{A_{ij}}{r_{ik}^{12}} - \frac{B_{ij}}{r_{ij}^6} - \frac{q_i q_j}{r_{ij}} \right] \qquad (2.21)$$

The first four terms in this equation are sums that reflect the energy needed to:

- stretch bonds (b)
- bend angles (θ) away from their reference values
- rotate torsion angles (ϕ) by twisting atoms about the bond axis that determines the torsion angle
- distort planar atoms out of the plane formed by the atoms they are bonded to (X)

The next five terms are cross-terms that account for interactions between the four types of internal coordinates.

The final term represents the nonbond interactions as a sum of repulsive and attractive Lennard-Jones terms as well as Coulombic terms, all of which are a function of the distance (r_{ij}) between atom pairs.

The force field defines the functional form of each term in this equation as well as the parameters such as D_b, a, and b_0. It also defines internal coordinates as a function of the Cartesian atomic coordinates, although this is not explicit in Equation (2.20).

2.5 MOLECULAR DYNAMICS

Once an energy expression and, if necessary, an optimized structure have been defined for the system of interest, a dynamics simulation can be run. The basis of this simulation is the classical equations of motion, which are modified, where appropriate, to deal with the effects of temperature and pressure on the system. The main product of a dynamics run is a trajectory file that records the atomic configuration, atomic velocities, and other information at a sequence of time steps that can be analyzed subsequently.

In its simplest form, molecular dynamics solves Newton's familiar equation of motion:

$$F_i(t) = m_i a_i(t) \tag{2.22}$$

where F_i is the force, m_i is the mass, and a_i is the acceleration of atom i.

The force on atom i can be computed directly from the derivative of the potential energy V with respect to the coordinate r_i:

$$-\frac{\partial V}{\partial r_i} = m_i \frac{\partial^2 r_i}{\partial t^2} \tag{2.23}$$

The classical equations of motion are deterministic. This means that, once the initial coordinates and velocities are known, the coordinates and velocities at a later time can be determined.

The standard method of solving an ordinary differential equation such as Equation (2.23) numerically is the finite-difference method. The general idea is as follows: Given the initial coordinates and velocities and other dynamic information at time t, the positions and velocities at time $t + \Delta t$ are calculated. The time step Δt depends on the integration method as well as the system itself.

Integrating Newton's equations of motion allows you to explore the constant-energy surface of a system. However, most natural phenomena occur under conditions where the system is exposed to external pressure and/or heat exchanges with the environment. Under these conditions, the total energy of the system is no longer conserved and extended forms of molecular dynamics are required.

Several methods are available for controlling temperature and pressure. Depending on which state variables (the energy E, enthalpy H [that is, $E + PV$], number of particles N, pressure P, stress S, temperature T, and volume V) are kept fixed, different statistical ensembles can be generated. A variety of structural, energetic, and dynamic properties can then be calculated from the averages or the fluctuations of these quantities over the ensemble generated. The constant-energy, constant-volume ensemble (NVE), also known as the *microcanonical ensemble*, is obtained by solving the standard Newton equation without any temperature and pressure control. Energy is conserved when this (adiabatic) ensemble is generated. However, because of rounding and truncation errors during the integration process, there is always a slight fluctuation, or drift in energy. Although the temperature is not controlled during true NVE dynamics, you might want to use NVE conditions during the equilibration phase of your simulation. Materials Studio (Accelrys Inc.) allows you to hold the temperature within specified tolerances by periodic scaling of the velocities. True constant-energy conditions (that is, without temperature control) are not recommended for equilibration because, without the energy flow facilitated by temperature control, the desired temperature cannot be achieved.

During the data collection phase, if you are interested in exploring the constant-energy surface of the conformational space or for other reasons do not want the perturbation introduced by temperature and pressure-bath coupling, this is a useful ensemble. The results can be used to calculate the thermodynamic response function [6]. The constant-temperature, constant-pressure ensemble (NPT) allows control over both the temperature and pressure. The unit cell vectors are allowed to change, and the pressure is adjusted by adjusting the volume (that is, the size and, in some programs, the shape of the unit cell). NPT is the ensemble of choice when the correct pressure, volume, and densities are important in the simulation. This ensemble can also be used during equilibration to achieve the desired temperature and pressure before changing to the constant-volume or constant-energy ensemble when data collection starts.

If the force field being used yields a high pressure at the experimental volume, it may be more realistic to simulate at the experimental pressure rather than the experimental volume. High simulated pressure is a sign that the system is unduly compressed, which restricts atomic motions, slowing down the dynamic relaxations. The constant-pressure, constant-enthalpy ensemble [7] is the analogue of constant-volume, constant-energy ensemble, where the size of the unit cell is allowed to vary. Enthalpy H, which is the sum of E and PV, is constant when the pressure is kept fixed without any temperature control. Although the temperature is not controlled during true (adiabatic) NPH dynamics, you might want to use these conditions during the equilibration phase of your simulation. The natural response functions (specific heat at constant pressure, thermal expansion, adiabatic compressibility, and adiabatic compliance tensor) are obtained from the proper statistical fluctuation expressions of kinetic energy, volume, and strain.

Because the ensembles are artificial constructs, they produce averages that are consistent with one another when they represent the same state of the structure. Nevertheless, the fluctuations vary in different ensembles. Some of the fluctuations are related to thermodynamic derivatives, such as the specific heat or the isothermal compressibility. The transformation and relation between different ensembles has been discussed in greater detail by Allen and Tildesley [8].

One of the objectives of MD is to obtain the equilibrium thermodynamic properties of a structure. If a microscopic dynamic variable A takes on values $A(t)$ along a trajectory, then the following time average yields the thermodynamic value for the selected variable.

$$A = \lim_{T \to \infty} \frac{1}{T} \int_0^T A(t)\, dt \qquad (2.24)$$

This dynamic variable can be any function of the coordinates and moment of the particles in the structure. Through time averaging, calculation of the first-order properties of a system (such as the internal energy, kinetic energy, pressure, and virial) is achievable. Similarly, using microscopic expressions in the form of fluctuations of these first-order properties, you can calculate thermodynamic properties of a system. These include the specific heat, thermal expansion, and bulk modulus. In the thermodynamic limit, the first-order properties obtained in one ensemble are equivalent to those obtained in other ensembles (differences are in the order of $1/N$). However, second-order properties such as specific heats, compressibility, and elastic constants differ between ensembles. For example, the specific heat at constant pressure differs from the specific heat at constant volume. Therefore, it is important to use the appropriate ensemble when performing simulations to obtain these properties.

Temperature is a state variable that specifies the thermodynamic state of the system. It is also an important concept in dynamics simulations. This macroscopic quantity is related to the microscopic description of simulations through the kinetic energy, which is calculated from the atomic velocities. The temperature and the distribution of atomic velocities in a system are related through the Maxwell-Boltzmann equation:

$$F(v) = \left(\frac{m}{2\pi k_B T}\right)^{3/2} \exp\left(-\frac{mv^2}{2k_B T}\right) 4\pi v^2 \tag{2.25}$$

This well known formula expresses the probability $f(v)$ that a molecule of mass m has a velocity of v when it is at temperature T.

The x, y, z components of the velocities have Gaussian distributions, which is a continuous probability distribution that describes data for cluster around mean or average.

$$g(v_x)dv_x = \left(\frac{m}{2\pi k_B T}\right)^{1/2} e^{\frac{mv_x^2}{2k_B T}} dv_x \tag{2.26}$$

The initial velocities are generated from the Gaussian distribution of v_x, v_y, and v_z. The Gaussian distribution is in turn generated using a random number generator and a random number seed.

As mentioned before, temperature is a thermodynamic quantity, which is meaningful only at equilibrium. It is related to the average kinetic energy of the system through the equipartition principle. This principle states that every degree of freedom (either in moment or in coordinates), which appears as a squared term in the Hamiltonian, has an average energy of $\frac{kT}{2}$ associated with it. This is true for momentum p_i, which appears as $\frac{p_i^2}{2m}$ in the Hamiltonian.

Hence we have:

$$\left\langle \sum_i^N \frac{p_i^2}{2m} \right\rangle = \langle K \rangle = \frac{N_j k_B T}{2} \tag{2.27}$$

The left side of this equation is also called the *average kinetic energy* of the system, N_f is the number of degrees of freedom, and T is the thermodynamic temperature. In an unrestricted system with N atoms, N_f is $3N$ because each atom has three velocity components (i.e., v_x, v_y, and v_z).

It is convenient to define an instantaneous kinetic temperature function:

$$T_{ins\,tan} = \frac{2K}{N_j k_B} \tag{2.28}$$

The average of the instantaneous temperature T_{instan} is the thermodynamic temperature T.

Temperature is calculated from the total kinetic energy and the total number of degrees of freedom. For a nonperiodic system:

$$\frac{(3N-6)k_B T}{2} = \sum_{i=1}^{N} \frac{miv_i^2}{2} \tag{2.29}$$

Six degrees of freedom are subtracted because both the translation and rotation of the center of mass are ignored. And for a periodic system:

$$\frac{(3N-3)k_B T}{2} = \sum_{i=1}^{N} \frac{miv_i^2}{2} \tag{2.30}$$

Only the three degrees of freedom corresponding to translational motion can be ignored, because rotation of a central cell imposes a torque on its neighboring cells.

Although the initial velocities are generated to produce a Maxwell-Boltzmann distribution at the desired temperature, the distribution does not remain constant as the simulation continues. This is especially true when the system does not start at a minimum-energy configuration of the structure. This situation often occurs because structures are commonly minimized only enough to eliminate any hot spots.

During dynamics, kinetic and potential energy are exchanged and the temperature changes as a consequence. To maintain the correct temperature, the computed velocities have to be adjusted appropriately. In addition to maintaining the desired temperature, the temperature-control mechanism must produce the correct statistical ensemble. This means that the probability of occurrence of a certain configuration obeys the laws of statistical mechanics.

For example, in order for constant-temperature, constant-volume dynamics to generate the canonical ensemble, $P(E)$ (i.e., the probability that a configuration with energy E will occur) must be proportional to $\exp(-E/k_B T)$, the Boltzmann factor.

Pressure is another basic thermodynamic variable that defines the state of a system. Standard atmospheric pressure is 1.013 bar, where 1 bar = 10^5 Pa. A single number for the pressure implies that pressure is a scalar quantity but, in fact, pressure is a tensor of the more general form [9]:

$$P = \begin{pmatrix} P_{xx} & P_{xy} & P_{xz} \\ P_{yx} & P_{yy} & P_{yz} \\ P_{zx} & P_{zy} & P_{zz} \end{pmatrix} \tag{2.31}$$

Each element of the tensor is the force that acts on the surface of an infinitesimal cubic volume that has edges parallel to the x, y, and z axes. The first subscript refers to the direction of the normal to the plane on which the force acts, and the second subscript refers to the direction of the force. Sometimes, especially in materials science, the stress tensor, or stress, is used in preference to the pressure tensor (its negative). The diagonal elements are known as the *tensile stress* and the nondiagonal elements are the *shear stress*. The changes in unit cell lattice parameters and volume resulting from the stress can be obtained from an analysis of dynamics trajectory data. Multiple dynamic runs can be performed at varying stress or pressure values and the strains obtained can be used to plot a stress–strain curve. Pressure is calculated through the use of the virial theorem [8]. It can be stated in terms of generalized coordinate. The virial theorem can be expressed as $\langle q_k \frac{\partial H}{\partial q_k} \rangle = kT$, so it can be written as $PV = Nk_B T + \langle W \rangle$, where $\langle W \rangle$ is the internal virial as defined later. Like temperature, pressure is also a thermodynamic quantity and, strictly speaking, meaningful only at equilibrium.

Thermodynamic pressure, thermodynamic temperature, volume, and internal virial can be related in the following way:

$$PV = Nk_B T + \frac{2}{3} \langle W \rangle \tag{2.32}$$

where W is defined as:

$$W = \frac{1}{2} \sum_{i=1}^{N} r_i - f_i \tag{2.33}$$

Pressure is defined only when the system is placed in a container having a definite volume. In a computer simulation, the unit cell under periodic boundary conditions is viewed as the container. Volume, pressure, and density can be calculated only when the structure is recognized as periodic. An instantaneous pressure function P can be defined, which is analogous to the temperature, so that thermodynamic pressure is the average of the instantaneous values:

$$P = \frac{Nk_B T}{V} + \frac{2}{3} \frac{W}{V} \tag{2.34}$$

where P is the instantaneous pressure and T is the instantaneous kinetic temperature, which is related to the instantaneous kinetic energy K of the system by:

$$T = \frac{2}{3Nk_B} K \tag{2.35}$$

The instantaneous pressure function can be written as:

$$P = \frac{2}{3V}(K + W) \tag{2.36}$$

As mentioned above, pressure is a tensor and its components can also be expressed in tensorial form. Equation (2.34) can be recast in the form of:

$$P = \frac{1}{V}\left[\sum_{l-1}^{N} m_i v_i v_i^T + \sum_{i=1}^{N} r_i f_i^T\right] \tag{2.37}$$

In detail, the two terms on the right-hand side of the equation are

$$\sum_{i=1}^{N} m_i v_i v_i^T = \begin{bmatrix} \sum_i m_i v_{ix} v_{ix} & \sum_i m_i v_{ix} v_{iy} & \sum_i m_i v_{ix} v_{iz} \\ \sum_i m_i v_{iy} v_{ix} & \sum_i m_i v_{iy} v_{iy} & \sum_i m_i v_{iy} v_{iz} \\ \sum_i m_i v_{iz} v_{ix} & \sum_i m_i v_{iz} v_{iy} & \sum_i m_i v_{iz} v_{iz} \end{bmatrix} \tag{2.38}$$

$$\sum_{i=1}^{N} r_i f_i^T = \begin{bmatrix} \sum_i r_{ix} f_{ix} & \sum_i r_{ix} f_{iy} & \sum_i r_{ix} f_{iz} \\ \sum_i r_{iy} f_{ix} & \sum_i r_{iy} f_{iy} & \sum_i r_{iy} f_{iz} \\ \sum_i r_{iz} f_{ix} & \sum_i r_{ix} f_{iy} & \sum_i r_{iz} f_{iz} \end{bmatrix} \tag{2.39}$$

where r_{ix}, v_{ix}, and f_{ix} indicate the x components of the position, velocity, and force vectors of the ith atom, respectively. From the definition of the instantaneous pressure tensor, the instantaneous hydrostatic pressure is calculated as one third the trace of the pressure tensor ($P_{xx} + P_{yy} + P_{zz}$). When periodic boundary conditions are used, atoms in the unit cell interact not only with the other atoms in the unit cell but also with their translated images. Forces on the images in the virial W must be included correctly. If the interaction is in pair, using Newton's third law, W can be written as

$$W = \frac{1}{2}\sum_{i>j} r_{ij} f_{ij} \tag{2.40}$$

instead of

$$W = \frac{1}{2}\sum_i r_i f_i \tag{2.41}$$

Berendsen et al. [10] use the $r_{ij} \ldots f_{ij}$ formalism by evaluating the virial and the kinetic energy tensor based on the centers of mass, which is valid because the internal contribution to the virial is canceled (on the average) by the internal kinetic energy. Because of the way forces are evaluated, rescaling of coordinates is also done on the basis of the centers of mass of the structures.

Molecular dynamics simulations are limited largely by the speed and storage constrains of available computers. Hence, simulations were performed on systems containing 100–1,000 particles, which is about 1,000,000 atoms. Because of the size limitation, simulations are confined to systems of particles that interact with small, relatively short-range forces; that is, intermolecular forces should be small when molecules are separated by a distance equal to half of the smallest overall dimensions of the system. Because of the speed limitation, simulations are confined to studies of relatively short-lived phenomena, roughly those occurring in less than 100–1,000 ps. The characteristic relaxation time for the phenomenon under investigation must be small enough so that one simulation generates several relaxation times.

Molecular dynamics simulations of flexible molecules such as alkanes and protein molecules are more complex and time consuming than simulations of atomic and rigid molecular systems. Not only are the equations of motion more complicated because of internal degrees of freedom, but the internal modes, such as bond vibration and rotation, tend to relax on time scales very different from those of molecular collisions, which dominate external translational and rotational modes. Consequently, lengthy molecular dynamics simulations must be performed to capture many relaxation processes typical of flexible molecules.

The important part of the dynamics is similar to any simulation technique the analysis part. The analysis is divided into two broad parts: (1) To assess the thermodynamic property including temperature and pressure dependent properties and (2) To look into the static structural aspect of the results involving geometry and displacement related functions. Static structure of a material can be measured by the radial distribution function $g(r)$, which describes the spatial organization of molecules about a central molecule. The function $g(r)$ plays a central role in the distribution function theory of dense fluids and provides a signature for identifying the lattice structure for crystalline solids. The radial distribution function $g(r)$ measures how atoms organize themselves around one another — "local structure." Specifically, it is proportional to the probability of finding two atoms separated by a distance $r \pm \Delta r$. It plays a central role in statistical mechanical theories of dense substances, and for atomic substances, it can be extracted from X-ray and neutron diffraction experiments. Because molecular dynamics provides positions of individual atoms as function of time, $g(r)$ can be readily computed from molecular dynamics trajectories. The radial distribution function depends on density and temperature, and therefore, in computer simulation studies, $g(r)$ serves as a helpful indicator of the nature of the phase assumed by the simulated system. The behavior of $g(r)$ in crystalline solids is very different form that for the gases at low density. In a crystal, the positions are more constrained, whereas in case of gas, the atoms move freely and interact primarily through binary collisions, and only weak structures form around one atom. For liquids and amorphous solids, the behavior of $g(r)$ is intermediate between crystal and gas: liquids exhibit short-range order similar to that in crystals but long-range disorder like that of gases. Apart from static properties, one can analyze the dynamic properties like time autocorrelation functions, thermal transport coefficients, and dynamic structures. Time correlation functions measure how the value of some dynamic quantity $A(t)$ may be related to the value of some other quantity $B(t)$. Certain time correlation functions can be used to compute thermal transport coefficients such as viscosity, thermal conductivity, and diffusion coefficients. The computations can be done either by invoking Green-Kubo formulas, in which a correlation function is integrated over time, or by invoking Einstein relations, in which mean square displacement is differentiated with respect to time. In either case, transport coefficients are equilibrium properties of a substance and can be obtained from equilibrium molecular dynamics. Self-diffusion coefficients can be calculated in many ways; one of the simplest is through the Einstein equation:

$$D = \frac{1}{6N} \lim_{t \to \infty} \frac{d}{dt} \left\langle \sum_i^N [r_i(t) - r_j(0)]^2 \right\rangle \tag{2.42}$$

which shows that D is proportional to the slope of the mean square displacement at long times. If one chooses to evaluate D from the mean square displacement, then the atomic positions must be unbounded; otherwise, periodic boundaries will prevent development of the proper linear behavior. One may also be interested in the velocity autocorrelation function — which leads to the self-diffusion coefficient — and one collective function — the stress autocorrelation function — which leads to the shear viscosity. The space–time correlation function $G(r,t)$ characterizes the dissipation of local density fluctuations; as such, it measures correlations in both space and time, correlations that molecular dynamics is able to address easily.

One can conclude that Monte Carlo is easier to implement for systems in which it is difficult to extract the intermolecular force law from partition functions. Systems having this difficulty include those composed of molecules that interact through discontinuous forces. Similar difficulties may arise for which the potential function is a complicated multidimensional surface, such as might be generated by *ab initio* calculations.

For determination of simple equilibrium properties such as the pressure in atomic fluids, Monte Carlo and molecular dynamics are equally effective; both require about the same amount of computer time to attain similar levels of statistical precision. However, molecular dynamics more efficiently evaluates properties like heat capacity, compressibility, and interfacial properties. Besides configurational properties, molecular dynamics provides access to dynamic quantities such as transport coefficients and time correlation functions. Molecular dynamics also offers certain computational advantages because of the deterministic way in which it generates trajectories. The presence of an explicit time variable allows one to estimate the length needed for a run; the duration must be at least several multiples of the relaxation time for the slowest phenomenon to be studied. No such guide is available for Monte Carlo simulation.

2.6 CHALLENGES

These methods have been well applied to address many critical issues and a wide domain of matrices. Applications include modeling reaction and dissociation on classical potential energy surfaces [11–14], studies of equilibrium crystal properties (e.g., density, packing, specific heats) [15–21], dynamic investigations of shock interactions with crystals and defects [22, 23], and simulating detonation in molecular crystals [24–27]. We will discuss the specific examples with these methods in the relevant chapters to cover different issues in nanoporous materials.

The major challenges of any simulation technique are its accuracy in comparison to experiments and how many experimental parameters can be implemented in one set of calculations. The answer varies from one method to another. The force field–based methods have the challenge of parameters, whole range of materials cannot be covered still with one set of parameters with highest accuracy, say of *ab initio* type. There are experimental parameters that may vary with the type of experiment and lack of experimentation for specific interactions. The challenges remain with inorganic materials and with hybrid materials with inorganic and organic components included. One has to focus on the functional form of the energy; once the functional form of the energy expression is acceptable, then one can take the charges from, say, DFT, and use it and look into the other bonding and bending parameters to tune for own application. The advantage of molecular mechanics is immense because it allows looking at things in a wider perspective with a minimum CPU cost and one therefore can devote time to tuning the force field. The coverage of MM encouraged people to focus on code like QM+MM, where MM will be used for the larger domain and QM for the specific interaction.

The Monte Carlo method has its own limitation in terms of its approach along with the force field. There has been some progress in the methodology, but this still remains the cheapest way of looking at primitive interaction among different materials of interest.

The challenge for dynamics remains to explore the liquid phase, people are more interested to see the drying process in a membrane, and it is a challenge to model that behavior as well. The challenge remains to generate a rational and general force field to address the issues. Force field fitting technology is needed where one can get a GUI to tune the force field. One initiative was developed by Accelrys for Drieding and in GULP type of MD engine, which can be a pioneer in this field. At this point, it does not allow changing the functional form in case of Drieding type of force field, which is plausible in GULP. One has to remember that force field editing is not a simple process and must be handled with knowledge on the system, the interactions possible, and, most importantly, the functional form. There is still a long way for the force field to reach to compare with the accuracy level of the experiment.

Molecular modeling has overcome immense challenges so far in its reasonably long journey of more than twenty years. Those challenges were focused on improving its accuracy and hence in the development of new algorithms and their validation for different materials. Once a new methodology is identified, it generates from a relentless effort with the theory to consolidate a mechanism to rationalize the physical chemistry or the chemical physics behind the process. As you have explored in this chapter, the simpler the method, the lower the accuracy level. Also, in most cases, modeling is a qualitative guide of a process, and its advantage lies with handling larger systems. The large in molecular model mindset is still smaller in the experimental domain and hence the philosophy of multiscale simulation is expanding. Accelrys is a software company with a wide variety of algorithms within its fold is taking lead here to offer a multiscale environment with a combination of MC, MD at the MM level with its expansion in mesoscale simulation, which is beyond the scope of this book. In nanoporous materials with a size expansion with the discovery of mesoscale and hybrid materials, meso can be the future of these material designs.

Designing materials with only composition and property known is the hardest challenge. Simulation is possible for systems where the structure of the material is known or can be modeled, but if one only knows the property he wants to make and knows the general composition of the matrix, which will expose the person with hundreds of choices. This needs a process that will help the selection process by intelligent questioning and will result in plausible composition, which one will try and test with the structures and compare the physical properties to validate the composition one has to design the real material of interest.

REFERENCES

1. D. Frenkel, B. Smit, *Understanding Molecular Simulation: From Algorithms to Applications* (2nd ed.), Academic Press, San Diego (2002).
2. N. Metropolis, A. W. Rosenbluth, M. N. Rosenbluth, A. H. Teller, E. Teller, *J. Chem. Phys.*, 21 (1953) 1087.
3. J. L. Siepmann, D. Frenkel, *Mol. Phys.*, 75 (1992) 59.
4. (a) S. Kirkpatrick, C. D. Gelatt, M. P. Vecchi, *Science*, 220 (1983) 671; (b) V. Cerný, *J. Optim. Theor. Appl.*, 45 (1985) 4.
5. M. Born, J. R. Oppenheimer, *Ann. Phys.*, 84 (1927) 457.
6. J. R. Ray, *Comput. Phys. Rep.*, 8 (1988) 109, and references therein.
7. H. C. Andersen, *J. Chem. Phys.*, 72 (1980) 2384.
8. M. P. Allen, D. J. Tildesley, *Comput. Simul. Liquids*, Clarendon Press/Oxford Science Publications, London (1987).
9. D. A. McQuarrie, *Statistical Mechanics*, Harper & Row, New York (1976).
10. H. J. C. Berendsen, J. P. M. Postma, W. F. van Gunsteren, A. DiNola, J. R. Haak, *J. Chem. Phys.*, 81 (1984) 3684.
11. (a) T. D. Sewell, D. L. Thompson, *J. Phys. Chem.*, 95 (1991) 6228; (b) C. C. Chambers, D. L. Thompson, *J. Phys. Chem.*, 99 (1995) 1588; (c) D. V. Shalashilin, D. L. Thompson, *J. Phys. Chem. A*, 101 (1997) 961.
12. E. P. Wallis, D. L. Thompson, *J. Chem. Phys.*, 99 (1993) 2661.
13. B. M. Rice, J. Grosh, D. L. Thompson, *J. Chem. Phys.*, 102 (1995) 8790.
14. Y. Kohno, K. Ueda, A. Imamura, *J. Phys. Chem.*, 100 (1996) 4701.
15. D. C. Sorescu, D. L. Thompson, *J. Phys. Chem. B.*, 101 (1997) 3605.
16. (a) D. C. Sorescu, B. M. Rice, D. L. Thompson, *J. Phys. Chem. B*, 101 (1997) 798;
    (b) D. C. Sorescu, B. M. Rice, D. L. Thompson, *J. Phys. Chem. A*, 102 (1998) 8386;
    (c) D. C. Sorescu, B. M. Rice, D. L. Thompson, *J. Phys. Chem. B*, 102 (1998) 948;
    (d) D. C. Sorescu, B. M. Rice, D. L. Thompson, *J. Phys. Chem. B*, 102 (1998) 6692.

17. (a) T. D. Sewell in *Decomposition, Combustion and Detonation Chemistry of Energetic Materials, Materials Research Society Symposium Proceedings*, Boston, MA, November 1995, T. B. Brill, T. P. Russell, W. C. Tao, R. B. Wardle (eds.), Materials Research Society, Pittsburgh, PA (1996); (b) T. D. Sewell, *Preprints of the Proceedings of the 11th Symposium (International) on Detonation*, Snowmass, CO, Office of the Chief of Naval Research, Arlington, VA (1998).

18. J. P. Ritchie, *Proceedings of the 10th Symposium (International) on Detonation*, Boston, MA, Office of the Chief of Naval Research, Arlington, VA (1989); J. P. Ritchie, *Structure and Properties of Energetic Materials, Materials Research Society Symposium Proceedings*, Boston, MA, November 1992, D. H. Liebenberg, R. W. Armstrong, J. J. Gilman (eds.), Materials Research Society, Pittsburgh, PA (1993).

19. (a) A. B. Kunz, *J. Phys. Condens. Matter*, 6 (1994) L233; (b) A. B. Kunz, *Phys. Rev. B*, 53 (1996) 9733; (c) A. B. Kunz, *Decomposition, Combustion and Detonation Chemistry of Energetic Materials, Materials Research Society Symposium Proceedings*, Boston, MA, November 1995, T. B. Brill, T. P. Russell, W. C. Tao, R. B. Wardle (eds.), Materials Research Society, Pittsburgh, PA (1996).

20. A. V. Dzyabchenko, T. S. Pivina, E. A. Arnautova, *J. Mol. Struct.*, 378 (1996) 67.

21. G. Filippini, A. Gavezzotti, *Chem. Phys. Lett.*, 231 (1994) 86.

22. (a) L. Phillips, R. S. Sinkovits, E. S. Oran, J. P. Boris, *J. Phys. Condens. Matter*, 5 (1993) 6357; (b) L. Phillips, *J. Phys. Condens. Matter*, 7 (1995) 7813.

23. (a) D. H. Tsai, *Structure and Properties of Energetic Materials, Materials Research Society Symposium Proceedings*, Boston, MA, November 1992, D. H. Liebenberg, R. W. Armstrong, J. J. Gilman (eds.), Materials Research Society, Pittsburgh, PA (1993); (b) D. H. Tsai, *Decomposition, Combustion and Detonation Chemistry of Energetic Materials, Materials Research Society Symposium Proceedings*, Boston, MA, November 1995, T. B. Brill, T. P. Russell, W. C. Tao, R. B. Wardle (eds.), Materials Research Society, Pittsburgh, PA (1996); (c) D. H. Tsai, *J. Chem. Phys.*, 95 (1991) 7497; (d) D. H. Tsai, S. F. Trevino, *J. Chem. Phys.*, 81 (1991) 5636; (e) D. H. Tsai, R. W. Armstrong, *J. Phys. Chem.*, 98 (1994) L10997.

24. P. J. Haskins, M. D. Cook, *Preprints of Proceedings of the 11th Symposium (International) on Detonation*, Snowmass, CO, Office of the Chief of Naval Research, Arlington, VA (1998), and references therein.

25. M. L. Elert, D. H. Robertson, C. T. White, *Decomposition, Combustion and Detonation Chemistry of Energetic Materials, Materials Research Society Symposium Proceedings*, Boston, MA, November 1995, T. B. Brill, T. P. Russell, W. C. Tao, R. B. Wardle (eds.), Materials Research Society, Pittsburgh, PA (1996).

26. J. C. Barrett, D. H. Robertson, D. W. Brenner, C. T. White, *Decomposition, Combustion and Detonation Chemistry of Energetic Materials, Materials Research Society Symposium Proceedings*, Boston, MA, November 1995, T. B. Brill, T. P. Russell, W. C. Tao, R. B. Wardle (eds.), Materials Research Society, Pittsburgh, PA (1996).

27. L. Soulard, *Decomposition, Combustion and Detonation Chemistry of Energetic Materials, Materials Research Society Symposium Proceedings*, Boston, MA, November 1995, T. B. Brill, T. P. Russell, W. C. Tao, R. B. Wardle (eds.), Materials Research Society, Pittsburgh, PA (1996).

3 Density Functional Theory

The microscopic description of the physical and chemical properties of matter is a complex problem. In general, we deal with a collection of interacting atoms, which may also be affected by some external field. This ensemble of particles may be in the gas phase (molecules and clusters) or in a condensed phase (solids, surfaces, wires); they may be solids, liquids, or amorphous, homogeneous, or heterogeneous (molecules in solution, interfaces, adsorbates on surfaces). However, in all cases we can unambiguously describe the system by a number of nuclei and electrons interacting through Coulombic (electrostatic) forces.

The key problem in the structure of matter is to solve the Schrödinger equation for a system of N interacting electrons in the external Coulombic field created by a collection of atomic nuclei (and maybe some other external field). It is a very difficult problem in the many-body theory and, in fact, the exact solution is known only in the case of the uniform electron gas, for atoms with a small number of electrons, and for a few small molecules. These exact solutions are always numerical. At the analytic level, one always has to resort to approximations. However, the effort of devising schemes to solve this problem is really worthwhile because the knowledge of the electronic ground state of a system gives access to many of its properties; for example, relative stability of different structures/isomers, equilibrium structural information, mechanical stability and elastic properties, pressure–temperature (P–T) phase diagrams, dielectric properties, dynamical (molecular or lattice) properties such as vibrational frequencies and spectral functions, (nonelectronic) transport properties such as diffusivity, viscosity, ionic conductivity, and so forth. Excited electronic states (or the explicit time dependence) also give access to another wealth of measurable phenomena such as electronic transport and optical properties.

Density functional is a remarkable theory that allows one to replace the complicated N-electron wave function Ψ ($x1$, $x2$, …, xn) and the associated Schrödinger equation by the much simpler electron density $\rho(r)$ and its associated calculational scheme. The principles of density functional theory are conveniently expounded by making reference to conventional wave function theory.

In this chapter the focus is to introduce the density functional theory (DFT) technique to the readers. The methodology has been introduced with a context of the basic wave mechanics, followed by an explanation of density matrix. The chemical potential will then be explained to address the issue with chemical bonds and the issue related to address the modeling of chemical bond. The challenges and the future of the DFT technology are described at the end of this chapter. The book aims to guide experimentalists to design material in combination with molecular modeling and DFT, which is one of the key technologies in the process.

3.1   BASIC WAVE MECHANICS

We state the basic problems of quantum physics in such a way as to emphasize the fact that what they all have in common is that they are all problems in the theory of the electromagnetic interaction. The issues are the following:

1. The structure of the source charge distributions and fields of static elementary charges.
2. The field of a moving elementary charge, especially an accelerated one.
3. The structure of the bound states of systems of elementary charges.

The first problem was generally treated under the heading of "the theory of the electron," because most of its definitive efforts preceded the discovery of other charged particles besides electrons and protons; in particular, the discovery of antiparticles was still far in the future. The main contributors to that theory were Abraham, Lorentz, and Rohrlich [1–3]. The second and third problems are both concerned with the radiative field modes of elementary charges; in one case, the scattering states, and in the other, the bound states. Although it was agreed that the motion of an accelerated charge should be accompanied by an emitted radiation field (i.e., a photon) — for instance, photons are produced by the oscillation of electrons in antennas and in the form of the Bremsstrahlung* emitted by decelerating charges — nevertheless, the precise manner by which this happened was incompletely understood. Dirac [4] proposed an extension to the Lorentz force law that included the radiation reaction associated with the acceleration of a charge, but the fact that it also included the third derivative of position, and thus raised the order of the dynamical equation above the customary two, led to various unacceptable pathological solutions. For completeness, we briefly summarize wave mechanics and its statistical interpretation. The turning point in quantum theory seems to have been de Broglie's suggestion that, just as Einstein had shown that the photon had particle-like properties, conversely, massive particles were also associated with a wavelike nature. Any problem in the electronic structure of matter is covered by Schrödinger's equation including the time. In most cases, however, one is concerned with atoms and molecules without time-dependent interactions, so at the beginning let us consider a time-independent Schrödinger equation. For an isolated N-electron atomic or molecular system in the Born-Oppenheimer nonrelativistic approximation, this is given by

$$\hat{H}\psi = E\psi$$

where E is the electronic energy, $\Psi = \Psi(x1, x2, x3, \ldots, xn)$ is the wave function, and $\hat{H}$ is the Hamiltonian operator. When a system is in the state Ψ, which may or may

---

* Bremsstrahlung denotes electromagnetic radiation produced by a sudden slowing down or deflection of charged particles, especially electrons, passing through matter in the vicinity of the strong electric fields of atomic nuclei.

not satisfy the simplest Schrödinger wave equation stated above, the average of many measurements of the energy is given by the formula

$$E[\psi] = \frac{\langle \psi | \hat{H} | \psi \rangle}{\psi | \psi}$$

where

$$\langle \psi | \hat{H} | \psi \rangle = \int \psi^* \hat{H} \psi dx$$

Because each particular measurement of the energy gives one of the eigenvalues of $\hat{H}$, we then have

$$E[\psi] \geq E_0$$

The energy computed from a guessed Ψ is an upper bound to the true ground state energy E_0. Full minimization of the functional $E[\psi]$ with respect to all allowed N-electron wave functions would give the true ground state Ψ and energy $E[\psi_0] = E_0$; that is,

$$E_0 = \min_{\psi} E[\psi]$$

In an electronic system, the number of electrons per unit volume in a given state is the electron density for that state. This quantity is designated by $\rho(r)$. It can be represented in terms of the wave function Ψ as follows:

$$\rho(r_1) = N \int \ldots |\psi(x_1, x_2, \ldots x_N)|^2 dx_1 dx_2 \ldots dx_N$$

This is a nonnegative simple function of three variables x, y, and z integrating to the total number of electrons,

$$\int \rho(r) dr = N$$

At any atomic nucleus in an atom, molecule, or solid, the electron density has a finite value.

A largely heuristic process by Schrödinger in the form obtained the equation for the nonrelativistic time evolution of these waves:

$$\frac{h}{i} \frac{\partial \psi}{\partial t} = -\frac{h^2}{2m} \Delta \psi + U$$

in which Ψ is the complex-valued wave function that describes a particle of mass m, and U describes the force potential that acts upon it. Klein and Gordon [5] extended this to a relativistic form, which described particles with integer spin:

$$\Delta\Psi + k_c^2\psi = 0$$

in which $k_c = m_0 c/h$ is the Compton wave number for a particle of rest mass m_0 and ultimately was given a relativistic form for half-odd-integer particles by Dirac [4]:

$$\gamma^\mu \partial^\mu \psi + k_c^2 \psi = 0$$

in which the γ^μ are the Dirac matrices that define the generators of a representation of the Clifford algebra of Minkowski space in the algebra of 4×4 complex matrices, and ψ is a wave function that takes its values in $\mathbb{C}^4$, this time. The main problem with Schrödinger's equation was finding some fundamental physical basis for explaining its empirical success in describing quantum phenomena. Bohr, Born, and Heisenberg [6] proposed the statistical interpretation, which gives the complex-valued quantum wave function Ψ a sort of "prephysical" character, like the potential 1-form A of the electromagnetic field F, and resolved this by saying that all of the physical meaning is in the modulus-squared $|\Psi|^2$, which represents the probability density function for the position of the particle. The fact that Ψ is complex is then attributed to the association of a phase factor with the wave/particle, as well.

The early hope of de Broglie [7] and Schrödinger [8] was that they could construct an exact analogy between the relationship of wave mechanics to particle mechanics and the relationship between wave optics and geometrical (ray) optics. The common element was the Hamilton-Jacobi equation [9]:

$$\begin{cases} \dfrac{\partial S}{\partial t} + H(t, x^i, p_i) = 0 \\ pi = \dfrac{\partial S}{\partial x^i} \end{cases}$$

From the standpoint of wave motion, the level surfaces of the Jacobi principal function $S = S(t, xi)$ describe isophase hypersurfaces, and one can recover the particle trajectories from the characteristic equations of this first-order partial differential equation, namely, Hamilton's equations:

$$\begin{cases} \dfrac{\partial x^i}{dt} = \dfrac{\partial H}{\partial p_i} \\ \dfrac{\partial p_i}{dt} = \dfrac{\partial H}{\partial x^i} + \dfrac{\partial p_i}{\partial x_i} \dfrac{\partial H}{\partial p_j} \end{cases}$$

By expressing these equations in the latter form, we are, of course, anticipating the possibility that the matter whose momentum is described by p_i is spatially extended, rather than point like.

The similarity between the first equation of nonrelativistic time evolution and Schrödinger's wave equation is undeniable, except for the missing factor of h/i, which, of course, accounts for the difference between classical and quantum mechanics, as well. Ultimately, this analogy proved to be useful mostly in the geometrical optical approximation, in which one makes the limiting assumption that the wavelength of the wave is vanishingly small; i.e., $h \to 0$, in order that the wave phenomena associated with S, such as interference and diffraction, do not contradict the particle motion described earlier. An interesting optical aspect of the Klein-Gordon equation that defined one of the early attempts at giving a relativistic form to Schrödinger's equation is something that Klein himself observed [5].

The following equation can be obtained from the five-dimensional linear wave equation by separating the fifth variable in the same way as that of Helmholtz's equation:

$$\Delta\Psi + k2\Psi = 0$$

which is fundamental to the time-invariant formulation of wave optics, can be obtained from the four-dimensional wave equation. This carries with the problem of physically interpreting the fifth dimension that one has introduced. Because $kC = h/m_0c$ is the Compton wavelength associated with a particle of *total* rest mass m_0, it can be suggested that mass appears like a separation constant; i.e., an eigenvalue of something. One can make a good case for the notion that the fifth dimension that one has introduced is the proper time parameter, in such a way that the fifth component of the velocity becomes c. Another interesting aspect of this wave of representing matter waves is that although massive matter waves can propagate with any speed between 0 and c, noninclusive, nevertheless, the five-dimensional wave that represents it must always propagate with the same characteristic velocity, like a massless particle. Further work on the physical details of five-optics was also carried out by Rumer [10], as well. Of course, Klein, along with Kaluza, Jordan, and Thirry, examined the possibility that one could unify the theories of gravitation and electromagnetism by means of the geometry of five-dimensional Lorentzian manifolds [11, 12].

3.2 DENSITY MATRIX

In quantum mechanics, a density matrix is a self-adjoin (or Hermitian) positive, semidefinite matrix (possibly infinite dimensional) of trace one that describes the statistical state of a quantum system. The formalism was introduced by John von Neumann (and according to other sources, independently by Lev Landau and Felix Bloch) in 1927. The details have been described elsewhere [13]. It is the quantum mechanical analogue to a phase-space probability measure (probability distribution of position and momentum) in classical statistical mechanics. The need for a statistical description via density matrices arises when one considers either an ensemble of systems or one system when its preparation history is uncertain and one does not know with 100% certainty which pure quantum state the system is in.

Situations in which a density matrix is used include the following: a quantum system in thermal equilibrium (at finite temperatures); nonequilibrium time-evolution that starts out of a mixed equilibrium state; entanglement between two subsystems, where each individual system must be described, via the partial trace operation, by a density matrix even though the complete system may be in a pure state; and in analysis of quantum decoherence, which is the mechanism by which quantum systems interact with their environments to exhibit probabilistically additive behavior.

3.2.1   The Density Matrix and Density Operator

In general, the many-body wave function $\psi(q1,\cdots,q3N,t)$ is far too large to calculate for a macroscopic system. If we wish to represent it on a grid with just ten points along each coordinate direction, then for $N = 10^{23}$, we would need $10^{10^{28}}$ total points, which is clearly enormous.

We wish, therefore, to use the concept of ensembles in order to express expectation values of observables $\langle A \rangle$ without requiring direct computation of the wave function. Let us introduce an ensemble of systems, with a total of Z members, and each having a state vector $|\psi^{(\alpha)}\rangle$, $\alpha = 1,\dots Z$. Furthermore, introduce an orthonormal set of vectors $|\phi_k\rangle$ $(\langle \phi k| \phi j\rangle = \delta_{ij})$ and expand the state vector for each member of the ensemble in this orthonormal set to get the following:

$$|\psi^{(\alpha)}\rangle = \sum_k C_k (\alpha | \phi_k\rangle$$

The expectation value of an observable, averaged over the ensemble of systems, is given by the average of the expectation value of the observable computed with respect to each member of the ensemble:

$$\langle A \rangle = \frac{1}{Z}\sum_{\alpha=1}^{z} \langle \psi^{(\alpha)} | A | \psi^{(\alpha)}\rangle$$

Substituting in the expansion for $|\psi^{(\alpha)}\rangle$, we obtain

$$\langle A \rangle = \frac{1}{Z}\sum_{k,l} C_k^{(\alpha)*} C_l^{(\alpha)} \langle \phi_k |A|\phi_l\rangle = \sum_{k,l}\left(\frac{1}{Z}\sum_{\alpha=1}^{z} C_l^{(\alpha)} C_k^{(\alpha)*}\right)\langle \phi_k |A|\phi_l\rangle$$

Let us define a matrix

$$\rho lk = \sum_{\alpha=1}^{z} C_l^{(\alpha)} C_k^{(\alpha)*}$$

and a similar matrix

$$\bar{\rho}lk = \sum_{\alpha=1}^{z} C_l^{(\alpha)} C_k^{(\alpha)*}$$

Thus, ρlk is a sum over the ensemble members of a product of expansion coefficients, whereas $\bar{\rho}lk$ is an average over the ensemble of this product. Also, let $A_{kl} = \langle \phi_k | A | \phi_l \rangle$. Then, the expectation value can be written as follows:

$$\langle A \rangle = \frac{1}{Z} \sum_{k,l} \rho_{lk} A_{kl} = \frac{1}{Z} \sum (\rho A)_{kk} = \frac{1}{Z} Tr(\rho A) = Tr(\bar{\rho} A)$$

where ρ and A represent the matrices with elements ρ_{lk} and A_{kl} in the basis of vectors $\{|\phi_k\rangle\}$. The matrix ρ_{lk} is known as the *density matrix*. There is an abstract operator corresponding to this matrix that is basis independent. It can be seen that the operator

$$\rho = \sum_{\alpha=1}^{z} |\psi^{(\alpha)}\rangle\langle\psi^{(\alpha)}|$$

and, similarly,

$$\bar{\rho} = \sum_{\alpha=1}^{z} |\psi^{(\alpha)}\rangle\langle\psi^{(\alpha)}|$$

have matrix elements ρ_{lk} when evaluated in the basis set of vectors $\{|\phi_k\rangle\}$.

$$\langle \phi_l | \rho | \phi_k \rangle = \sum_{\alpha=1}^{z} \langle \phi_l | \psi^{(\alpha)}\rangle\langle\psi^{(\alpha)} | \phi_k \rangle = \sum_{\alpha=1}^{z} C_l^{(\alpha)} C_k^{(\alpha)*} = \rho lk$$

Note that ρ is a Hermitian operator, and

$$\rho^f = \rho$$

so that its eigenvectors form a complete orthonormal set of vectors that span the Hilbert space. If w_k and $|w_k\rangle$ represent the eigenvalues and eigenvectors of the operator $\bar{\rho}$, respectively, then several important properties they must satisfy can be deduced.

Firstly, let A be the identity operator I. Then, because $\langle I \rangle = 1$, it follows that

$$1 = \frac{1}{Z} Tr(\rho) = Tr(\bar{\rho}) = \sum_k w_k$$

Thus, the eigenvalues of $\bar{\rho}$ must sum to 1. Next, let A be a projector onto an eigenstate of $\bar{\rho}$,

$$A = |w_k\rangle_k| \equiv P_k.$$

Then $\langle P_k \rangle = Tr(\bar{\rho}|w_k\rangle_k)|$

But, because $\bar{\rho}$ can be expressed as

$$\bar{\rho} = \sum_k w_k |w_k\rangle\langle w_k|$$

and the trace, being basis set independent, can be therefore be evaluated in the basis of eigenvectors of $\bar{\rho}$, the expectation value becomes

$$\langle P_k \rangle = \sum_j \left\langle w_j \middle| \sum_i w_i |w_i\rangle \middle\rangle \langle w_i|w_k\rangle\langle w_k|w_j\rangle = \sum_{i,j} w_i \delta_{ij}\delta_{ik}\delta_{kj} = w_k$$

However,

$$\langle P_k \rangle = \frac{1}{Z}\sum_{\alpha=1}^{z} \langle \psi^{(\alpha)}|w_k\rangle\langle w_k|\psi^{(\alpha)}\rangle = \frac{1}{Z}\sum_{\alpha=1}^{z} |\langle \psi^{(\alpha)}|w_k\rangle|^2 \geq 0$$

thus, $w_k \geq 0$. Combining these two results, we see that, because $\sum_k w_k = 1$ and $w_k \geq 0$, $0 \leq w_k \leq 1$, so that w_k satisfies the properties of probabilities.

With this in mind, we can develop a physical meaning for the density matrix. Let us now consider the expectation value of a projector $|a_i\rangle_i \equiv P_{a_1}$ onto one of the eigenstates of the operator A. The expectation value of this operator is given by

$$|a_i\rangle_i \equiv P_{a_1}|$$

But $|\langle a_i|\psi^{(\alpha)}\rangle|^2 \equiv P_{a1}^{(\alpha)}$ is just probability that a measurement of the operator A in the αth member of the ensemble will yield the result a_i. Thus,

$$\langle P_{a1} \rangle = \frac{1}{Z}\sum_{\alpha=1}^{P} P_{ai}^{(\alpha)}$$

or the expectation P_{a1} value is just the ensemble averaged probability of obtaining the value a_i within each member of the ensemble. However, note that the expectation value of P_{a1} can also be written as

$$\langle P_{a1} \rangle = Tr(\bar{\rho}P_{a1}) = Tr\left(\sum_k w_k |w_k\rangle k |a_i\rangle\langle a_i|\right) = \sum_{k,l} \langle w_l | w_k | w_k \rangle_k |a_i\rangle\langle a_i | w_i\rangle$$

$$= \sum_{k,l} w_k \delta_{kl} \langle w_k a_i\rangle\langle a_i w_i\rangle$$

$$= \sum_k w_k |\langle a_i | w_k \rangle|^2$$

Equating the two expressions gives

$$\frac{1}{Z}\sum_{\alpha=1}^{z}\langle P_{ai}^{(\alpha)}\rangle = \sum_k w_k |\langle a_i | w_k \rangle|^2$$

The interpretation of this equation is that the ensemble averaged probability of obtaining the value a_i if A is measured is equal to the probability of obtaining the value a_i in a measurement of A if the state of the system under consideration were the state $|w_k\rangle$, weighted by the average probability w_k that the system in the ensemble is in that state. Therefore, the density operator ρ (or $\bar{\rho}$) plays the same role in quantum systems that the phase space distribution functions $f(T)$ play in classical systems.

3.3 CHEMICAL POTENTIAL

The chemical potential of a thermodynamic system is the amount by which the energy of the system would change if an additional particle is introduced, keeping the entropy and volume fixed. If a system contains more than one species of particle, there is a separate chemical potential associated with each species, defined as the change in energy when the number of particles of that species is increased by one. The chemical potential is a fundamental parameter in thermodynamics and it is a conjugate to the particle number.

Consider a thermodynamic system containing n constituent species. Its total internal energy U is postulated to be a function of the entropy S, the volume V, and the number of particles of each species $N_1, \ldots, N_n$.

$$U = U(S, V, N_1, \ldots, N_n)$$

By referring to U as the *internal energy*, it is emphasized that the energy contributions resulting from the interactions between the system and external objects are excluded. For example, the gravitational potential energy of the system with the Earth is not included in U.

The chemical potential of ith species, μ_i is defined as the partial derivative

$$\mu_i = \left(\frac{\partial U}{\partial N_i} \right)_{S,V,Nj \neq i}$$

where the subscripts simply emphasize that the entropy, volume, and the other particle numbers are to be kept constant.

The electronic chemical potential is the functional derivative of the density functional with respect to the electron density, where μ is the chemical potential varying for the density functional for a variation of density for a specific range within a limit of the density variance between ρ and ρ reference as shown below:

$$\mu(r) = \left[\frac{\partial E[\rho]}{\partial \rho(r)} \right]_{\rho = \rho_{ref}}$$

Formally, a functional derivative yields many functions. Just as a derivative yields a function, and a particular function as a number is being evaluated with respect to a reference point, a particular function when evaluated about a reference electron density is important.

The density functional is written as:

$$E[\rho] = \int \rho(r)\, v(r) d^3r + F[\rho]$$

where $v(r)$ is the external potential — e.g., the electrostatic potential of the nuclei and applied fields — and F is the universal functional, which describes the electron–electron interactions; e.g., electron Coulomb repulsion, kinetic energy, and the nonclassical effects of exchange and correlation. With this general definition of the density functional, the chemical potential is written as

$$\mu(r) = v(r) + \left[\frac{\partial F[\rho]}{\partial \rho(r)} \right]_{\rho = \rho_{ref}}$$

Thus, the electronic chemical potential is the effective electrostatic potential experienced by the electron density.

The ground-state electron density is determined by a constrained variational optimization of the electronic energy. The Lagrange multiplier enforcing the density normalization constraint is also called the *chemical potential*; i.e.,

$$\partial \left\{ E[\rho] - \mu \left(\int \rho(r) d^3r - N \right) \right\} = 0$$

where N is the number of electrons in the system and μ is the Lagrange multiplier enforcing the constraint. When this variational statement is satisfied, the terms within the second brackets obey the property:

$$\left[\frac{\partial E[\rho]}{\partial \rho(r)}\right]_{\rho=\rho 0} - \mu \left[\frac{\partial N[\rho]}{\partial \rho(r)}\right]_{\rho=\rho 0} = 0$$

where the reference density is the density that minimizes the energy. This expression simplifies to

$$\left[\frac{\partial E[\rho]}{\partial \rho(r)}\right]_{\rho=\rho 0} = \mu$$

The Lagrange multiplier enforcing the constraint, by constructing, a constant; however, the functional derivative is, formally, a function. Therefore, when the density minimizes the electronic energy, the chemical potential has the same value at every point in space. The gradient of the chemical potential is an effective electric field. An electric field describes the force per unit charge as a function of space. Therefore, when the density is the ground-state density, the electron density is stationary, because the gradient of the chemical potential (which is invariant with respect to position) is zero everywhere; i.e., all forces are balanced. As the density undergoes a change from a non-ground-state density to the ground-state density, it is said to undergo a process of chemical potential equalization.

The chemical potential of an atom is sometimes said to be the negative of the atom's electronegativity. Similarly, the process of chemical potential equalization is sometimes referred to as the process of *electronegativity equalization*. This connection comes from the Mulliken definition of electronegativity. By inserting the energetic definitions of the ionization potential and electron affinity into the Mulliken electronegativity, it is possible to show that the Mulliken chemical potential is a finite difference approximation of the electronic energy with respect to the number of electrons; i.e.,

$$\mu_{\text{Mulliken}} = -\chi_{\text{Mulliken}} = \frac{IP - EA}{2} = \left[\frac{\partial E[N]}{\partial N}\right]_{N=N0}$$

where IP and EA are the ionization potential and electron affinity of the atom, respectively.

According to the Hohenberg and Kohn (HK) theorems, the ground-state energy functional [14] of an N-electron system with density $\rho(r)$ in an external potential υ is given by

$$E[\rho(r)] = F[\rho(r)] + \int \upsilon(r)\rho(r)dr$$

where $F[\rho(r)]$ is called the universal HK-functional containing the contribution of the kinetic energy (T) and the electron–electron interaction (V_{ee}) of the system. The usual minimization of the energy functional of (1) using the method of Lagrange multipliers subject to the constraint

$$N = \int \rho(r)dr$$

leads to the Euler-Lagrange equation,

$$\mu = \left[\frac{\delta E}{\delta \rho}\right] \upsilon = \upsilon(r) + \frac{\delta F[\rho(r)]}{\delta \rho(r)}$$

where the constant μ has been identified, in the grand canonical ensemble at 0 K, as the electronic chemical potential [14]. This quantity arising within the DFT measures the escaping tendency of an electronic cloud in the ground-state system. Being a constant over all space for the ground state of an atom or molecule μ is recognized as a global reactivity index. It has also been shown that the chemical potential is the slope of the curve E versus N at a fixed external potential.

$$\mu = \left(\frac{\partial E}{\partial N}\right) \upsilon$$

Within the finite difference approximation this slope can be written in terms of the ionization potential I and the electron affinity A. In this way, the DFT chemical potential can be associated with the negative of the Mulliken electronegativity (χ), as [15]:

$$\mu \approx -\frac{(I + A)}{2} = -\chi$$

The chemical potential can be considered as a function of N and υ and describes a change in the system from

$$[N, \upsilon] \rightarrow [N + \delta N, \upsilon + \delta \upsilon]$$

according to the following

$$\delta \mu = \eta \delta N + \int \partial \upsilon(r) f(r) dr$$

This consideration leads to the definition of another two important indices, which have been used to study chemical reactivity: the global hardness, η, and the local reactivity index, f, called the Fukui function [16, 17].

3.4 MODELING OF CHEMICAL BONDS

Chemical bonds form when electrons can be simultaneously close to two or more nuclei, but beyond this, there is no simple, easily understood theory that would not only explain why atoms bind together to form molecules but would also predict the three-dimensional structures of the resulting compounds as well as the energies and other properties of the bonds themselves. Unfortunately, no one theory exists that accomplishes these goals in a satisfactory way for all of the many categories of compounds that are known. Moreover, it seems likely that if such a theory does ever come into being, it will be far from simple.

When we are facing a scientific problem of this complexity, experience has shown that it is often more useful to concentrate instead on developing models. A scientific model is something like a theory, in that it should be able to explain observed phenomena and to make useful predictions. Whereas a theory can be discredited by a single contradictory case, a model can be useful even if it does not encompass all instances of the phenomena it attempts to explain. We do not even require that a model be a credible representation of reality; all we ask is that be able to explain the behavior of those cases to which it is applicable in terms that are consistent with the model itself. An example of a model that you may already know about is the kinetic molecular theory of gases. Despite its name, this is really a model (at least at the level that beginners use it) because it does not even try to explain the observed behavior of real gases. Nevertheless, it serves as a tool for developing our understanding of gases and as a starting point for more elaborate treatments. Given the extraordinary variety of ways in which atoms combine into aggregates, it should come as no surprise that a number of useful bonding models have been developed. Most of them apply only to certain classes of compounds or attempt to explain only a restricted range of phenomena.

The bonding model can even be explained using a classical model. By *classical*, we mean models that do not take into account the quantum behavior of small particles, notably the electron. These models generally assume that electrons and ions behave as point charges, which attract and repel according to the laws of electrostatics. Because the quantum theory was developed in the 1920s, it completely ignores what has been learned about the nature of the electron, although these classical models have not only proven extremely useful, but the major ones also serve as the basis for chemists' general classifications of compounds into covalent and ionic categories. These models of bonding take into account the fact that a particle as light as an electron cannot really be said to be in any single location. The best we can do is to define a region of space in which the probability of finding the electron has some arbitrary value and will always be less than unity. The shape of this volume of space is called an *orbital* and is defined by a mathematical function that relates the probability to the (x, y, z) coordinates of the molecule. Like other models of bonding, the quantum models attempt to show how more electrons can be simultaneously close to more nuclei. Instead of doing so through purely geometrical arguments, they attempt this by predicting the nature of the orbital that the valence electrons occupy in joined atoms.

3.4.1 THE HYBRID ORBITAL MODEL

The hybrid orbital model was developed by Linus Pauling in 1931 [18] and was the first quantum-based model of bonding. It is based on the premise that if the atomic *s*, *p*, and *d* orbitals occupied by the valence electrons of adjacent atoms are combined in a suitable way, the ***hybrid orbitals*** that result will have the character and directional properties that are consistent with the bonding pattern in the molecule. The rules for bringing about these combinations turn out to be remarkably simple, so once they were worked out it became possible to use this model to predict the bonding behavior in a wide variety of molecules. The hybrid orbital model is most usefully applied to the *p*-block elements the first two rows of the periodic table and is especially important in organic chemistry.

3.4.2 THE MOLECULAR ORBITAL MODEL

This model takes a more fundamental approach by considering molecule as a collection of valence electrons and positive cores. Just as the nature of atomic orbitals derives from the spherical symmetry of the atom, so will the properties of these new ***molecular orbitals*** be controlled by the interaction of the valence electrons with the multiple positive centers of these atomic cores. These new orbitals, unlike those of the hybrid model, are ***delocalized***; that is, they do not belong to any one atom but extend over the entire region of space that encompasses the bonded atoms. The available (valence) electrons then fill these orbitals from the lowest to the highest, very much as in the ***Aufbau principle*** for working out atomic electron configurations. For small molecules, there are simple rules that govern the way that atomic orbitals transform themselves into molecular orbitals as the separate atoms are brought together. The real power of the molecular orbital theory, however, comes from its mathematical formation, which lends itself to detailed predictions of bond energies and other properties.

3.4.3 THE ELECTRON TUNNELING MODEL

A common theme of uniting all models discussed is that the bonding depends on the fall in potential energy that occurs when opposite charges are brought together. In the case of covalent bonds, the shared electron pair acts as a kind of "electron glue" between the joined nuclei. In 1962, however, it was shown by Loudin et al. [19] that this assumption is not strictly correct, and that instead of being concentrated in the space between the nuclei, the electron orbitals become even more concentrated around the bonded nuclei. At the same time, however, they are free to move between the two nuclei by a process known as *tunneling*. This refers to a well-known quantum mechanical effect that allows electrons (or other particles small enough to exhibit wavelike properties) to pass ("tunnel") through a barrier separating two closely adjacent regions of low potential energy. As a result, the effective volume of space available to the electron is increased, and according to the ***uncertainty principle*** it will reduce the kinetic energy of the electron.

According to this model, the bonding electrons act as a kind of fluid that concentrates within different regions of the nucleus, lowering the potential energy, and at the same time are able to freely flow between them, reducing the kinetic energy. Despite its conceptual simplicity and full implementation of the laws of quantum mechanics, this model is not widely known and is rarely taught.

The density functional theory begins with a theorem by Hohenberg and Kohn [20], later generalized by Levy [21], which states that all ground-state properties are functionals of the charge density ρ and, specifically, the total energy, t, may be written as:

$$E_t[\rho] = T[\rho]) + U[\rho] + E_{xc}[\rho]$$

where $T[\rho]$ is the kinetic energy of a system of noninteracting particles of density ρ. $U[\rho]$ is the classical electrostatic energy due to Coulombic interactions. $E_{xc}[\rho]$ includes all many-body contributions to the total energy, in particular, the exchange and correlation energies.

The total energy can now be written as:

$$E_t[\rho] = \sum_i \left\langle \phi_i + \left| \frac{-\nabla 2}{2} \right| \phi_i \right\rangle + \left\langle \rho(r_1) \left[\varepsilon_{xc} \left[\rho(r_1) + \frac{V_e(r_1)}{2} - V_N \right] \right] \right\rangle + V_{NN}$$

To determine the actual energy, variations in E_t must be optimized with respect to variations in ρ, subject to the orthonormality constraints [19]:

$$\frac{\partial Et}{\partial \rho} - \sum_i \sum_i \varepsilon ij \langle \phi i | \phi j \rangle = 0$$

This process leads to a set of coupled equations first proposed by Kohn and Sham [19]:

$$\left\{ \frac{-\nabla^2}{2} - V_n + V_e + \mu_{xc}[\rho] \right\} \phi_i = \varepsilon_i \phi_i$$

The term μ_{xc} is the exchange-correlation potential, which results from differentiating E_{xc}. The potential μ_{xc} for the local spin-density approximation is

$$\mu_{xc} = \frac{\partial}{\partial \rho}(\rho \varepsilon_{xc})$$

Use of the eigenvalues of an earlier equation leads to a reformulation of the energy expression:

$$E_t = \sum_i \varepsilon_i + \left\langle \rho(r_1) \left[\varepsilon_{xc}[\rho] - \mu_{xc}[\rho] - \frac{V_e(r_1)}{2} \right] \right\rangle + V_{NN}$$

To describe this in a little detail, in analogy with the Hohenberg-Kohn definition of the universal functional $F_{HK}[\rho]$, Kohn and Sham [22] invoked a corresponding noninteracting reference system, with the Hamiltonian

$$\hat{H}_s = \sum_i^N \left(-\frac{1}{2} \nabla_i^2 \right) + \sum_i^N v_s(r_i)$$

in which there are no electron–electron repulsion terms, and for the ground-state electron density is exactly ρ. For this system there will be an exact determinantal ground-state wave function and therefore the kinetic energy is $T_s[\rho]$ and is given by

$$T_s[\rho] = \left\langle \psi_s \middle| \sum_i^N \left(-\frac{1}{2} \nabla_i^2 \right) \middle| \psi_s \right\rangle$$

The foregoing definition of $T_s[\rho]$, leaves an undesirable restriction on the density — it needs to be noninteracting v-representable; that is, there must exist a noninteracting ground state with the given $\rho(r)$. To produce this separation, Kohn and Sham intelligently brought the definition of exchange correlation energy containing the difference between T and T_s and the nonclassical part of $V_{ee}[\rho]$.

Thus, interpreting Kohn-Sham equations, one can see that through the introduction of N orbitals, the equations can handle the kinetic energy exactly. The price for this gain in accuracy is that there are now N equations to solve as opposed to only one equation for the total density derived from direct approximation of $T_s[\rho]$ of the Thomas Fermi type.

The ability to evaluate the derivative of the total energy with respect to geometric changes is critical for the study of chemical systems. Without the first derivatives, a laborious point-by-point procedure is required, which is taxing to both computer and human resources. The availability of analytic energy derivatives for methods like Hartree-Fock [23], configuration interaction [24], and many-body perturbation theory [25] are few of the theories that have made these remarkably successful methods for predicting chemical structures.

In practice, it is convenient to expand the MOs in terms of atomic orbitals (AOs):

$$\phi_i = \sum_\mu C_{i\mu} \chi_\mu$$

The atomic orbitals χ_μ are called the *atomic basis functions* and the $C_{i\mu}$ are the *MO expansion coefficients*. Several choices are possible for the basis set, including Gaussian functions [26], Slater functions [27], and plane waves [28]. The energy gradient formulas for the Hartree-Fock-Slater method were first derived by Satako [29] and later implemented practically using Slater basis sets [30]. Others have used Gaussian basis sets to compute derivatives of the DFT energy [26].

The concept of bond order and valence indices is well established in chemistry. It allows for interpretation and deeper understanding of the results of DFT calculations using ideas familiar to chemists. First, we have to define a density matrix or, as it is sometimes called, a *charge-density bond-order matrix*. If ϕ is a molecular orbital and $C_{i\mu}$ are the SCF expansion coefficients, then:

$$P_{\mu v} = \sum_i C_{i\mu} C_{iv}$$

and matrix $P_{\mu v}$ and a set of atomic orbitals completely specify the charge density The trace of matrix P and the overlap S is equal to the total number of electrons in the molecule:

$$N = TrPS = \sum_\mu (PS)_\mu$$

Summing $(PS)_{\mu v}$ contributions over all $\mu \in A$, $v \in B$, where A and B are centers, we can obtain P_{AB}, which can be interpreted as the number of electrons associated with the bond A–B. This is the so-called Mulliken population analysis [31]. The net charge associated with the atom is then given by:

$$q_A = Z_A - \sum_{\mu=A} (PS)_{\mu\mu}$$

where Z_A is the charge on the atomic nucleus A.
    Mayer [32] has defined the following quantities.

• Bond order between atoms A and B:

$$B_{AB} = 2\sum_{\mu=A}\sum_{v=B}\left[(P^\alpha S)_{\mu v}(P^\alpha S)_{v\mu} + (P^\beta S)_{\mu v}(P^\beta S)_{v\mu}\right]$$

where P^α, P^β are the density matrices for spin α and β.

• Actual total valence of atom A in the molecule:

$$V_A = \sum_{B(B\neq A)} B_{AB} + F_A$$

• Actual free valence of atom A in the molecule:

$$F_A = \sum_{\mu,v \subset A} (P^S S)_{\mu v}(P^S S)_{v\mu}$$

where $P^S = P^\alpha - P^\beta$.

The Mayer bond orders and valence indices have several useful properties:

- The values of bond orders are close to the corresponding classical values. This means that the double bond in H_2CO would have a C–O Mayer bond order close to 2.0.
- The total valence indicates how many single bonds are associated with the atom. For example, in the methane molecule, the C atom would have a total valence close to 4.0.
- The free valence index is zero for closed-shell systems. For open-shell radicals, it is a measure of the reactivity. The free valence index indicates whether free electrons are available for bonding on a particular atom.
- Unlike Mulliken bond orders, Mayer quantities are less dependent on the basis set choice and they are transferable, so they can be used to describe similar molecules.
- For similar molecules, the trends in Mayer quantities can be correlated well with electronic and geometrical changes due to the substituent.

This has been well introduced in Material Studio of Accelrys.

3.5   CHALLENGES

Though DFT itself does not provide any hints on how to construct approximate exchange-correlation functionals, it holds both the promise and the challenge that the true E_{xc} is a universal functional of the density; i.e., it has the same functional form for all systems. On one hand, this is a promise because an approximate functional, once constructed, may be applied to any system of interest. On the other hand, this is a challenge because a good approximation should perform equally well for very different physical situations. Both the promise and the challenge are reflected by the fact that the simplest of all functionals, the so-called local density approximation (LDA), has remained the approximation of choice for many years after the formulation of the Kohn-Sham theorem. In LDA, the exchange correlation energy is given by:

$$E_{xc}^{LDA}[n] = \int d^3 rn(r) e_{xc}^{unif}(n(r))$$

where $e_{xc}^{unif}(n)$ is the exchange-correlation energy per particle of an electron gas with spatially uniform density n. It can be obtained from quantum Monte Carlo calculations and simple parameterizations are available. By its very construction, the LDA is expected to be a good approximation for spatially slowly varying densities. Although this condition is hardly ever met for real electronic systems, LDA has proved to be remarkably accurate for a wide variety of systems. In the quest for improved functionals, an important breakthrough was achieved with the emergence of the so-called generalized gradient approximations (GGAs). Within GGA, the exchange-correlation energy for spin unpolarized systems is written as follows:

$$E_{xc}^{GGA}[n] = \int d^3 rf(n(r), \nabla n(r))$$

Though the input $e_{xc}^{unif}(n)$ in LDA is unique, the function f in GGA is not, and many different forms have been suggested. The construction of a GGA usually incorporates a number of known properties of the exact functional into the restricted functional form of the approximation. The impact of GGAs has been quite dramatic, especially in quantum chemistry where DFT is now competitive in accuracy with more traditional methods while being computationally less expensive. In recent years, still following the lines of GGA development, a new class of "meta-GGA" functionals has been suggested. The additional flexibility in the functional form gained by the introduction of the new variable can be used to incorporate more of the exact properties into the approximation. In this way it has been possible to improve upon the accuracy of GGA for some physical properties without worsening the results for others. Unlike LDA or GGA, which are explicit functionals of the density, meta-GGAs also depend explicitly on the Kohn-Sham orbitals. In order to go beyond exact exchange DFT, one might be tempted to use the exact exchange functional in combination with a traditional approximation such as LDA or GGA for correlation. Because both LDA and GGA benefit from a cancellation of errors between their exchange and correlation pieces, an approach using LDA or GGA for only one of these energy components is bound to fail. The obvious alternatives are approximate, orbital-dependent correlation energy functionals. One systematic way to construct such functionals is known as Görling-Levy perturbation theory. Structurally, this is similar to what is known as Møller-Plesset perturbation theory in quantum chemistry. However, it is only practical for low orders. A more promising route uses the fluctuation–dissipation theorem, which establishes a connection to linear response theory in time-dependent DFT. A final class of approximations to the exchange-correlation energy is the so-called hybrid functionals, which mix a fraction of exact exchange with GGA exchange,

$$E_x^{HYB}[n] = aE_x^{EXX}[n] + (1-a)E_x^{GGA}[n]$$

where a is the (empirical) mixing parameter. This exchange functional is then combined with some GGA for correlation. Hybrid functionals are tremendously popular and successful in quantum chemistry but much less so in solid-state physics. This last fact highlights a problem pertinent to the construction of improved functionals: available approximations are already very accurate and hard to improve upon. In addition, and this makes it a very difficult problem, one would like to have improved performance not only for just one particular property or one particular class of systems but for as many properties and systems as possible. After all, the true exchange-correlation energy is a universal functional of the density.

3.6 FUTURE

Within only twenty-five years, quantum mechanical simulation techniques based on density functional theory have progressed from the modeling of two atoms of silicon to hundreds of atoms of any species. Using such calculations, we can now make reliable predictions about the structure and properties of many atomistic systems. There have been numerous applications to study surfaces, point defects such as vacancies and impurities, extended defects such as grain boundaries and dislocations, catalysis, and many other applications. In recent years, these techniques have been extended to predict a wider range of physical and chemical properties and to predict theoretical values for experimentally measured spectra, such as optical, nuclear magnetic resonance (NMR), electron paramagnetic resonance (EPR), and many others. The numerous codes based on DFT provide us with an incredibly powerful scientific tool. However, in most of these approaches the cubic scaling of the computational time with the number of atoms in the system ultimately limits the useful system size that can be addressed. It has long been known that for nonmetallic systems it is possible, in principle, to reformulate the quantum mechanical problem so that the computational time scales linearly with the number of atoms in the system. A number of groups worldwide have developed linear scaling techniques, though these do not appear to offer the combination of the accuracy of the best conventional codes with the efficiency and robust convergence that has allowed such codes to be widely disseminated to nonexperts. A novel code, ONETEP, an *ab initio* linear scaling code, that offers state-of-the-art accuracy, robust and efficient convergence of the electronic structure to its ground state, has been developed by Skylaris et al. [33]. This code presents all the advantages of conventional DFT codes but allows calculations to be performed routinely on systems containing many thousands of atoms. Skylaris has published a second paper in that series to present a detailed comparison between ONETEP, [34] our linear-scaling density functional method, and the conventional pseudo potential plane wave approach in order to demonstrate its high accuracy. Further comparison with all electron calculations shows that only the largest available Gaussian basis sets can match the accuracy of routine ONETEP calculations. Results indicate that our minimization procedure is not ill conditioned and that convergence to self-consistency is achieved efficiently. Finally, the authors present calculations with ONETEP, on systems of about 1,000 atoms, of electronic, structural, and chemical properties of a wide variety of materials such as metallic and semiconducting carbon nanotubes, crystalline silicon, and a protein complex. Among the factors that make this possible is the fact that in ONETEP the calculated properties converge rapidly with the radii of the localization spheres of nonorthogonal generalized Wannier functions (NGWFs) and the rate of self-consistent convergence is affected neither by the size of these regions nor the number of atoms. In all these cases we have managed to obtain excellent agreement with CASTEP [35] in comparing either smaller systems of the same material or, where possible by the use of k-points, systems of equivalent size. These results confirm that ONETEP is a robust, highly accurate linear-scaling density functional approach, which makes possible a whole new level of large-scale simulation in systems of interest to nanotechnology, biophysics, and condensed matter physics.

The density functional theory of electronic structure should expand in the future and should become increasingly useful. Its calculation methods are in need of developmental as well as fundamental theoretical work, but they offer a special economy with respect to difficulty as the number of electron increases. Density functional concepts offer a compellingly appealing language for discussion of molecular structure and behavior: almost pictorial, almost intuitive, yet quantitative.

Dirac's familiar pronouncement is history, which was the laws of chemistry were now known, things have started, rather than ended, the development of quantum chemistry [36]. The contemporary successful and accurate calculations of molecular properties signal the end of the first phase of Dirac's era, as Robert G. Parr quotes in his book [37]:

> The next phase, just beginning, comprises the shakedown, codification, and unification of the basic ideas in the science of chemistry itself (as distinct from Physics). We expect the language of chemistry of the future to be replete with the idiom of density-functional theory. (p. 245)

REFERENCES

1. M. Abraham, R. Becker, *Theorie der Elektrizität*, vol. II, Teubner, Leipzig (1933).
2. H. A. Lorentz, *Theory of Electrons*, Dover, NY (1952).
3. F. Rohrlich, *Classical Charged Particles*, Addison-Wesley, Reading, MA (1965).
4. P. A. M. Dirac, *Proc. Roy. Soc. A*, 167 (1938) 148.
5. O. Klein, *Z. Phys.*, 37 (1926) 895.
6. M. Born, *Science*, 14 October 122. (1995) pp. 675–679.
7. L. de Broglie, *Journal de Physique et le Radium*, 7 (1926) 1.
8. D. Schrödinger, *Ann. Phys.*, 4 (1926) 79.
9. H. Rund, *The Hamilton-Jacobi Method in Mathematical Physics*, Krieger, NY (1973).
10. Y. B. Rumer, *Investigations into 5-Optics*, Gostekh-Teorizdat, Moscow (1956) (in Russian).
11. V. P. Vizgin, *Unified Field Theories in the First Third of the Twentieth Century*, Birkhäuser, Boston (1994).
12. A. Lichnerowicz, *Théorie relativiste de la gravitational et de l'electromagnetisme*, Masson and Co., Paris (1955).
13. J. Kohanoff, N. I. Gidopoulos, Density functional theory: Basics, new trends and applications, in *Handbook of Molecular Physics and Quantum Chemistry*, Vol. 2: *Molecular Electronic Structure*, S. Wilson (ed.), John Wiley & Sons (2003).
14. P. Hohenberg, W. Kohn, *Phys. Rev.*, 136 (1964) B864.
15. R. G. Parr, W. Yang, *Density-Functional Theory of Atoms and Molecules*, Oxford University Press, New York (1989).
16. R. G. Parr, R. A. Donelly, M. Levy, W. E. Palke, *J. Chem. Phys.*, 72 (1978) 3669.
17. R. G. Parr, W. Yang, *J. Am. Chem. Soc.*, 106 (1984) 4049.
18. L. Pauling, *J. Am. Chem. Soc.*, 53 (1931) 1367.
19. P. O. Loudin, *J. Mech. Phys. Solid*, 3 (1962) 969.
20. P. Hohenberg, W. Kohn, *Phys. Rev. B*, 136 (1964) 864.
21. (a) M. Levy, *Proc. Natl. Acad. Sci. USA*, 76 (1979) 6062; (b) M. Levy, J. A. McCammon, M. Karplus, *Chem. Phys. Lett.*, 64 (1979) 4.
22. W. Kohn, L. Sham, *Phys. Rev. A*, 140 (1965) 1133.
23. P. Pulay, *Mol. Phys.*, 17 (1969) 197.
24. B. R. Brooks, W. D. Laidig, P. Saxe, J. D. Goddard, Y. Yamaguchi, H. F. Schaefer, *J. Chem. Phys.*, 72 (1980) 4652.
25. J. A. Pople, R. Krishnan, H. B. Schlegel, J. S. Binkley, *Internet J. Quantum Chem. Symp.*, 13 (1979) 225.
26. J. Andzelm, E. Wimmer, D. R. Salahub, Spin density functional approach to the chemistry of transition metal clusters: Gaussian-type orbital implementation, in *The Challenge of d- and f-Electrons: Theory and Computation*, D. R. Salahub, M. C. Zerner (eds.), (1989).
27. L. Versluis, T. Ziegler, *J. Chem. Phys.*, 88 (1988) 322.

28. (a) N. W. Ashcroft, N. D. Mermin, *Solid State Phys.*, Holt Saunders, Philadelphia (1976);
    (b) M. C. Payne, M. P. Teter, D. C. Allan, T. A. Arias, J. D. Joannopoulos, *Rev. Mod. Phys.*, 64 (1992) 1045.

29. C. Satako, *Chem. Phys. Lett.*, **83** (1981) 111.

30. L. Versluis, T. Ziegler, *J. Chem. Phys.*, 88 (1988) 3322.

31. R. S. Mulliken, *J. Chem. Phys.*, 23 (1955) 1833.

32. I. Mayer, *Int. J. Quant. Chem.*, 29 (1986) 477.

33. C. K. Skylaris, A. A. Mosto, P. D. Haynes, O. Dieguez, M. C. Payne, *Phys. Rev. B*, 66 (2002) 035119.

34. C.-K. Skylaris, P. D Haynes, A. A. Mostofi, M. C. Payne, *J. Phys. Condens. Matter*, 17 (2005) 5757–5769.

35. M. C. Pyne, M. P. Teter, D. C. Allan, T. A. Arias, J.P Johnnopolus, *Rev. Mod. Phys.*, 64 (1992) 1045.

36. P. A. M. Dirac, *Physics Today*, 21 (1968) 52.

37. R. G. Parr, W. Y. Anag, in *Density Functional Theory of Atoms and Molecules*, Oxford University Press (1994).

4 Local Reactivity Descriptors

Chemical reactivity is the best success story of density functional theory (DFT). In a recent article, Geerlings and De Proft provide an extensive description of the successful pathway of DFT [1]. The step from quantum mechanics into what quantum chemistry can, in principle, be situated in the pioneering work by Heitler and London [2] on hydrogen molecules in 1927, which endows with the insight into the nature of the chemical bond [3]. In the years between 1930 and 1950, Pauling [3], Huckel [4], and Coulson [5] combined quantum mechanical principles with their chemical intuition to create a new discipline, called *quantum chemistry*. The valence bond approach (after Heitler and London) was prominent in those days, to the detriment of Hund's and Mulliken's molecular orbital (MO) method [6]. A revolution was provoked by Roothaan's matrix formulation of the MO method in 1951 [7]. Its elegance together with the increasing computer power paved the way for the large-scale introduction of the molecular orbital–lineal combination of atomic orbital (MO-LCAO) method within the framework of the Hartree Fock self-consistent field (SCF) approach as detailed in Pople's comprehensive treatise [8]. The late 1970s, 1980s, and 1990s saw the development and/or the adaptation for systematic use of methods beyond SCF including electron correlation: Møller Plesset perturbation theory [9], the method of configuration interaction [10], and various types of coupled cluster theory [11]. The introduction of initially freely distributed, later on commercially available, computer programs that became increasingly user friendly (Pople et al.'s Gaussian series) [12], Delley's DMol^3 [13], and Payne's CASTEP [14]) definitely promoted quantum chemistry, from a branch of theoretical chemistry almost exclusively reserved for "pure-sang" theoreticians and concentrating on diatomic and small polyatomic molecules, to a field also creating tools for nonspecialists, in many other subfields of chemistry (inorganic, organic, biochemistry).

Both wave functions quantum chemistry (QC) and density functional theory (DFT), when used to compute atomic and molecular properties, results, which often produced, are not always directly exploitable. The numbers they produced in many cases are translated or casted into a language or formalism pointing out their chemical relevance. This is simply stated by Parr [15]: "Accurate calculation is not synonymous with useful interpretation. To calculate a molecule is not to understand it." Quite often this translation involves terms going back to the early days of theoretical chemistry but still in use as a guideline for chemists in the interpretation of experimental data: hybridization, electronegativity, aromaticity, etc.

This chapter aims to discuss local reactivity descriptors including the definition, origin explanation, and possible application areas. The terms *electronic, nuclei Fukui functions* will be described. The local softness and hardness paradigm along with philicity concepts will be explained to have a basic idea of the concept, to have a background to discuss and rationale for the applications of these methodologies to understand the experimental postulates.

Rapid advances are taking place in the application of DFT to describe complex chemical reactions. Researchers in different fields working in the domain of quantum chemistry tend to have different perspectives and to use different computational approaches. DFT owes its popularity to recent developments in predictive powers for physical and chemical properties and its ability to accurately treat large systems. Both theoretical content and computational methodology are developing at a pace that offers a tremendous opportunity for scientists working in diverse fields of quantum chemistry, cluster science, and solid-state physics. In recent years, DFT offered a perspective for the interpretation/prediction of experimental/theoretical reactivity data on the basis of a series of reactivity functions to perturbations in the number of electrons and/or external potential. This approach has enabled the sharp definition and computation, from first principles, of a series of well-known chemical concepts such as electronegativity and hardness. This chapter focuses on the elucidation of the reactivity/philicity concept derived from the great work of Fukui [16] famously known as the Fukui function. The second-order approach, well known until now, introduces the hardness/softness and Fukui function concepts related to polarizability and frontier MO theory, respectively. The introduction of polarizability/softness is also considered as a historical perspective in which polarizability (with some exceptions) mainly put forward in noncovalent interactions. DFT is situated within the evolution of QC as a facilitator of computations and a provider of new chemical insights. The importance of the latter branch of DFT, conceptual DFT is highlighted following Parr's dictum [15] "to calculate a molecule is not to understand it." The hard–soft acid–base (HSAB) principles classify the interaction between acids and bases in terms of global softness. Pearson proposed the global HSAB principle [17]. The global hardness was defined as the second derivative of energy with respect to the number of electrons at constant temperature and external potential, which includes the nuclear field. The global softness is the inverse of global hardness. Parr and Pearson also suggested a principle of maximum hardness (PMH) [18], which states that, for a constant external potential, the system with the maximum global hardness is most stable. In recent days, DFT has gained widespread use in quantum chemistry [19, 20]. This concept of HSAB gains impetus with the observation that DFT provides a necessary and efficient framework for a quantitative description of many global and local parameters (including hardness and softness) directly related to the inherent reactivity of chemical species. There are different forms of describing the reactivity index, where the idea is to find the donor acceptor capability of an atom present in a molecule interacting with another molecule or the interaction within itself. This is the main concept, now depending on the interaction that is taking place; one can look into local softness of the atom, approaching the other interacting species or the group of atoms together to the active site. It has also been mentioned that it is possible to describe the interaction between atoms for an intermolecular interaction through the concept of an equilibrium using the idea of reactivity index. Hence, the concept of reactivity index attributes to the activity of atom center and its capability to interact with other species of comparable activity in its localized/nonlocalized neighbor.

4.1 ELECTRONIC FUKUI FUNCTION

Electronic and nuclear reactivity descriptors have been defined on the basis of Taylor series expansions of the energy functional within the four ensembles of DFT, providing a formal basis for a deeper understanding of chemical reactivity and chemical reaction processes within a perturbative approximation.

As mentioned earlier, computational DFT is founded on a variational principle, more precisely for the energy functional

$$E = E[\rho]$$

Looking for an optimal ρ — i.e., the one that minimizes E — is thereby subjected to the constraint that ρ should at all times integrate to N, the number of electrons.

$$N = \int \rho(\vec{r})\,d\vec{r}$$

Within a variational calculation this constraint is introduced via the method of Lagrangian multipliers, yielding the variational condition

$$\partial[E - \mu\rho] = 0$$

where μ is the Lagrangian multiplier, a constant gaining its physical significance in the differential equation (the Euler equation) resulting from the previous equation.

$$v(r) + \frac{\partial F_{HK}}{\partial \rho} = \mu$$

Here, $v(r)$ is the external potential (i.e., due to the nuclei) and FHK is the Hohenberg Kohn functional containing the electronic kinetic energy and the electron–electron interaction operators [21]. It has been Parr's impressive contribution to identify this abstract as Lagrange multiplier [22].

DFT [22] is mainly based on the existence of an universal functional $F[\rho(r)]$, where $\rho(r)$ is the electron density, which allows us to write the total energy, E, as a functional of the density:

$$E_\upsilon[\rho(\vec{r})] = F[\rho(\vec{r})] + \int \upsilon(\vec{r})\rho(\vec{r})\,d\vec{r}$$

where $v(r)$ is the local external potential. The second important ingredient is a variational principle, which gives the Euler equation:

$$\mu = \frac{\delta E \upsilon[\rho(\vec{r})]}{\delta \rho(\vec{r})}$$

where m is a Lagrange parameter due to the normalization condition of the electron density. It has been proven [18, 22, 24] that:

$$\mu = \left(\frac{\partial E}{\partial N}\right)_{\nu_0}$$

Therefore, μ is called the electronic chemical potential.

It is a global property of a ground state, constant from point to point in the atom or molecule (or solid), in principle calculable for each N electrons and v in terms of the fundamental constants of physics as $\mu[n, v]$. The important Sanderson principle of electronegativity [25, 26] states that when atoms (or other combining groups) of different chemical potentials unite to form a molecule with its own characteristic chemical potential, to the extent that the atoms (groups) retain their identity, their chemical potential must equalize.

The electronegativity χ, defined by Mulliken as the average of ionization potential I and electron affinity A, $\chi = 1/2(I + A)$, is such a parameter; it is a useful measure of the tendency of a species to attract electrons. Thus, the electronegativity χ has been identified as the negative of the chemical potential μ, which is the Lagrange multiplier in the Euler-Lagrange equation in density functional theory, and where E is the total electronic energy, N is the number of electrons, and $v(r)$ is the external electrostatic potential an electron at r feels due to the nuclei. Mulliken's formula — $1/2(I + A)$ — is no more than the finite-difference approximation for the equation. The Mulliken values, the arithmetic average of ionization energy (IE) and electron affinity (EA), were already shown to correlate with the Pauling values [27, 28] and have received increasing importance in recent years on the basis of their simpler foundation. It can easily be seen that they correspond to the following:

$$\mu = \left(\frac{\partial E}{\partial N}\right) v(r) = -\chi$$

Similarly, the natural definition of the hardness, called η, has been shown to be

$$\eta = \frac{\partial^2 E}{\partial N^2} v(r) = \frac{\partial \mu}{\partial N} v(r)$$

which has the finite-difference approximation $I - A$. The hardness/softness/acid/base principle has been derived using the equation above as the definition of hardness. The physical meaning of hardness becomes clear if one considers the disproportionation reaction in which an electron is taken from one S and given to another S

$$\dot{S} + \dot{S} \rightarrow S^+ + \ddot{S}^-$$

The energy change is

$$\Delta E_s = I_s - A_s$$

This is hardness. Small or zero hardness means that it is easy for electrons to go from S to S'; that is, S is a soft species. Just as with electronegativity, and for the same reasons, there will be a state dependence and an environmental dependence of the hardness of a chemical species, which will have to be elucidated in any particular circumstance. There is a consensus with the meaning of *hard* and *soft* from the concept of Pearson [29] and Klopman [30].

- Soft base: Donor atom has high polarizability and low electronegativity, is easily oxidized, and is associated with empty low-lying orbitals.
- Hard base: Donor atom has low polarizability and high electronegativity, is hard to oxidize, and is associated with empty orbitals of high energy.
- Soft acid: Acceptor atom is of low positive charge, is of large size, and has easily excited other electrons.
- Hard acid: Acceptor atom has high positive charge and small size and does not have easily excited outer electrons.

On the other hand, the frontier electron densities proposed by Fukui are local properties that depend on ρ; they differentiate one part of a molecule from another and serve as reactivity indices.

In density functional theory, hardness (η) is defined as

$$\eta = \frac{1}{2}\left(\frac{\partial^2 E}{\partial N^2}\right)_{v(r)} = \frac{1}{2}\left(\frac{\partial \mu}{\partial N}\right)_v$$

where E is the total energy and N is the number of electrons of the chemical species and the chemical potential.

The global softness, S, is defined as the inverse of the global hardness, η.

$$S = \frac{1}{2}\eta = \left(\frac{\partial N}{\partial \mu}\right)_v$$

Using the finite difference approximation, S can be approximated as

$$S = \frac{1}{(IE - EA)}$$

where IE and EA are the first ionization energy and electron affinity of the molecule, respectively.

With the relationship between chemical potential and electronegativity, one can therefore say that the electronegativity (chemical potential) is a property of the state of the system, calculable in terms of the constants of physics from density functional theory, or from experiment by statistical ensemble or else. Electronegativity differences (chemical potential differences) drive the electron transfer. Electrons trend to flow from a region of low electronegativity (high chemical potential) to a region of high electronegativity (low chemical potential). In the simplest case, the number of electrons that flow is first-order proportional to the electronegativity (chemical potential) difference; the concurrent energy stabilization is proportional to its square. On formation of a molecule, electronegativities (chemical potentials) of constituent atoms or groups equalize (neutralize), all becoming equal to the electronegativity of the final molecule.

The concept of electronegativity has been well known for years, and density functional theory brings it back in a new way. The density functional formulation of the quantum theory leads rigorously to the concept of electronegativity and the electronegativity equalization.

One therefore can say that electronegativity or chemical potential of the ground state of a microscopic species (atom, ion, molecule, or solid) is a property of the system that behaves much like the chemical potential of macroscopic thermodynamics. It measures the escaping tendency of the electrons from the system and is constant through a system in equilibrium.

4.2 NUCLEAR FUKUI FUNCTION

In the absence of other fields, the external potential $n(r)$ arises only from the spatial configuration of the nuclei in the system [31],

$$v(r) = \sum_\alpha \frac{Z_\alpha}{|r - R_\alpha|}$$

$F[r]$ stands for the kinetic energy and the electron–electron repulsion functional. The responses to changes in the external potential, $v(r)$, are associated with changes in nuclear configuration, which is the most obvious variable in a chemical reaction [32–34]. Although the electron density determines all ground-state properties of a molecular system, the response of the nuclei to a perturbation in N remains unknown, and a complicated response kernel translates electron density changes into external potential changes. Cohen and Ganduglia-Pirovano [35, 36] introduced an alternative to this problem defining a nuclear Fukui function (NFF) as the change of the Hellman-Feynman force on the nucleus a, F_a, due to a perturbation in the number of electrons at a constant external potential,

$$\Phi_\alpha = \left(\frac{\partial F_\alpha}{\partial N} \right)_{v(r)}$$

where

$$F_\alpha = \frac{\partial E}{\partial R_\alpha} = -Z_\alpha \frac{\delta}{\delta R_\alpha} \int \frac{\rho(r)}{|r - R_\alpha|} dr + Z_\alpha \frac{\delta}{\delta R_\alpha} \sum_\beta \frac{Z_\beta}{|R_\alpha - R_\beta|}$$

φ_a, a vectorial quantity, does not directly measure the change in the external potential $dv(r)$; instead, it describes the onset of that change [37]. In the limit of very low temperature, the density matrix in the grand canonical ensemble is a piecewise function of the number of electrons [35]; consequently, any derivative with respect to N will be a discontinuous function of N. Therefore, the derivative respect to N must be taken just above or below the correct integral number of electrons of the system, which permits us to define left-hand (−) and right-hand (+) nuclear Fukui functions:

$$\Phi_{\alpha}^{+/-} = \left(\frac{\partial F_{\alpha}^{+/-}}{\partial N} \right)_{v(r)'}$$

In the case of diatomic molecules and using internal coordinates as the reference system, the sign of the NFF has a simple meaning. In the case of $\varphi_{\alpha}+$, a positive sign indicates a bond lengthening with ionization, whereas the opposite sign in φ_{α} predicts a shortening of the bond length. This tendency to change the bond length can be rationalized in the light of the Berlin division of the molecular space [38]. The density rearrangement due to changes in the number of electrons implies an intramolecular charge transfer. The bond length changes depend on whether charge is arriving or leaving the bonding regions. Using Maxwell relations, Baekelandt [39] showed that the nuclear Fukui function might also be identified as the configurational contributions to the change in the chemical potential:

$$\Phi_{\alpha} = \left(\frac{\partial F_{\alpha}}{\partial N} \right)_{v(r)} = -\left(\frac{\delta \mu}{\delta R_{\alpha}} \right)_{N'}$$

where R_a designates the position of the nucleus. Further nuclear-related reactivity indexes, based on a nuclear kernel softness quantity, have been defined and its relations with electronic descriptors explored [40–47]. An extension of this formalism to the spin-polarized DFT has been recently presented [48, 49]. It is clear that these nuclear reactivity hierarchies complement and extend the usefulness of DFT as applied to gain insight into the reactivity within molecular chemical systems, despite its remaining a perturbative approximation.

4.3 LOCAL SOFTNESS AND LOCAL HARDNESS

The HSAB principle has also been invoked in a local sense to explain the response of a chemical system to different kinds of reagents [50–54]. In particular, Fukui functions [55] and local softness [56] may be used to determine the nature of reactive sites of a given molecule. Generally, it is believed that the larger the value of the Fukui function, the greater the reactivity. However, current understanding has been changing with the recent proposal of Gazquez and Méndez [57]. They propose that when two molecules A and B interact, a bond is likely to form between an atom of A and an atom of B whose Fukui function values are similar. They also propose that local softness may play the same role as the Fukui function when the softness of the two interacting molecules is different. This can be taken as a local version of the HSAB principle [58]. The proposition was subsequently verified by Geerlings and coworkers [59]. The determination of the specific sites of interaction between two chemical species is of fundamental importance to detecting the products of a given reaction without actual calculations of the corresponding potential energy surface [60].

The Fukui function $f(r)$ is defined by:

$$f(r) = \left[\frac{\partial \mu}{\partial v(r)} \right] N = \left[\frac{\partial \rho(r)}{\partial N} \right] v$$

The function f is thus a local quantity, which has different values at different points in the species, N is the total number of electrons, μ is the chemical potential, and v is the potential acting on an electron due to all nuclei present. This simultaneously generates the frontier-electron theory. Of the two different sites with generally similar dispositions for reacting with a given reagent, the moiety prefers the one with the maximum response of the system's chemical potential. In short, large $|d\mu|$ is good. This is perhaps an oversimplified view of chemical reactivity, but it is useful. And in any case, $f(r)$ is established as an index of considerable importance for understanding molecular behavior — the natural reactivity index of density functional theory. Note that $f(r)$ is defined independently of any model, whereas the concepts of classical frontier theories are couched in the language of the independent-particle model.

The last equation immediately gives us a very important condition of the Fukui function. It is normalized to one:

$$\int f(\vec{r})d\vec{r} = 1$$

Owing to the discontinuity of the chemical potential at integer N, the derivative of the equation will be different if taken from the right or the left side. One has $m+$ when the derivative is taken from above $(N + d; d > 0)$, $m-$ when the derivative is taken from below $(N - d; d > 0)$ and for the cases where there is not a net charge exchange a good approximation is to use the average of

$$\mu^0 = (\mu^+ + \mu^-) = 2$$

The local softness $s(r)$ can be defined as

$$s(r) = \left(\frac{\partial \rho(r)}{\partial \mu} \right)_v$$

This equation can also be written as

$$s(r) = \left[\frac{\partial \rho(r)}{\partial N} \right]_v \left[\frac{\partial N}{\partial \mu} \right]_v = f(r)S$$

Thus, local softness contains the same information as the Fukui function $f(r)$ plus additional information about the total molecular softness, which is related to the global reactivity with respect to a reaction partner, as stated in the HSAB principle. Atomic softness values can easily be calculated by using the following equation:

$$s_x^+ = [q_x(N+1) - q_x(N)S]$$

$$s_x^- = [q_x(N) - q_x(N-1)S]$$

$$s_x^0 = \frac{S[q_x(N+1) - q_x(N-1)]}{2}$$

In this way, the frontier-orbital theory of reactivity of Fukui [61] can be easily incorporated into the theory. The function $f_{(r)}^+$ is associated with the lowest unoccupied molecular orbital LUMO and measures reactivity toward a donor reagent; the function $f_{(r)}^-$ is associated with the highest occupied molecular orbital HOMO and measures reactivity toward an acceptor reagent; and finally, the average of both, $f_o(r)$, measures reactivity toward a radical. One of the important characteristics of the Fukui function is that it is a local index; e.g., a function of the position. Therefore, it is able to give us information about site reactivity and about what region of a molecule is better prepared to accept or donate charge. Suppose that the reagent R donates a charge to the substrate S, the electrons flow from R to S. This will give one an ample indication for the preferential bonding and will allow experimentalists to use this simple concept to illustrate and design new materials of interest.

4.4 PHILICITY

Since the introduction of the concepts of electrophilicity and nucleophilicity by Ingold [62] in 1929, there has been a growing interest in classifying atoms and molecules within empirical scales of electrophilicity and nucleophilicity. The main idea behind this objective was the search of absolute scales that could be independent of the reactivity of the nucleophile/electrophile partners. This objective is ambitious if one considers that a universal scale should accommodate a wide diversity of chemical species presenting quite different structural and bonding properties. For instance, one of the first attempts to classify electron donors within a single nucleophilicity scale was reported by Swain and Scott [63], who defined a nucleophilicity number as an intrinsic property of nucleophiles, using rate coefficients for a series of SN2 reactions. Other attempts to quantitatively rank the nucleophilic power of molecules were proposed by Edwards, using a four-parameter scheme [64], and by Edwards and Pearson [65], using the HSAB empirical rule. Recently, in a series of articles, Mayr et al. persuasively argued in favor of nucleophilicity and electrophilicity parameters that are independent of the reaction partner [66–71]. They proposed that the rate coefficients for the reactions of carbocations with uncharged nucleophiles obey the linear free energy relationship:

$$\log k = z(E + N)$$

where E and N are the electrophilicity and nucleophilicity parameters, respectively, and z is the nucleophilic-specific slope parameter. This sensitivity parameter is usually close to unity, so that it may be neglected for the purpose of qualitative comparisons. These authors have clearly emphasized and illustrated the usefulness of the nucleophilicity/electrophilicity scales to quantitatively discuss reactivity as well as intermolecular selectivity [66–71].

The concept of electrophilicity viewed as a reactivity index is based on a second-order expansion of the electronic energy with respect to the charge transfer dN at fixed geometry. Because electrophiles are species that stabilize upon receiving an additional amount of electronic charge from the environment, there exists a minimum of energy for a particular dN* value. Using this simple idea, Parr [15] performed a variational calculation that led to the definition of the global electrophilicity index as:

$$w = -KDE\,(dN^*)$$

which may be recast into the more familiar form, using global electrophilicity index (ω), which was introduced by Parr et al. [72] as:

$$\omega = \frac{\mu^2}{\eta}$$

where μ and η are chemical potential [73] and chemical hardness [74], respectively. A local variant of ω has been proposed [75] via the resolution of the identity associated with the normalization of the Fukui function [76], $f(r)$, as:

$$\omega(r) = \omega f^\alpha(r)$$

where $f^\alpha(r)$ is the Fukui function [76] associated with $\alpha = +$, $-$, and 0 referring to nucleophilic, electrophilic, and radical reactions, respectively. Corresponding condensed-to-atom variants may be written for the kth atomic site in a molecule as:

$$\omega^\alpha_k = \omega f^\alpha_k$$

A special case of this general treatment is provided in the literature [77, 78]. Similarly, local softness is defined as [79]:

$$S^\alpha(r) = Sf^\alpha(r)$$
$$S^\alpha_k = Sf^\alpha_k$$

where $S = \frac{1}{2}\eta$ is the global softness [73].

The local softness is obtained through the decomposition of the global softness and the former indicates that the soft–soft interactions are preferred in comparison to the hard–soft ones. On the other hand, the decomposition of global electrophilicity provides the local philicity, which is capable of showing the preference of electrophile–nucleophile interactions over electrophile–electrophile interactions. It may, however, be noted that the local philicity and the local softness will provide the same trend of reactivity if μ^2 remains more or less constant or varies more slowly than S or has a similar variation as that of S in a group of molecules, because $\omega = \mu^2 S$.

A molecule with a high global electrophilicity value would be more reactive toward that with a corresponding low value. For two such molecules the reaction would be through the atomic center having the largest $\omega - k$ in one molecule, with the atomic center having the largest $\omega + k$ of the other molecule. The electrophilic or nucleophilic power is distributed over all atomic sites in a molecule, keeping the overall philicity conserved. When an electrophile interacts with a nucleophile from a large distance, their global electrophilicities (ω) decide their behavior, viz. ω electrophile > ω nucleophile. The local variants, in general, remain spectators until they come very close, when the most electrophilic site of an electrophile will attack the most nucleophilic site of the nucleophile. It may not be true in all cases (especially when two or more strong electro[nucleo]philic centers of comparable strength are present in a molecule) that the most electrophilic site in the electrophile has larger ω_k^+ value than that of the most electrophilic site of the nucleophile. Similarly, global and local HSAB principles may be at variance with each other in some occasions. Of course the HSAB principle is not violated in that case and also during strong interactions between a hard and a soft species with a substantial electronegativity difference. Charge-based descriptors should be used in analyzing charge-controlled hard–hard interactions [80] where the Fukui function and the related descriptors (S_k^α, ω_k^α) may provide wrong reactivity trends.

The principle "soft likes soft" has to do with high polarizabilities enhancing covalent bond formation, whereas the principle "hard likes hard" relates to factors enhancing purely electrostatic interactions. The fact is that when A approaches B, not only do N_A and N_B change but so also do the effective external potentials v_A and v_B acting on the electrons in A and B, respectively. Nalewajski [81] proposed the following. Let ΔN_A and ΔN_A be the changes in going from the separate atom A, having chemical potential μ_A^0, to the atom A in the molecule, having chemical potential μ_A, and similarity with B. Letting $\Delta N = \Delta N_A = -\Delta N_B$, one will have,

$$\mu_A = \mu_A^0 + \left(\frac{\partial \mu_A}{\partial N_A}\right)\Delta N_A + \int\left[\frac{\partial \mu}{\partial v_A(r)}\right]_{n,A}\Delta v_A(r)dr$$

$$= \mu_A^0 + 2\eta_A\Delta N + \int f_A(r)\Delta v_A(r)dr$$

$$\mu_B = \mu_B^0 - 2\eta_B\Delta N + \int f_B(r)\Delta v_B(r)dr$$

where η_A, η_B are hardnesses of atoms A and B and f_A and f_B are their Fukui functions. The equilibrium condition $\mu_A = \mu_B$ then yields the first-order formula, where the Fukui function terms moderate the chemical potential difference in driving the charge transfer [82].

$$\Delta N = \frac{\left(\mu_B^0 - \mu_A^0\right) + \int f_B(r)\Delta v_B(r)\,dr - \int f_A(r)\Delta v_A(r)\,dr}{2(\eta_A + \eta_B)}$$

This is a modified version of the concept that electronegativity differences should be determinative for a dipole moment [3], which is based on a simpler relation to say that charge transfer is first-order proportional to the original chemical potential difference. The equation is

$$\Delta N = \frac{\mu_B^0 - \mu_A^0}{2(\eta_A + \eta_B)}$$

A different viewpoint was put forward by Berkowitz [83]. When A and B are at a large distance apart but already in equilibrium, to satisfy the situation, their chemical potentials must be equal:

$$\mu_A = \mu_B$$

As the reaction coordinate changes by a small amount, this condition will remain unsatisfied, but there may be a flow of electrons from B to A and therefore a change in external potential. One will then have

$$d\mu_A = 2\eta_A dN_A + \int f_A(r)\,dv_A(r)\,dr$$

$$d\mu_B = 2\eta_B dN_B + \int f_B(r)\,dv_B(r)\,dr$$

As we know, $dN_A = -dN_B = dN$ and $d\mu_A = d\mu_B$. So one can see a change along a reaction coordinate

$$dN = \frac{\int f_B(r)\,dv_B(r)\,dr - \int f_A(r)\,dv_A(r)\,dr}{2\eta_A + 2\eta_B}$$

With this, Berkowitz says that for small change along the reaction coordinate, as parameterized by a change $dv(r)$ in the potential due to nuclei, the induced electron flow is greater the softer that A and B (small $\eta_A + \eta_B$) are, the larger the frontier quantity $f_B - f_A$ (differences in local softness) is, and the larger the overlap between Fukui functions is. For an essentially homogeneous system divided into two parts, A and B, from the concept of Nalewajski and Konisiski [81], one sees that the flow of electrons from B to A could nevertheless be extremely large if the hardness η were extremely small. This situation, of very large or infinite softness, may characterize the superconducting state [73].

Two different finite-difference approaches can be followed as detailed by Pal et al. [84]. Separate calculations of $(N + 1)$ and $(N - 1)$ electrons imply relaxation of orbital from the neutral system. This approach is known as the *relaxed orbital* approach. The practical difficulty of the Fukui function, especially of $f^+(r)$, using this approach suffers the technical difficulties of computing the density of anionic system at the same level of accuracy as the neutral system. At this stage it is difficult to get the correlation effect. A manner of taking into account, in an approximate way, the orbital relaxation effects doing a single point calculation on the neutral system and avoiding the complexity of the anionic species has been presented and implemented in the literature. On the other hand, assuming that the shape of molecular orbital does not change when a small amount of charge is added or subtracted is known as a *frozen orbital* approximation [85].

For determining site selectivity or site reactivity, one usually calculates atom-condensed Fukui functions, first introduced by Yang et al. based on the idea of integrating the Fukui function over atomic regions [86], similar to the procedure followed in population analysis technique [56]. Combined with finite difference approximation, this yields working equations of the type

$$f_a^+ = q_{A,N+1} - q_{A,N}$$

$$f_a^- = q_{A,N} - q_{A,N_{+1}}$$

where $q_{A,N}$ denotes the electronic population of atom A of reference system. In *ab initio* calculations these numbers are obtained mostly by a Mulliken and Lowdin population analysis [87]. Under frozen orbital approximation, atom-condensed Fukui functions will be nothing but the respective atomic population of HOMO or LUMO orbital.

In addition to the Fukui function, other reactivity descriptors based on Fukui functions have also been used and we have discussed two major philicity index uncovered by Parr and Yang [73] and Chattaraj et al. [75]. Earlier, Krishnamurthy et al. [88] proposed *group softness* to describe intermolecular reactivity trends in carbonyl compounds and organic acids. They defined group softness as the sum of softness of atoms present in an appropriately defined group. This has been further validated by the current author [89, 90] and applied it to different larger system to propose the interaction between heterocyclics and clay materials. Based on this concept, Partha and coworkers [91] proposed *group philicity* as the summed condensed-philicity of atoms present in the group.

$$w_g = \sum_{k=1}^{n} W_k$$

where n is the number of atoms bonded to the reactive atom, wk is the atom-condensed philicity of the atom k, and wg is the group philicity. We have applied this methodology extensively to many different materials of interest [92–100]. The interaction energy scheme [99, 100] can be explained as follows. It is known that A and B interact in two steps: (1) interaction will take place through the equalization of chemical potential at constant external potential and (2) A and B approach the equilibrium state through changes in the electron density of a global system generated by making changes in the external potential at constant chemical potential. That means within DFT we can write

$$\Delta E_{inter} = E[\rho_{AB}] - E[\rho_A] - E[\rho_B]$$

where ρ_{AB}, ρ_A, ρ_B are the electron densities of the systems AB at equilibrium and of the isolated systems A and B, respectively or in terms of the potentials we can write

$$\Delta E \ inter = \Delta E v + \Delta E \mu$$

where

$$\Delta E v = -1/2 \left[(\mu A - \mu B)^2 \Big/ (S_A + S_B) \right]$$

$$\Delta E \mu = -1/2 N A B^2 k \left[1 \Big/ (S_A + S_B) \right]$$

N_{AB} total number of electrons, k is the proportionality constant between S_{AB} and S_A + S_B, and the product of N and K is λ.

$$\Delta E \mu = (-1/2) \ \lambda / (S_A + S_B)$$

If the interaction is taking place through j site of A,

$$\lambda Aj = qAj - qAj$$

qAj is the density of jth atom of A in complex AB and qAj is the density in an isolated system.

In recent years, conceptual density functional theory has offered a perspective for the interpretation/prediction of experimental/theoretical reactivity data on the basis of a series of response functions to perturbations in the number of electrons and/or external potential as described in a recent review by Geerlings and De Proft [101]. In this contribution, a short overview of the shortcomings of the simplest first-order response functions is illustrated leading to a description of chemical bonding in a covalent interaction in terms of interacting atoms or groups, governed by electrostatics with the tendency to polarize bonds on the basis of electronegativity differences. The second-order approach introduces the hardness/softness and Fukui function concepts related to polarizability and frontier MO theory, respectively. The introduction of polarizability/softness is also considered in a historical perspective in which polarizability is, with some exceptions, mainly put forward in noncovalent interactions. A particular series of response functions, arising when the changes in the external potential are solely provoked by changes in nuclear configurations (the "R analogues"), is also systematically considered. The main part of the contribution is devoted to third-order response functions apparently may be expected not to yield chemically significant information, as turns out to be for the hyper hardness. A counterexample is the dual descriptor and its R analogue, the initial hardness response, on a density-only basis; i.e., without involving the phase, sign, and symmetry of the wave function. Even the second-order nonlinear response functions are shown to possibly bear interesting information; e.g., on the local and global polarizability. Its derivatives may govern the influence of charge on the polarizability, the R analogues being the nuclear Fukui function and the quadratic and cubic force constants. Although some of the higher order derivatives may be difficult to evaluate, a comparison with the energy expansion used in spectroscopy in terms of nuclear displacements, nuclear magnetic moments, and electric and magnetic fields leads to the conjecture that, certainly, cross-terms may contain new and intricate information for understanding chemical reactivity. The application of the Fukui function formalism in chemistry has been reviewed. The basic equations have been presented and its exact properties were discussed. The electronic Fukui function and its associated response function have been formulated within a framework that makes it the natural descriptor of selectivity and pair site reactivity. Approximated expressions have been developed to deal with regional reactivity condensed to atoms and functional groups in a molecule. The nuclear counterpart and the important effect associated to the nuclear configuration motion during a chemical reaction were expected to introduce into the theory. A relationship between the nuclear and the electronic counterpart has been discussed. It is clear that a generalized formalism including both electronic and nuclear Fukui functions should be of great potential for giving more complete and complementary information regarding the concepts of selectivity and site activation for the molecular system in its different conformations defined by the stationary points in a potential energy surface. We have illustrated the usefulness of this local descriptor of chemical reactivity and shown how the static reactivity picture described by the electrophilic Fukui functions helps in the prediction of the potential site of protonation in a polyfunctional system. At the same time, this reactivity index has been implemented within an energy-density formalism to obtain useful Hammett-like relationships to describe substituent effects in some gas-phase acid–base equilibrium. The theory was applied intensively in various domains of materials and applications will be discussed with some examples in the following chapters.

This is a simple concept of electron–donor acceptor; if there is a chemical bond formation, it needs some affinity between the atoms, and if there is affinity, there will be either donation of electrons or an acceptance. Based on this theory, one can simply calculate the charges of the system one is considering and can look into the population analysis on the atom centers. This will guide about the highest occupied state orbital within the atom and the lowest unoccupied orbital state present in the atom. This therefore will propose a feasible electron donation or acceptance. Now, this can be intermolecular or intramolecular. The caution is that this is predictive behavior and one will be mimicking this donor acceptor behavior by removing one electron from its filled state and to see the change in its donor capability or acceptor capability. Hence, we recommended considering a ratio of the donor acceptor capability of each center to scale the site's activity. This method has been applied for a range of molecules from organic to inorganic and it is able to propose a selective order as well as being a useful tool to scale a chemisorption phenomenon in the reaction process, where the primary need is to find an active site and then whether or not the reactant proposed have some affinity to the surface or not. So the procedure is to first look at the arbitrary position by Monte Carlo simulation and then see a potential surface by molecular mechanics and perform a reactivity index study with DFT to reconfirm the activity and hence the adsorption. This can go in other way and can be performed at the cluster level; the challenge remains still to see the effect in a periodic boundary condition.

REFERENCES

1. P. Geerlings, F. De Proft, *Int. J. Mol. Sci.*, 3 (2002) 276.
2. W. Heitler, F. London, *Z. Phys.*, 44 (1927) 455.
3. L. Pauling, *The Nature of the Chemical Bond* (3rd ed.), Cornell University Press, Ithaca (1960).
4. K. L. Huckel, *Zetsrift fur Physik*, 70 (1931) 204; 76 (1932) 6211; 83 (1933) 632.
5. C. A. Coulson, Oxford, Clendon Press, 1952, Latest Edn. R. McWeeny, *Coulson's Valence*; Third Edition, Oxford, 1979.
6. (a) F. Hund, *Z. Phys.*, 51 (1928) 759; (b) R. S. Mulliken, *Phys Rev.*, 32 (1928) 186.
7. C. C. J. Roothaan, *Rev. Mod. Phys.*, 23 (1951) 69.
8. W. J. Hehre, L. Radom, P. V. R. Schleyer, J. A. Pople, *Ab Initio Molecular Orbital Theory*, Wiley, New York (1986).
9. C. Møller, M. S. Plesset, *Phys. Rev.*, 46 (1934) 618.
10. I. Shavitt, *The Method of Configuration Interaction in Modern Theoretical Chemistry*, Vol. 3, *Methods of Electronic Structure Theory*, H. F.Schaefer, III (ed.), Plenum Press, New York (1977).
11. R. J. Bartlett, *J. Phys. Chem.*, 93 (1989) 1697.
12. (a) J. A. Pople, et al., *Gaussian 98*, and previous releases (Gaussian 94, Gaussian 92, Gaussian 70), Gaussian Inc., Pittsburgh, PA (1998); (b) M. J. Frisch, G. W. Trucks, H. B. Schlegel, G. E. Scuseria, M. A. Robb, J. R. Cheeseman, G. Scalmani, et al. Gaussian, Inc., Wallingford CT, 1995.
13. B. Delley, *J. Chem. Phys.*, 113 (2000) 7756.
14. M. P. Teter, M. C. Payne, D. C. Allen, *Phys. Rev. B*, 40 (1989) 12255.
15. R. G. Parr, in *Density Functional Methods in Physics*, R. M. Dreizler, J. da Providencia (eds.), Plenum (1985).
16. K. Fukui, *Science*, 217 (1962) 747.
17. R. G. Pearson, *J. Chem. Educ.*, 64 (1987) 561.
18. R. G. Parr, R. G. Pearson, *J. Am. Chem. Soc.*, 105 (1983) 7512.
19. P. Geerlings, F. De Proft, *Int. J. Quant. Chem.*, 80 (2000) 227.
20. A. Chatterjee, Lecture Notes in Computer Science (**LNCS**), 3993, 77-81. Volume editor V.N. Alexandrov et al for ICCS 2006, Part III ; 39930077_LNCS.pdf, 2006 Computational Chemistry and its application –Application of the reactivity indices to propose intra and inter molecular reactivity in catalytic materials.
21. R. G. Parr, W. Yang, *Density Functional Theory of Atoms and Molecules*, Oxford (1989).
22. R. G. Parr, R. A. Donelly, M. Levy, W. E. Palke, *J. Chem. Phys.*, 68 (1978) 3801.
23. P. Hohenberg, W. Kohn, *Phys. Rev.*, 136 (1964) B864.

24. W. Yang, R. G. Parr *Proc. Nat. Acad. Sci. USA*, 82 (1985) 6723.
25. R. T. Sanderson, *Science*, 114 (1951) 670.
26. R. T. Sanderson, *Chemical Bonds and Bond Energy* (2nd ed.), (1976) Academic Press, New York.
27. L. Pauling, *The Nature of Chemical Bond* (3rd ed.), (1960) Cornell, Ithaca, NY.
28. J. G. Malone, *J. Chem. Phys.*, 1 (1933) 197.
29. R. G. Pearson, *Science*, 151 (1966) 172.
30. G. Klopman, *J. Am. Chem. Soc.*, 90 (1968) 223.
31. R. S. Mulliken, *J. Chem. Phys.*, 2 (1934) 782.
32. For a series of papers covering various aspects of electronegativity see: K. D. Sen, C. K. Jørgensen (eds.), *Structure and Bonding*, Vol. 66, Springer Verlag, Berlin (1987).
33. C. Cárdenas, E. Chamorro, M. Gálvan, P. Fuentealba, *Int. J. Quant. Chem.*, 107 (2007) 807.
34. M. H. Cohen, *Top. Curr. Chem.*, 183 (1996) 143.
35. M. H. Cohen, M. V. Ganduglia-Pirovano, *J. Chem. Phys.*, 101 (1994) 8988.
36. M. H. Cohen, M. V. Ganduglia-Pirovano, J. Kudrnovsky, *J. Chem. Phys.*, 103 (1995) 3543.
37. P. Geerlings, F. De Proft, R. Balawender, *Rev. Mod. Quant. Chem.*, 2 (2002) 1053.
38. T. Berlin, *J. Chem. Phys.*, 19 (1951) 208.
39. B. G. Baekelandt, A. Cedillo, R. G. Parr, *J. Chem. Phys.*, 103 (1995) 8548.
40. B. G. Baekelandt, *J. Chem. Phys.*, 105 (1996) 4664.
41. R. Balawender, F. De Proft, P. Geerlings, *J. Chem. Phys.*, 114 (2001) 4441.
42. R. Balawender, P. Geerlings, *J. Chem. Phys.*, 114 (2001) 682.
43. E. Chamorro, P. Fuentealba, R. Contreras, *J. Chem. Phys.*, 115 (2001) 6822.
44. F. De Proft, S. Liu, P. Geerlings, *J. Chem. Phys.*, 108 (1998) 7549.
45. P. Ordon, L. Komorowski, *Chem. Phys. Lett.*, 292 (1998) 22.
46. P. Ordon, L. Komorowski, *Int. J. Quant. Chem.*, 101 (2005) 703.
47. M. Torrent-Sucarrat, J. M. Luis, M. Duran, A. Toro-Labbe, M. Sola, *J. Chem. Phys.*, 119 (2003) 9393.
48. E. Chamorro, F. De Proft, P. Geerlings, *J. Chem. Phys.*, 123 (2005) 84104.
49. C. Cárdenas, A. M. Lamsabhi, P. Fuentealba, *Chem. Phys.*, 322 (2006) 303.
50. W. Yang, W. J. Mortier, *J. Am. Chem. Soc.*, 108 (1986) 5708.
51. W. Langenaeker, M. De Decker, P. Geerlings, *J. Mol. Struct.*, 207 (1990) 115.
52. W. Langenaeker, K. Demel, P. Geerlings, *J. Mol. Struct.*, 234 (1991) 329.
53. M. Galvan, A. Dal Pino, J. D. Joannopoulos, *Phys. Rev. Lett.*, 70 (1993) 21.
54. F. Méndez, M. Galvan, A. Garritz, A. Vela, J. L. Gazquez, *J. Mol. Struct.*, 277 (1992) 81.
55. R. G. Parr, W. Yang, *J. Am. Chem. Soc.*, 106 (1984) 4049.
56. W. Yang, R. G. Parr, *Proc. Natl. Acad. Sci. USA*, 82 (1985) 6723.
57. J. L. Gazquez, F. Méndez, *J. Phys. Chem.*, 98 (1994) 4591.
58. R. G. Pearson, *Hard and Soft Acids and Bases*, Dowden, Hutchinson and Ross, Stroudsville, PA (1973).
59. S. Damoun, G. Van de Wounde, F. Méndez, P. Geerlings, *J. Phys. Chem.*, 101 (1997) 886.
60. A. K. Chandra, M. Nguyen, *J. Chem. Soc. Perkin Trans.*, 2 (1997) 1415.
61. K. Fukui, T. Yonezawa, H. Shingu, *J. Chem. Phys.*, 20 (1952) 722.
62. C. K. Ingold, *Recl. Trav. Chim.*, 48 (1929) 797.
63. C. G. Swain, C. B. Scott, *J. Am. Chem. Soc.*, 75 (1953) 141.
64. J. O. Edwards, *J. Am. Chem. Soc.*, 76 (1954) 1540.
65. J. O. Edwards, R. G. Pearson, *J. Am. Chem. Soc.*, 84 (1962) 16.
66. H. Mayr, M. Patz, *Angew. Chem. Int. Ed.*, 33 (1994) 938.
67. H. Mayr, K. H., A. R. Ofial, M. Buhl, *J. Am. Chem. Soc.*, 121 (1999) 2418.
68. H. Mayr, T. Bug, M. F. Gotta, N. Hering, B. Irrgang, B. Janker, B. Kempf, R. Loos, A. R. Ofial, G. Remennikov, H. Schimmel, *J. Am. Chem. Soc.*, 123 (2001) 9500.
69. H. Mayr, B. Kempf, A. R. Ofial, *Accounts Chem. Res.*, 36 (2003) 66.
70. S. Minegishi, H. Mayr, *J. Am. Chem. Soc.*, 125 (2003) 286.

71. S. Minegishi, S. Kobayashi, H. Mayr, *J. Am. Chem. Soc.*, 126 (2004) 5174.

72. R. G. Parr, L. V. Szentpaly, S. Liu, *J. Am. Chem. Soc.*, 121 (1999) 1922.

73. R. G. Parr, W. Yang, *Density Functional Theory of Atoms and Molecules*, Oxford University Press, Oxford (1989).

74. (a) R. G. Pearson, *J. Am. Chem. Soc.*, 85 (1963) 3533; (b) R. G. Parr, R. G. Pearson, *J. Am. Chem. Soc.*, 105 (1983) 7512.

75. P. K. Chattaraj, H. Lee, R. G. Parr, *J. Am. Chem. Soc.*, 113 (1991) 1855.

76. (a) P. K. Chattaraj, P. V. R. Schleyer, *J. Am. Chem. Soc.*, 116 (1994) 1067; (b) P. K. Chattaraj, B. Maiti, *J. Am. Chem. Soc.*, 125 (2003) 2705.

77. P. K. Chattaraj, B. Maiti, U. Sarkar, *J. Phys. Chem. A*, 107 (2003) 4973.

78. (a) R. G. Parr, W. Yang, *J. Am. Chem. Soc.*, 106 (1984) 4049; (b) K. Fukui, *Science*, 218 (1987) 747; (c) P. W. Ayers, M. Levy, *Theor. Chem. Accounts*, 103 (2000) 353.

79. L. R. Domingo, M. J. Aurell, P. Pérez, R. Contreras, *J. Phys. Chem. A*, 106 (2002) 6871. In the abstract of this paper the major problems associated with relative electrophilicity over local electrophilicity are highlighted.

80. E. Chamorro, P. K. Chattaraj, P. Fuentealba, *J. Phys. Chem. A*, 107 (2003) 7068.

81. N. F. Nalewajski, *J. Am. Chem. Soc.*, 106 (1984) 944.

82. N. F. Nalewajski, T. Konisiski, *Naturforsch*, 42a (1987) 451.

83. M. Berkowitz, *J. Am. Chem. Soc.*, 109 (1987) 4823.

84. B. S. Kulkarni, A. Tanwar, S. Pal, *J. Chem. Sci.*, 119 (2007) 489.

85. (a) P. K. Chattaraj, *J. Phys. Chem. A*, 105 (2001) 511; (b) J. Melin, F. Aparicio, V. Subramanian, M. Galvan, P. K. Chattaraj, *J. Phys. Chem. A*, 108 (2004) 2487; (c) A. Hocquet, A. Toro-Labbé, H. Chermette, *J. Mol. Struct.*, 686 (2004) 213.

86. (a) R. G. Parr, W. Yang, *J. Am. Chem. Soc.*, 106 (1984) 4049; (b) Y. Yang, R. G. Parr, *Proc. Natl. Acad. Sci. USA*, 821 (1985) 6723.

87. (a) P. O. Lowdin, *J. Chem. Phys.*, 21 (1953) 374; (b) P. O. Lowdin, *J. Chem. Phys.*, 18 (1950) 365; (c) R. S. Mulliken, *J. Chem. Phys.*, 23 (1955) 1833.

88. S. Krishnamurthy, S. Pal, *J. Phys. Chem. A*, 104 (2000) 7639;

89. A. Chatterjee, T. Iwasaki, T. Ebina, *J. Phys. Chem. A*, 105 (2001) 10694.

90. A. Chatterjee, T. Iwasaki, T. Ebina, *J. Phys. Chem. A*, 106 (2002) 641.

91. R. Parthasarathi, J. Padmanabhan, M. Elango, V. Subramania, P. K. Chattaraj, *Chem. Phys. Lett.*, 394 (2004) 225.

92. A. Chatterjee, T. Ebina, T. Iwasaki, *Stud. Surf. Sci. Catal.*, 145 (2003) 371.

93. A. Chatterjee, T. Ebina, T. Iwasaki, F. Mizukami, *J. Chem. Phys.*, 118 (2003) 10212.

94. A. Chatterjee, T. Ebina, Y. Onodera, F. Mizukami, *J. Mol. Graph. Model.*, 22 (2003) 93.

95. A. Chatterjee, T. Ebina, T. Iwasaki, F. Mizukami, *J. Mol. Struct.*, 630 (2003) 233.

96. A. Chatterjee, T. Suzuki, Y. Takahashi, D. A. P. Tanaka, *Chem. Eur. J.*, 9 (2003) 3920.

97. A. Chatterjee, *J. Mol. Graph. Model.*, 24 (2006) 262.

98. A. Chatterjee, T. Balaji, H. Matsunaga, F. Mizukami, *J. Mol. Graph. Model.*, 25 (2006) 208.

99. A. Chatterjee, A. Kawazoe, *Mater. Trans.*, 48 (2007) 2152.

100. A. Chatterjee, M. Chatterjee, *Mol. Simat.*, 34 (2008) 1091.

101. P. Geerlings, F. De Proft, *Phys. Chem. Chem. Phys.*, 10 (2008) 3028.

5 Synthesis of Nanoporous Materials

In recent years considerable progress has been made to design and characterize nanoporous materials to make them obtainable for variety of applications from medicine to catalysts. However, it is evident that nanoporous materials are complex in character due to their disordered pore structure and different surface characteristics; in addition, the methods used for their analysis are sometimes old and based on oversimplified concepts. Furthermore, considerable caution should be exercised prior to the interpretation of the experimental data and their deviations. Hence, there is a need for molecular modeling to resolve issues before the synthesis. Thus, more advanced simulation procedure is now available for the designing of the nanoporous solid comprising with their application to the specific field. Again, simulations now face serious challenges. In particular, contemporary materials are becoming so complex that it may soon be impossible to generate models that are sufficiently realistic to describe them adequately. Structural complexity evolves during synthesis, and therefore one way of capturing such complexity within atomistic models is to "simulate synthesis." The whole idea of this book is to show how simulation can help support the bench chemist in the experimental process starting from synthesis, characterization, and reaction to establish a smart, tailor-made material for a specific application.

5.1 NUCLEATION AND GROWTH

It has been mentioned before that zeolites are the most widely used nanoporous aluminosilicates and contribute significantly to catalysis and separation in an industrial scale [1]. One of the most studied areas of computer simulation is that of the nucleation and growth of its structure based on the concept that such understanding could be used to optimize catalysis and separations by tailoring zeolite crystallite size and shape. Depending on the structure of the zeolite, its application can be extended to some new fields like optical electronics [2], bio-implants [3], and enantioselective separation [4]. Thus, it is necessary to have a clear idea on the nucleation and growth of zeolites to obtain a tailor-made structure for specific field of applications [5]. Some fascinating experiments and theoretical studies were conducted at the beginning of the nanoporous era that commented on the crystallization process and variables [6–8]. Detailed aspects of the numerous events in zeolite crystallization are elaborated on in an article by Serrano and Grieken [9]. When zeolite synthesis starts from a solid-containing gel, the participation of the amorphous raw phase in the formation of the zeolite nuclei has been clearly established. Moreover, in some cases the zeolite crystals grow by reorganization of the hydrogel through solid–solid transformations. In zeolite synthesis starting from clear solutions, the crystallization mechanism has been studied by *in situ* light scattering techniques, showing the formation in the first stages of an X-ray amorphous gelatinous phase, consisting of particles with sizes below 10 nm. These nanoparticles seem to be involved directly in the crystallization, because their aggregation and subsequent densification have been proposed to lead to the formation of the zeolite crystals. The crystallization of zeolites through heterogeneous versus homogeneous pathways is favored at high solid concentrations or when the solubility of the silicate species is low, which occurs during zeolite synthesis through the fluoride route.

Derounae et al. [7, 8] postulated two extreme synthesis mechanisms governing the formation and growth of ZSM-5 crystallites in the presence of TPA as structure-directing agent, depending on the source of silica and the relative concentrations of the reactants. Syntheses starting from silica solutions (type A) with low Si/Al and Na/SiO$_2$ ratios were found to take place by liquid-phase ion transportation, whereas syntheses starting from aqueous silicate solutions (type B), with higher Si/Al and Na/SiO$_2$ ratios, appeared to be governed by a solid hydrogel reconstruction. Figure 5.1 shows a schematic representation of both types of mechanisms. Mechanism A essentially occurs in highly alkaline medium, where the solubility of polymeric silicate ions is increased. In this case, depolymerization of the silica source to yield the appropriate building blocks is the rate-limiting step. The resulting monomeric or oligomeric silicate anions can either condense with aluminate species to form aluminosilicate complexes or interact directly with TPA ions, the latter being capable of ordering around them (preferably Si-richer silicate units) to form stable nuclei. The preference to accommodate silica instead of alumina by the ZSM-5 nuclei leads to an increase in the crystal growth rate at the expense of the nucleation rate from dissolved silicate species in solution. The gel dissolution supplies the reactants for the process, balancing the thermodynamic equilibrium of the complex system. As the silicate species available in solution are exhausted, the gel continues to dissolve and bring progressively Al-rich soluble species to the outer layer of the growing particles. This mechanism yields large crystals with an inhomogeneous Al distribution. According to mechanism B, starting from an aqueous sodium silicate solution, ingredients are mixed at acidic pH, whereas the hydrogel formation at pH ~11 takes place through NaOH addition. In this case, the starting solution contains monomeric or low oligomeric silica species because the presence of silicate ions is not limited by any depolymerization process, and, as well, the solution composition is very similar to the reagent ratio. The high Si/Al and Na/SiO$_2$ molar ratios favor a rapid nucleation through the interaction of the structure directing TPA cations, present throughout the gel, with the reactive aluminosilicate anions. Under these conditions, a direct recrystallization process is postulated mainly involving a solid hydrogel transformation, as suggested by the absence of low-molecular-weight species in the analysis of the solution by liquid ^{29}Si NMR. The analysis of the Si/Al molar ratios during the whole crystallization process shows a remarkably constant value, indicating that both the growing crystallites and their gel precursors must continuously keep the same composition, supporting the direct hydrogel formation.

Several experimental techniques have been established to shed light on the early stage of zeolite nucleation such as X-ray and neutron scattering, infrared (IR) spectroscopy, nuclear magnetic resonance (NMR), and dynamic light scattering (DLS), but these have encountered a number of difficulties because relevant length scales

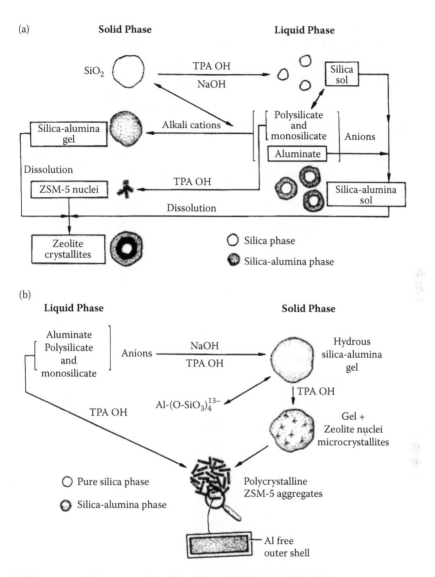

FIGURE 5.1 Steps of zeolite crystallization: (a) solution-mediated phase and (b) hydrogel phase. [7] With permission.

of the zeolite formation process lie just in between the accessible length scale of NMR and diffraction techniques [10]. Thus, it has been found that molecular simulation has the potential to go deeper into the process of nucleation and zeolite growth [11–16]. For zeolite scientists, it is well known that the synthesis of zeolites is comprised of many experimental parameters like the nature of the template, temperature, the nature of the alkali metal cation, silica and alumina source, gelation time, and crystallization time. On the other hand, simulation parameters lead to the sizes and structure of the critical zeolite nuclei, the role in zeolite growth of secondary building units such as rings or partial cages, structure directing agents such as organic templates or hydrated alkali cations, and precursor silica nanoparticles. Indeed, a considerable body of work has been done to simulate synthesis of zeolites to provide predictive value to experiments. The most significant advantage of the molecular simulation method is that it has potential to choose a particular synthetic parameter without affecting the others and more meticulous information can be obtained on any particular phase of synthesis. For example, detailed work by Hamad et al. [17] provides insight into the embryonic cluster leading to the formation of the nuclear seeds. Again the simulation of dissolution shows an influence of structural imperfection on the morphology. For the nanoparticle phase, small-angle X-ray and neutron scattering pair distance distribution functions have been modeled using a simulated annealing Monte Carlo (MC) algorithm to determine the silica structure as well as the cation amount and location (particle tested with tetramethylammonium hydroxide). To study the nucleation and growth of the zeolite structure, so far fully quantum mechanical calculations using density functional theory (DFT) have been applied to the silica polymerization [18]. Those studies have shown that polymerization is preferable over strong and stable silicate six rings and linear polymers compared to the smaller rings and branched polymers. These studies also include the effect of pH on the polymerization process. Based on *ab initio* calculations not including template molecules, Wu and Deem analyzed the free energy barriers and critical cluster sizes as a function of pH and Si-monomer concentration at ambient conditions using a series of advanced MC techniques [19]. They found that the critical clusters for the polymerization contained relatively few ($\approx$30–40) Si atoms. There have been several other attempts to model silica polymerization using molecular simulation [20]. One can really simulate polymerization by building the single building unit of the silica source — say, the SiO_4 unit, which is Si terminated at the next O-Si distance as mentioned in our previous work [21, 22]. It is possible to build the subsequent polymerization unit in the form of dimer, trimer, tetramer, and finally the secondary building unit (SBU) in terms of a pentamer. These exemplary building blocks are shown in Figure 5.2a through Figure 5.2e. The zeolite structures are a fascinating piece of work where the internal bonding may play a crucial role with their local architecture. Our work is the first study [21] to choose the best template for a particular zeolite synthesis by estimating the individual activity of templating molecules and the framework cluster in terms of reactivity index and then validating the results through interaction energy calculations. Here, we are successful in justifying the charge-compensating role of templates, which is observable through the favorable interaction of the templates with the framework containing aluminum. Thus, the role of aluminum has also been monitored. This paves a novel qualitative way for estimating the activity of interacting template molecules to choose a matrix for their usage, which is a prime need for tailoring. A few research groups in this field assumes that at the initial stage precursor particles are formed as a core shell structure of silica entrapping the template molecule, which then sticks together in the form of a nanoslab with the correct crystalline structure. These particles finally form the zeolite by a clicking mechanism when the solution is heated up. It should be mentioned that the formation of silica nanoparticles in the early stage is a debatable issue and opinions largely differ; for example, the nanoparticles are added one by

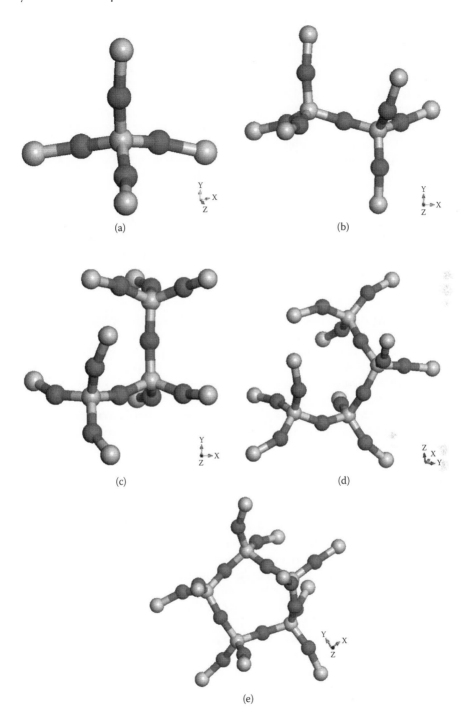

FIGURE 5.2 (For color version see accompanying CD) Polymerization unit: (a) monomer, (b) dimmer, (c) trimer, (d) tetramer, and (e) secondary building unit (SBU) pentamer.

one to the growing crystals [23], or the particles are regarded as monomer reservoirs: monomer dissolves into the solution and attaches to the growing nucleus or in other words an aggregative growth mechanism of discrete nanoparticles may dominate the early stage of the growth process [24]. Jorge et al. presented a lattice model describing the formation of silica nanoparticles in the early stages of the clear-solution templated synthesis of silicalite-1 [25]. Using this simplified model, they identified the nanoparticles in a metastable state, stabilized by electrostatic interactions between the negatively charged silica surface and a layer of organic cations. The nanoparticle size is controlled mainly by the solution pH, through nanoparticle surface charge. The size and concentration of the charge-balancing cation are found to have a negligible effect on nanoparticle size and it has been suggested that this mechanism may play a role in the growth of zeolite crystals [25]. Such understanding could be used to optimize catalysis and separations by tailoring zeolite crystallite size and shape.

High-resolution transmission electron microscopy (HRTEM) can resolve surface and bulk crystallographic features and surface structural features to compare with DFT simulation results and in combination prove the best tools to address the growth issue, especially with zeolites and related nanoporous materials. In a recent article [26], the growth of the double four ring is found to be a crucial and rate-determining step in the surface-mediated, postnucleation crystal growth mechanism of zeolite Beta C. Growth of four rings is found to be more favorable on fast-growing rather than slow-growing faces, explaining the relative growth rate of crystal faces in these materials. Similarly, the terminating structures of zeolite Y/faujasite can be partly explained by considering the condensation of six-ring and double six-ring species at the crystal surface. Though four rings and double four rings are known to be present as solution species, six rings and double six rings are not, and hence it is speculated that postnucleation crystal growth may involve competition between primary building unit– and secondary building unit–mediated crystal growth mechanisms. Invoking the SBU-mediated growth mechanism, it seems that unless there is a very large concentration of D4Rs, diffusion-limited growth would occur, caused by the low density of D4Rs in a viscous medium occasionally reaching the nucleation sites on the surface. Because the conditions at the interface are very distinct from the conditions probed by NMR and mass spectroscopic studies, we cannot discount the possibility that the gel interface is abnormally rich in oligomers not found in clear solution conditions sampled by various spectroscopic techniques [27].

The stability of FAU, EMT, and the intergrowth can be obtained from relatively simple interatomic potential lattice energy calculations using the potential set described previously with a cutoff of 12 Å [28]. The work of Henson et al. [29] has demonstrated the remarkable accuracy of the Sanders and Catlow force field [30] in comparison to calorimetric data and indeed the mean error in geometry and energy relative to quartz is comparable to *ab initio* methods within the General Utility Lattice Program (GULP) [31]. This is a program created to deal with inorganic matrixes with a special emphasis on the growth process or nucleation as well as the high temperature–pressure hydrothermal condition.

Mora-Fonz et al. [32] reported a fascinating work of polymerization and nucleation using DFT calculations with salvation and found that there is pH dependence for polymerization and the formation of small rings. Their work suggested a very important aspect that, even with the continuum model they have chosen to address the pH by considering anionic silicate species, can give a reasonable description of the system. Secondly, the formation of cyclic fragments is clearly favored in the process they mimicked through the polymerization process of the small chains — in agreement with experiment and with the expectation of the formation of zeolite-like nucleation species. With this methodology they have also proven that it is very unlikely for larger noncyclic oligomers to play a significant role in either nucleation or crystal growth. Hence, growth is much more likely to occur by condensation of relatively small units, particularly those with lead to rings, in either nucleation species or in subsequent surface growth. They have also validated the proposition of Slater et al, that surfaces with complete (small) rings are prevalent from both high-resolution transmission electron microscopy and computational studies of siliceous zeolite surfaces [33]. Thirdly, it is clear that large single (>5 and even 5) rings are not formed as free species in solution. Another study [34] looked into oligomerization through a similar mechanism and the initial steps of silica formation by quantum chemical techniques. The formation of various oligomers (from dimer to tetramer) was investigated using DFT and a continuum solvation model. The calculations show that the anionic pathway is kinetically preferred over the neutral route. The first step in the anionic mechanism is the formation of the Si–O–Si linkage between the reactants to form a five-coordinated silicon complex, which is an essential intermediate in the condensation reaction. The rate-limiting step is water removal, leading to the oligomer product. The activation energy for dimer and trimer formation is 80 kJ/mol, which is significantly higher than those of the subsequent oligermerization. Again, the activation energy for the ring closure reaction (100 kJ/mol) is even higher. The differences in activation energies can be related to the details in intra- and intermolecular hydrogen bonding of the oligomeric complexes.

However, truly realistic simulations of all stages in zeolite synthesis are still a long way off, because the typical system sizes and time scales at which zeolite formation takes place are generally beyond the capabilities of present computer resources.

A unique proposition has recently been tested [32] within the helm of many different calculations, which allows chemically accurate deprotonation and dimerization energetics to be calculated at conditions reflecting those of high pH hydrothermal zeolite synthesis. The free energies of condensation reactions leading to silicate species up to the linear tetramer are considered at room temperature and at 450 K. It is a well-known fact that the complexity of silicate chemistry limits the experimental studies to only a few species, specifically the monomer and dimer, because many different reactions occur once polymerization is initiated, the individual speciation of which proves almost impossible. Similarly, computational methods have also been restricted by their relative high cost and lack of incorporation of descriptions of solvent and pH effects.

The dimerization reaction is the fundamental reaction of silica chemistry. Again, the range of conditions (acidic, neutral, basic) and the difficulties in attempting to restrict further polymerization make it a challenging reaction to study experimentally. The experimental estimate of the ΔG is = -7 kJ mol/mol [35]. Most recently, Trinh et al. [34] reported values of $+9$ kJ mol/mol and -28 kJ/mol for the neutral dimerization and the reaction of a monomer with a deprotonated monomer, respectively, by B3LYP/6-31+G (d, p) with COSMO solvation. However, none of these studies consider explicit hydration. This is the only calculation in years that has reproduced the experimental data for the dimerization reaction, both under neutral and under basic conditions. The most important object is that these calculations correctly reflect how an increase in pH makes polymerization more favorable. Moreover, we find the reaction of two deprotonated monomers (with cations and explicit water) to be slightly less favorable (−3.4 kJ/mol). This validates the similar observation from experiments, which predicts that high pH does not necessarily result in increased polymerization. Thus, a combination of the DFT method used together with a suitably explicit representation of the inner solvation sphere of the key species and a continuum representation of the remainder of the solvent allows the aqueous chemistry of silicate oligomers to be modeled reliably. At this point it is the best technology to monitor the polymerization process beyond experiment and model a situation of the inside of the autoclave during hydrothermal conditions.

For the correct modeling of the nucleation process, the simulation box should at least be larger than the critical nucleus, a requirement that is out of limits for quantum mechanical calculations and demands the development of accurate reactive force fields.

In contrast to zeolite-like microporous material, theoretical studies on the nucleation of mesoporous materials are scant even in experiments. Few calorimetric and spectroscopic techniques have been developed to study the synthesis mechanisms of mesoporous materials. Calorimetric experiments show that the formation of MCM-41 mesoporous silica involves three steps. In the first step, rod-like inorganic/surfactant complexes are formed that spontaneously pack into an ordered arrangement. Second, with the help of the new template, a novel kind of pore structure was formed. Third, the interaction between the pore structures formed in the former steps 1 and 2, respectively results in a new pore structure. This new pore structure is very similar to the architecture formed in the first process [36]. EPR was also performed on the reaction mixtures leading to the formation of the hexagonal, MCM-41, and the lamellar, MCM-50, materials at room temperatures. These *in situ* measurements provide direct experimental evidence that micelles serve as precursors for mesoporous materials. In contrast to the calorimetric measurements, formation of the MCM-41 appears to form in two stages. The first stage, which starts immediately after the mixing of the silica source with the surfactant solution, involves the formation of domains with hexagonal order. These domains consist of micellar rods encapsulated with silicate ions oligomers. At room temperature this stage lasts 3–5 min. The second, much slower stage (1–1.5 h), involves polymerization of the silicate ion at the interface. It results in hardening of the inorganic phase and at the same time in restricting the motion of the surfactant molecules at the interface. The latter stage requires a minimum Si/surfactant ratio, below which the polymerization will not start [37].

The situation of metal organic framework (MOF) materials is similar to that of mesoporous materials. Most studies are mainly related to adsorption over MOF because of its high surface area. A simple mechanism for the formation of MOF was proposed by Ramanan et al. [38]. According to their prediction, as soon as a metal salt is dissolved in water or a nonaqueous solvent, a soluble metal complex is initially formed. Knowing how this complex organizes with the organic groups in the medium of the solvent is crucial to understanding the transformation of these molecules into the final solid through hydrolysis and condensation [38].

5.2 SCREENING RAW MATERIALS

There are three basic stages in the hydrothermal synthesis of zeolite and those are (1) the creation of an alkaline hydrogel of silica alumina species, (2) transformation of the hydrogel in to supersaturated solution, and, finally, (3) crystallization of aluminosilicate product. In the creation of silica alumina hydrogel, a range of silica sources such as silica gel, water glass fumed silica, and silicon alkoxide has been used. Aluminum sources are mainly gibbsite or soluble aluminum salt. Quantum chemical calculations are used to determine the relative stability and reactivity of the component of the reaction mixture. Pereira et al. reported the energies of the smallest aluminum-based cluster and suggested that $SiOAl\ (OH)_6^-$ and $AlOAl\ (OH)_6^{2-}$ have strong hydrogen bonds and the energies are controlled by the charge distribution and depend very little on the structure factor [39]. Ermoshin et al. reported that $Si(OH)_5^-$, $Al(OH)_4^-$, and $SiOAl(OH)_6^-$ ions in alkaline solution are much more stable than their neutral species $Si(OH)_4$, $Al(OH)_3$, and $SiOAl(OH)_5$ and the main mechanism of formation of the aluminosilicate structure is polycondensation using anionic species [40]. Determining the actual scenario behind the mechanism of the zeolite synthesis is still a challenge. Recently, the molecular mechanism of aluminosilicate dimer formation in basic solution has been discussed in detail using DFT calculations. Two different reaction pathways (concerted and stepwise) for condensation between $Al(OH)_4$ and $Si(OH)_4$ were proposed and suggested the formation of large oligomers via further condensation of the aluminosilicate anion [41]. Thus, basicity of the medium is an important criterion for condensation followed by crystallization of the aluminosilicate structure and in order to achieve alkalinity, a base such as sodium hydroxide is introduced into the medium. Now what kind of role could these cations play? Crystallization of zeolites has taken place in aqueous medium. It is well known that small, densely charged ions interact with water and rearrange it into an organized cluster. So, the larger cations like alkyl ammonium cations (acting as a template; details will be discussed later) can also order the water molecule. During crystallization, cations surrounding the water molecules are replaced by the aluminosilicate and those structures combine to produce aluminosilicate topology. The formation of this extended structure is controlled by Na^+ or K^+ ions present in the medium.

The most exciting advancement in zeolite synthesis is the prediction of the role of the template. By far the most successful strategy to obtain a new zeolite is the variation of the template because the charge distribution and the size and geometric shape of a template are believed to be the causes for structure directing. What kind of organic molecule could be used as a template for zeolite synthesis? Lewis and Freeman [42] answered the question by studying the interaction between the template and the zeolite framework based on a combined molecular dynamics, MC, and energy minimization technique. Results of their calculations shed some light on the aspects of organic molecule to act as template in the synthesis of zeolite. A nonbonding interaction has been predicted between the template and the zeolite framework, which have strong impact on the nature of the product. Again, the template molecule must be packed efficiently within the framework [42]. Direct experimental determination of the location and conformation of the organic molecule inside the zeolite host is difficult. Thus, a variety of molecular modeling methods were used to screen novel templates for zeolite synthesis. The activity of different representative templating molecules along with zeolite frameworks was investigated using a range of reactivity indexes using DFT. From the values of local softness and the charge on the hydrogen atom of the bridging hydroxyl, resulting from the presence of aluminum in the framework, it has been observed that the acidities of the aluminum-containing zeolite-type model systems are dependent on several characteristics of importance within the framework of the hard and soft acids and bases (HSAB) principle. The local softness of the interacting templates has also been calculated to compare their affinity with the zeolite framework cluster models. The cluster models are chosen to mimic the secondary building units of zeolite crystals for both silicalite and silica aluminates. Conformational flexibility was brought out as a common feature of those representative organic templates. The influence of the nature of the functional group and alkyl group on the electronic interaction has been studied systematically. An *a priori* rule was formulated for choosing the best template for a particular zeolite (e.g., ZSM-5) synthesis [43]. Results of computational studies on the structure-directing role of alkyl groups present in tetra alkyl ammonium (TAA) during the synthesis of ZSM-5 were reported [22]. To account for the extent of interaction, different cluster models were chosen. Propane and butane were chosen to interact with the framework clusters to understand the role of alkyl groups in TPA (tetrapropyl ammonium) and TBA (tetrabutyl ammonium), which are used in the synthesis of ZSM-5 as templates. It was observed that the location of the molecule inside the framework plays a decisive role in predicting the efficiency of template. This is due to the typical location of the additional methyl group of the butane molecule inside the framework and results in a stronger interaction with the framework. The additional methyl group of butane plays a crucial role in the polarization of the methylene hydrogen of C of the interacting butane molecule. The efficiency of the template depends on the variation in the number of alkyl groups in ammonium cations containing different alkyl groups. This study indicates that tetra-methyl ammonium cations give a more unfavorable interaction than mono-, di-, and tri-methyl ammonium cations do. These results indicate the trend that higher numbers of alkyl groups in the template produce more unfavorable interactions with the framework at a particular orientation of the molecule; i.e., keeping the position of the terminal methyl group fixed [22]. We show in Figure 5.3a some exemplary template molecules in the group of quaternary ammonium salts such as triethyl, tetraethyl, tetrapropyl, and tetrabutyl ammonium cations. These templating molecules direct the structure by helping the polymerization process in the presence of water. Those template molecules can be simulated either by atomistic or DFT methodologies. Finally, the structure-directing templates can be found at the pore of silicalite before calcination. Figure 5.3b is an example of TMA inside the straight channel, whereas TPA is at the channel intersection (Figure 5.3c) as confirmed by our continuous work on templates along with structural information

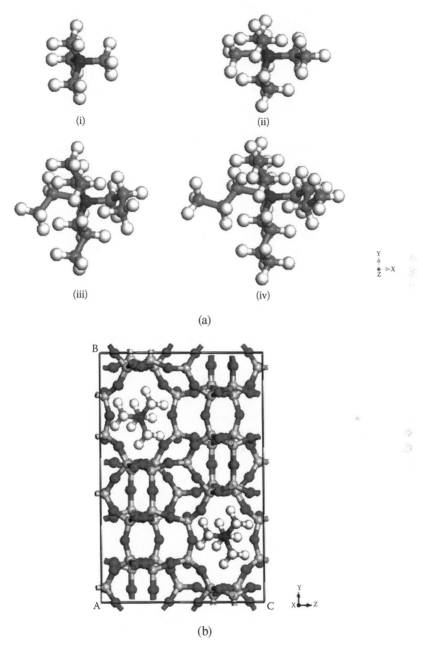

FIGURE 5.3 (For color version see accompanying CD) (a) Exemplary template molecules in the group of quaternary ammonium salts: (i) triethyl, (ii) tetraethyl, (iii) tetrapropyl, and (iv) tetrabutyl ammonium cation. (b) TMA inside the straight channel.

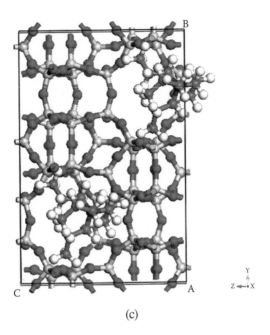

(c)

FIGURE 5.3   (Continued). (c) TPA at the channel intersection.

presented [21]. Structure analysis has also been carried out over several templating agents using molecular mechanical energy calculations and the molecular principle axes of inertia plotted to produce a three-dimensional shape-space diagram. A clear correlation is observed between the shape of the template and the zeolite product formed consequently, which explains the formation of the same product from the large and small templates [44]. The principal axes of inertia for each template were calculated using a quantitative structure property relationship (QSAR). The moment of inertia I of a polyatomic molecule is defined as

$$I = \sum_i m_i \vec{r}_i^2$$

where the summation includes all atoms I, and m is the mass of the ith atom, separated from the axis of gravity by a distance $\vec{r}_i$. The molecular moment of inertia is resolved using the orthogonal principal moments IX, IY, and IZ, described in g/cm². The principal axes of inertia RX, RY, and RZ are described in Å and are inversely proportional to the corresponding moments. The RX vector is aligned along the largest molecular moment IX, and RZ vector is aligned along the smallest moment IZ; the principal axes are scaled to define an internal ellipsoidal as shown in Figure 5.4. The axes have shown in the corner represents the three axes of inertia. An example with minimum and maximum values for templates forming two types of zeolites FER and MOR are shown in Table 5.1. This has proven that templates in terms of size and shape as shown in Figure 5.4 can be analyzed for screening purposes. Also, Monte Carlo docking algorithms based on a force field approach combined with crystallographic data were used by Toby et al. to study the effect of structure-directing agents

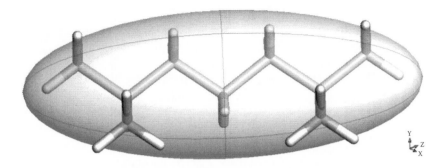

FIGURE 5.4 (For color version see accompanying CD) Domain of the principal axes of inertia for each template in terms of size and shape.

in the inhibition or formation of stacking faults in CIT-1 [45]. Sun et al. rationalized the efficiency of a template in terms of the energetics of the host-template nonbonding interaction energy and predicted that the positive match between di-quaternary cations and zeolite pair is a poor match and never appeared in experiments [46]. According to the templating theory, one template can give many structures and, similarly, many templates with different shapes and sizes can give the same structure. Force field calculations were utilized to study the structural and conformational properties, and semiempirical MNDO calculations were adopted to study the electronic properties more than fifty templating organic molecules that led to the synthesis of ZSM-5. The conformational flexibility and the charge-compensating function were brought out as common features of these organic molecules. The orientation of the template organic molecules and the actual conformation of the framework–template complex were derived, based on the X-ray crystal structure reports and the MNDO energetics. It has been suggested that quantum chemical calculations could be successfully utilized for the generalized description of the importance of several complex interactions, which are mutually present [47]. The role of structure-directing agents is also rationalized in terms of the energetic stabilization between the template and the microporous zeolite structure. An explanation is provided for the synthesis outcome in terms of a balance between kinetic and thermodynamic factors throughout the nucleation and crystallization stages. The energetic terms corresponding to the short-range interaction between the zeolite and the structure-directing agent play an important role during the nucleation stage, and zeolite synthesis may be mainly driven by this interaction. In such a case, the results can be explained in terms of the stability of the appropriate nuclei during the initial stages of zeolite formation. In addition, the conformation of the occluded organic cation inside the zeolite is an important parameter. Nevertheless, flexible templates may offer a wide range of conformations. The energy calculation shows that when the template molecule entered into the zeolite microcavities, the energy of the zeolite with respect to the equilibrium energy is increased. This energy enhancement has an important influence on the synthesis process, because during the nucleation process template cations must adopt a conformation that matches the channel/pore of the zeolite to be formed [48]. Sastre et al. has used the GULP program to investigate preferential location of aluminum and proton citing in zeolite type ITQ-7. They have shown that it is not the energy of the final structure that controls the aluminum distribution but rather the energetics during the synthesis process (zeolite+ template) needs to be considered. Hence, the accommodation of the template, which acts as the structure-directing agent, within the microporous space is detrimental. The template not only acts as a structure-directing agent but also as an aluminum distribution guider. This methodology allows us to compute the OH stretching frequencies and consequently a map of acid site frequencies was obtained. Moreover, relative population, which gives a simulation of the OH stretching IR bands in terms of frequencies and intensities, was found. The methodology has been validated by applying this general approach to the case of the newly synthesized ITQ-7 zeolite and the corresponding calculated IR spectrum shows a reasonable agreement with the experimental spectrum. Two bands centered at 3,577 and 3,624 cm^{-1} result from the simulation, which compare to the experimental values of 3,595 and 3,629 cm^{-1}. Preferential proton citing has also been studied and energy differences in each AlO_4 tetrahedra of about 30–40 kJ/mol were found, in qualitative agreement with ^{1}H NMR results.

Thus, based on *ab initio* calculation and experimental data, several force fields have been developed, which allows the study of larger systems including solvent and template molecules. The large number of experimental factors that influence the synthesis cannot be accurately simulated by any technique available today, and drastic approximations have to be made.

Hence, a combination of several computer modeling techniques is essential for better understanding of the structure-directing ability of the organic molecules used as templates. It has been observed that the worthiness of a template can be rationalized in terms of the energetics of the template–framework interactions. Those techniques can successfully apply to the design and the synthesis of new materials.

In addition to zeolites, mesoporous materials are an important part of nanoporous materials, where the surfactant plays an important role to tailor the pore size and the pore structure of the material. The generation of atomistic models of PMS by a standard molecular dynamics (MD) simulation of the synthesis is not feasible because of the size of the system and the time scale that has to be covered by the simulation. To circumvent this problem, several approaches have been taken. For example, MCM-41 has been modeled as isolated one-dimensional channels with regular and disordered walls [49], as oxygen atoms distributed randomly around the pore volume [50], or by relaxation of randomly generated amorphous silica walls by MD simulation at high temperature [51]. The oxygen atoms in a typical single unit cell of our model MCM-41 material are shown in Figure 5.5. The BET-type surface area of the pore shown in Figure 5.5 is therefore calculated from the diameter of the monolayer. All model surface areas presented here are taken from the center of the atoms in the monolayer and are therefore equivalent to experimentally derived BET surface areas. The polymerization of silicic acid monomers in the absence of templating micelles has been simulated in MD runs at high temperatures [52]. Wu and Deem studied the nucleation of silica clusters from acid monomers in a Monte Carlo scheme that takes the atomic positions of the atoms into account [53]. In addition, silica clusters and their interactions with other molecules present in the synthesis solution have been studied with *ab initio* methods [54]. A novel kinetic Monte Carlo method has

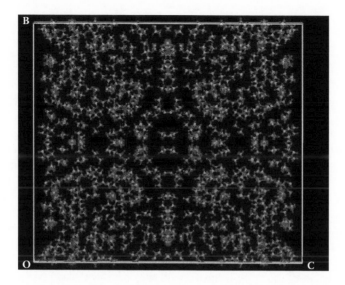

FIGURE 5.5 (For color version see accompanying CD) Single unit cell model MCM-41 material.

been developed to simulate the synthesis of templated PMS structures. First it was applied to silica without the presence of a templating surfactant. The simulation data showed reasonable agreement with experimentally observed kinetics of a similar system and with the properties of amorphous silica, which can in turn generate the model of MCM-41 materials generated with different pore diameters. The only input parameters in the simulations are the geometry of the micelle and the number of silicic acid monomers at the beginning of the simulation. The resulting amorphous structures seem to be realistic, but the degree of polymerization is lower than that of the synthesized material [55].

5.3 OPTIMIZING THE BULK STRUCTURE WITH VARIED COMPOSITION

The composition of the reaction mixture is one of the most important governing parameters to determine the zeolite structure. These include (1) silica/alumina ratio, (2) concentration of OH^- ion, and (3) inorganic cations. The silica/alumina ratio determines the candidate framework in the exploratory zeolite crystallization, hydrophobic and hydrophilic nature of the material, affinity for the polar adsorbent, thermal stability, acidity of the structure, and the cation exchange capacity. A vast experimental effort is needed to optimize the zeolite structure with varied composition. Gel compositions representing most areas of the silica-rich half-triangle of the crystallization field that yields open structure zeolites were prepared and crystallized under different conditions in a design of factorial experiment. The synthesized products were examined for their crystal structures and their Si/Al ratios. The data prove that the gel-alkalinity factor (a) increases the rate of crystallization to a point but a further increase reveals a decrease of the phase transformation rate, thus stabilizing metastable structures; and (b) decreases the Si/Al ratio of the zeolitic framework. The increase of the gel-aluminicity (a) generates new crystalline structures that grow simultaneously with those that are formed in low-aluminicity gels and (b) decreases the slope of the framework Si/Al ratio versus alkalinity. An increase of crystallization temperature produces structures that are denser and more stable. Structure stability sequences were established in experiments performed at relatively high gel alkalinity and aluminicity. The distributions of Brønsted acidic protons and their acid strengths in zeolite H-MCM-22 have been characterized by DFT calculations as well as magic angle spinning (MAS) NMR experiments. The embedded scheme (ONIOM) that combines the quantum mechanical (QM) description of active sites and semiempirical AM1 treatment of the neighboring environment was applied to predict the aluminum substitution mechanism and proton affinity (PA), as well as adsorption behaviors of acetone and trimethylphosphine oxide (TMPO) on the zeolite. Theoretical results indicate that the Al substitution takes place in the order of Al1–OH–Si2 > Al8–OH–Si8 > Al5–OH–Si7. The DFT and NMR results suggest that the acid strength of the three Brønsted acid sites in H–MCM–22 zeolite is slightly lower than that of H–ZSM–5 zeolite and the accessible Brønsted acidic protons most likely reside in both the supercages (at the Al8–OH–Si8 and Al1–OH–Si2 sites) and external surface pocket (at the Al8–OH–Si8 site) rather than in the sinusoidal channels (Al5–OH–Si7), with the Al1–OH–Si2 site having the strongest acid strength (as probed by TMPO) [56]. The cationic, structural, and compositional influences on the structure and bonding of zeolitic aluminosilicates have been investigated with the DFT method including local (VWN) and nonlocal spin density functionals (BLYP). Full optimization of structures has been carried out at the 6-31G*/VWN and 6-31G*/BLYP levels of theory for the different types of $\equiv$ Si–OH–Al $\equiv$ units in the secondary building unit of the zeolite cluster models $[(OH)_8 H_y Al_x Si_{8-x} O_{12}]^{(x-y)-}$ ($x,y = 0, 1, 2, 4$) and the silica model $(OH)_8 Si_8 O_{12}$. Changes in the environment of the silicon and aluminum framework atoms with a given Si/Al ratio generate new different acid sites. Proton affinities of Brønsted hydroxyl groups in H forms of zeolites associated with different Si/Al ratios indicate that the higher the ratio, the less the proton is constrained, which results in a stronger acid strength [57]. A first principle density functional procedure has been used to examine the influence of the silica/alumina ratio on the structure and proton affinity of zeolites. According to the study [58], low-alumina zeolites are more acidic than high-alumina zeolites due to the acidity of the Si–O–Al unit. It also suggested that large structural deformation occurs in the isomorphous substitution of silica by alumina [58]. The Si/Al ratio also has an effect on the concentration of the exchangeable cations, which have a profound effect on the physical properties of the material. This Si/Al ratio is also deciding the catalytic properties of the material and has been the subject of several experimental and theoretical studies. Experimentally, ^{29}Si and ^{27}Al NMR can provide ample evidence on the distribution of Si and Al and their particular sitting in zeolite. Derouane et al. [59–61] used nonempirical SCF–MO techniques to study monomer and dimer clusters to propose preferential sitting of aluminum in different zeolites. They performed charge distribution analysis to show that the anionic framework behaves as a weak but soft base. They also explained the high acid strength of zeolites as well as their strong affinity for large and polarizable cations. O'Malley and Dwyer [62] performed *ab initio* calculations on the acidic properties of boralite 4 and on the sitting of aluminum in the Theta-1 framework. The semiempirical quantum chemical MNDO technique has been adopted to calculate the substitution energy for boron at different possible framework sites in the ZSM-5 structure [63]. EHMO calculations on ferrisilicate zeolite models were used to report preferential sitting of iron in the ZSM-5 lattice [64]. Alvarado et al. used *ab initio* calculations for the monomeric cluster models to model the twelve different T sites of the ZSM-5 lattice [65]. They also proposed a model to explain the dependence of the energy cluster models on the geometry of T sites. Schroder et al. [66] used an energy minimization method to calculate the defect energy and to predict the preferential sitting of aluminum as well as protons over different oxygen sites of the ZSM-5 lattice. In recent years, there have been concerted efforts to calculate the sitting of aluminum in the ZSM-5 lattice by different quantum chemical methods [56–66]. The adopted models also vary from monomeric (with nine atoms) to pentameric (with thirty-three atoms) clusters. The MNDO calculation on larger pentameric cluster models predicted that the contributions of different geometric parameters for the relative energy and substitution energy are not uniform. Further, it is shown that the charge density on oxygen, which is an indication of acidity in zeolites, shows a better correlation to geometric parameters such as Si–O and Si–O–Si rather than to the total energy and the substitution energy [67]. Most of the studies regarding the Si/Al ordering were performed on the basis of Mott Littleton methodology [68] and consider aluminum as a defect site in all silica zeolites. The identification of the preferential aluminum sites is carried out by calculating the defect energies, which is the measure of the feasibility of the incorporation of aluminum at the framework site. Considering the location of aluminum atoms in zeolite, Catlow et al. found that in high-silica zeolites, the aluminum ions can be considered as dilute impurities in a siliceous matrix, and the framework will relax only locally without major distortions to the structure [28]. Thus, one can consider that the aluminum is distributed homogeneously. On the other hand, when the Si/Al ratio is equal to one, a strict alternation of silica and aluminum is expected, such as in the case of zeolite A [28]. However, the problem becomes very complicated when one considers zeolites having Si/Al ratios between the above two limiting cases; a prime example is the zeolite clinoptilolite, with an Si/Al ratio of ~5.

Another important parameter of zeolite synthesis strategy is the OH ion concentration. It functions as a structure director through the control of the degree of polymerization of the silicate solution. Hydroxide ions modify the nucleation time by influencing transport of silicates from the solid phase to solution. They enhance the crystal growth and control the phase purity. It was also found that the OH/Si ratio influences the pore size [69]. Wu and Deem [70] performed cluster simulations in the grand canonical ensemble calculation on the nucleation of zeolite at $T = 430$ K, a typical temperature for zeolite synthesis, and assumed that at pH = 12 the dominant monomeric and dimeric ions are $Si(OH)_2O_2^{2-}$ and $Si_2(OH)_4O_2^{3-}$, respectively. According to their calculations, both the concentration and pH value have an influence on the critical cluster size and the nucleation barrier. The pH value affects the critical cluster size and the nucleation barrier through the oxygen chemical potential, μ_O. The sensitivity of the nucleation barrier to the pH value is characteristic of the sensitivity of zeolite synthesis to solution conditions. A decrease of pH would shift the equilibrium of the following equation to the right and lead to lower nucleation barriers and smaller critical clusters.

$$2Si(OH)_2O_2^{2-} + H_2O \leftrightarrows 2Si_2(OH)_4O_2^{3-} + 2OH^-$$

Lowering the pH value does not always leads to faster crystallization, because other factors such as insufficient supply of monomers may prevent the formation of the nucleus of the crystal. Indeed, the mechanism of nucleation for zeolites seems to operate only at high pH, where the critical cluster size is on the order of fifty silicons [71]. Fan et al. also proposed a rough estimation of the nucleation rate depending on the classical nucleation theory. The nucleations of zeolites are mostly studied by atomistic models. However, Jorge et al. [25] simulated the silica-template nanoparticle formation comprising with the lattice model approach, which is much simpler than atomistic model, and found that nanoparticles are spontaneously formed when neutral silica are used. To address whether these nanoparticles are at true equilibrium states, Jorge et al. performed parallel tempering Monte Carlo simulations, which more efficiently surmount barriers that prevent equilibration. Jorge et al. found that the parallel tempering simulations always produce one single cluster, suggesting that silica-template nanoparticles observed experimentally are actually in metastable states. Their study again extended to the variation of metastable cluster size distribution with pH, silica concentration, and template size [25].

One of the most important roles that inorganic cations play is the balancing of the zeolite framework charge created by incorporation of the three valences aluminum in the lattice. In addition, they govern the morphology of the zeolite either by favoring the nucleation or by selectively enhancing the crystal growth along a suitable direction and they can also act as structure-directing agents. While in place, the cation mediates the acquisition of new T units from solution (through the condensation reactions given above), guiding them into more favorable coordination geometry and generating a periodically regular local structure. In due course, the cation may have sufficient of its original hydration shell replaced by lattice oxygen donors to stabilize the newly created site. Alkali metal cations Li^+, Na^+, and K^+ play a fundamental role in the processes of dissolution and recrystallization (depolymerization–polymerization of anionic moieties) of the amorphous phase, ensuring supersaturation of the solution followed by formation and growth of nuclei of zeolite crystals.

Similar to zeolites, the structure of $AlPO_4$ is also dependent on the chemical composition of the synthesis gel. Experimentally it was found that crystallization time and temperature, H_2O content, molar ratio of phosphorous to organic template, and the nature of the silica source dictate the structure formation. In addition, solvent also plays an effective role to the crystallinity of the resultant structure. For instance, ethanol, ethylene glycol, and glycerol seem to be the best solvents for the synthesis of Co substituted $AlPO_4$. By varying the type of solvent, Co content, and template in the synthesis gel, it was possible to increase the substitution of Co^{+2} in the structure [72]. Computationally derived virtual spectra of $AlPO_4$ confirm the effect of composition on the structural feature. The simulated spectra revealed the possibility to determine the effect of chemical composition; i.e., force constant from the nonuniform distribution of PO_4 and AlO_4. The chemical composition means one is focused to know the T site distribution depending on the chemical composition and loading to monitor the effect of the environment around. This dependence is more understandable in terms of internal interaction described by variation of force constant. The quantum chemical calculation is not enough to determine the force constant for this type of system. Thus, an indirect method like the determination of force constant from the infrared spectra is an alternative choice [72].

5.4 STABILITY TEST WITH HIGH TEMPERATURE AND PRESSURE

The structural stability of zeolites is very important to their application in different fields. The major factors affecting the stability are the temperature and external pressure. For over a decade, two powerful experimental techniques, atomic force microscopy (AFM) and HRTEM, have been used to check the external surface of zeolites. However, due to the lack of resolution to the desired level, it is not always possible to get an exact idea about the changes of the external surface. By using computer simulation methods one can assess the stability of zeolite structures related to temperature and pressure. It should be mentioned that mainly atomistic simulation has been used to predict the thermodynamic stability of zeolites. Ford et al. [73] studied the phase stability of silica zeolites based on low coordination and strong association to the calculation of SOD, LTA, MFI, and FAU. Applying the Frenkel and Ladd force field for calculating free energies of arbitrary solids, it was predicted that the all-silica MFI zeolite structure has a regime of thermodynamic stability relative to dense phases such as quartz and stable at low pressures and above ~1,400 K. On the other hand, less dense all-silica zeolites SOD, LTA, and FAU exhibit no regime of thermodynamic stability, which explains why silica MFI can be synthesized directly, whereas silica SOD, LTA, and FAU are typically prepared by postsynthesis dealumination. Moreover, their investigation also led to the fact that MFI may have regimes of thermodynamic stability but not because of template stabilization [73]. However, their model does not correlate well with experiments. DFT calculations have also been carried out to determine the thermodynamic stability of various Ga species in gallium-exchanged ZSM-5, the thermodynamics of H_2 adsorption, and the most favorable pathway for H_2/D_2 exchange. The thermodynamics of H_2 desorption from ZGa(H)(2) are favorable, though the process is projected to be slow because of a high activation barrier. The most favorable pathway for H_2/D_2 exchange over ZGa(H)(2) proceeds via Z(D)(Ga(H)(2)(D)) as an intermediate. Similar calculations have been carried out for H_2/D_2 exchange over H-ZSM-5.

Compared to the thermal stability of zeolites, the stability against pressure has been studied to a lesser extent. There are very few experimental reports dealing with this matter based on the high-pressure synchrotron X-ray diffraction study [74] and infrared spectroscopy [75]. Lee et al. reported that some of the zeolite expands rather than contracts when applied to the external pressure [76] and speculated that the application of this type of zeolite in the pollution control of water and removal of the larger radioactive ions from the environment would lighten up the corresponding field. To eliminate experimental difficulties, the effect of pressure on zeolite structure was checked by computer simulation. Several methods have been applied to get insight on the zeolite structure against the external pressure. Based on the classical

Born model of solids with interatomic potentials describing the forces between the atoms, White et al. revealed the pressure-induced hydration effects in zeolite lau-monite [77]. They have used energy minimizations [78] with the potential param-eters described by Jackson and Catlow [79] and de Leeuw et al. [80]. This clearly shows that when pressure is applied, the structural stability of the laumontite frame-work is enhanced by the presence of a fully occupied water network and the fully hydrated laumontite can be considered to be a pressure-induced overhydrated phase of the as-formed material. Another study [81] was performed on LTA structure using empirical potential model. This model correlates well with the IR spectra and their change with pressure. Simulations of ZK-4, Na-A, and Li-A zeolites strongly suggest the existence of a transition from a crystalline to an amorphous phase at pressures around 3 GPa. This transition can be monitored by checking the vibrational band, which experimentally falls at around 463 cm^{-1} related with the vibrations of the double four-membered rings (D4R). This band becomes very weak with increasing pressure. It significantly decreases at high pressure (3, 6 GPa). This fact has been experimentally reported and associated with the loss of the crystallinity of the ZK-4 framework. [81]. Another study was done with the pure surface structure to monitor effects of water environment on the geometries and stabilities of zeolite A [82]. The surface structure of the zeolite contained a large numbers of low-coordinated silica and oxygen atoms, so that the structure could be stabilized upon dissociative adsorp-tion of water. Using the relative thermodynamic stabilities of all possible termina-tions, Slater et al. have established that three terminations are possible for the {100} surface and suggested that the single four-member ring (s4r) termination, which is predicted to be of equal thermodynamic stability, is in fact a short-lived intermediate structure, is quickly evolved out of the growing crystal surface [82].

Metal organic frameworks (MOFs), consisting of metal atoms or clusters and organic ligands, are a relatively new class of nanoporous crystalline materials. Due to their unique framework structures similar to zeolitic materials, many novel proper-ties and potential applications have been identified [83–86]. In addition, they exhibit certain properties similar to those found in their inorganic counterparts, zeolite-like materials. Indeed, it was recently suggested that MOF-5 with the chemical formula $C_{192}H_{96}O_{104}Zn_{32}$ as shown in Figure 5.6 is one of the most well-known MOF com-pounds first identified by Kang et al. [87]. It exhibits an exceptional negative ther-mal expansion (NTE) behavior with its potential usage in thermoplastics [87–89] as found in some zeolites or zeolite-like materials. Experimentally, there exists some indirect evidence supporting the NTE of MOF-5. In particular, it was observed that MOF-5 with adsorbed N_2 or CO_2 expands upon cooling [88, 89]. However, no direct thermal expansion measurement on a clean MOF-5 sample has been reported so far. Theoretically, a recent molecular dynamics (MD) simulation [89] predicted that MOF-5 has a very large linear thermal expansion coefficient $\sim -20 \times 10^{-6}$ K^{-1}. Another MD study [90] suggested a linear thermal expansion coefficient of $\sim -8 \times 10^{-6}$ K^{-1} for MOF-5. Both simulations were carried out using a classical force field, and with the own limitation for the force field it has provided some insight on the mechanism of the thermal expansion and by providing a similar order of the thermal expansion coef-ficient in both the cases with the absolute number is varying. This is expected from the force field methodology, though one can safely use these types of methods to look into

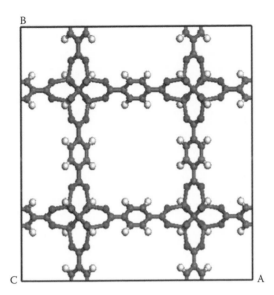

FIGURE 5.6 (For color version see accompanying CD) MOF-5 model with structural formula $C_{192}H_{96}O_{104}Cu_{32}$.

the surface stability in terms of temperature. But it is challenging to reveal the useful physics behind the process with simple MD, and in case of MOFs there is an issue of thermal expansion coefficient and its variation with the change in metal cation and its symmetrical environment. Snapshots of the structure equilibrated at high temperature often contain mixed lattice phonon characteristics, although they provide hints as to what kind of motions may be related to the NTE behavior investigated by using the lattice dynamics of MOF-5 at its equilibrium low-temperature structure [91].

Most relevant to the negative thermal expansion are those low-frequency phonon modes that involve the motion of the ZnO_4 clusters and benzene rings as rigid units [92]. This is an expensive calculation in terms of CPU, which needs to know the thermal expansion. To account for the effect of the models to the negative thermal expansion requires phonon calculations at various temperatures. On the other hand, thermal expansion in insulators is known to arise from anharmonic lattice vibrations; thus, thermal expansion can be calculated within quasi-harmonic approximation. In the quasi-harmonic approximation, the effect of the anharmonicity in the lattice energy is treated by allowing the phonon frequencies to depend on the lattice parameters. Hence, for a giving temperature T, one first takes a, minimizes the atomic positions by quantum mechanical calculations with plane wave, and calculates the phonon spectrum (a detailed description of phonon calculations can be found in Zhou and Yildirim [91]) and then the free energy. Repeating this for other values of a, one can find the optimum value of the lattice parameter that minimizes the free energy at a given temperature. The lowest energy phonon modes can easily turn "negative" in frequency with a large enough lattice contraction at high temperature and thus introduce structural instability; i.e., lowered crystal symmetry and then breakage between the zinc oxide cluster and the organic ligand. Indeed, under ambient pressure, the MOF-5 sample starts to decompose at 650 K and eventually turns into ZnO powders. Clearly, the chemical bonds bridging the benzene rings and the ZnO_4 units are the unstable part of the MOF-5 structure. This is further supported by the fact that the organic linker molecules in the precursor acids usually have higher thermal stability than their counterparts in the corresponding MOF structures. For example, the benzene-1,4-dicarboxylic acid (organic precursor of MOF-5) melts or sublimes at 700 K and decomposes; i.e., breaks covalent bonds at a much higher temperature (1000 K). MOFs with different symmetries and configurations that do not possess flexible structural units or chemical bonds that can serve as bridges between rigid units would not be expected to show much negative thermal expansion behavior. For example, zeolitic imidazolate framework-8 $Zn_6N_{24}C_{48}H_{60}$, space group $I4$-$3m$ consists of rigid units of ZnN_4 and a methylimidazolate ring [93]. The two rigid units are directly connected through the N atom without an additional buffering bridge. Not surprisingly, we found the experimental linear thermal expansion coefficient of zeolitic imidazolate framework-8 to be positive 10×10^{-6} K^{-1}.

The stability of MOF has another component to study, termed as the effect of hydration. Experimental terahertz (THz) spectroscopy investigations in the spectral range of 3–120 cm^{-1} can be used for detection and identification for a wide assortment of compounds, because the compounds have distinct spectra "fingerprints" in this region of the electromagnetic spectrum. THz spectroscopy is hence widely used for security applications [94] and more routinely in the pharmaceutical industry [95]. The assignment of the observed experimental features is often difficult. In some cases, the experimental interpretation of spectra does not yield a unique assignment; a spectrum may be consistent with more than one geometric structure. Computational chemistry plays a pivotal role: by computing the spectra, the correspondence between spectral peaks and molecular or crystal features may be assigned unambiguously. One can therefore explore and compare the structural

---

TABLE 5.1

Maximum and Minimum Values for Principal Axes of Inertia for Templates to Form FER- and MOR-Type Zeolites

Type of Framework	Template Range	Value	RX (Å)	RY (Å)	RZ (Å)	Surface Area (Å²)	Molecular Volume (Å³)
FER	Small	Max	20	2.8	1.6	133	81
		Min	3	1.6	1.2	99	57
	Diammoniums	Max	164	2.2	2.2	437	321
		Min	128	2.2	2.2	428	294
MOR	Small	Max	19	6.0	4.6	464	265
		Min	3	2.2	1.6	112	69
	Trioctylamine	Only	11	10.8	6.4	485	337

Note: Table 1 of Gale and Rohl [31] is reproduced with the permissible modification, with permission from the publisher.

features of MOFs by using quantum chemical calculations. They also help one to figure out the portion and effect of hydration in an MOF. Experimentally, one can the detect and characterize the adsorbed interfacial water within the cages of the metal–organic framework MOF-5 with a chemical composition of $(Zn_4O(BDC)_3)$ by terahertz time-domain spectroscopy (THz TDS) in the frequency range from 5 to 46 cm^{-1}. The experimental spectra suggest a coupling of the intermolecular motions of the water molecules adsorbed to the collective vibrations of the network at 4 wt% hydration. It is very difficult to probe the structural condition with this amount of hydration. This finding is supported by the results of MD simulations [96]. When the water content increases to 8 wt%, a nonreversible decomposition of MOF-5 is observed. The MD calculation was performed with classical dynamics with a shell model and the hydration level was varied by changing the water molecular content of the system. The message here is that there exists a challenge in MOF modeling due to the complicated structure and the structural flexibility but modeling remains the only tool that has a fair chance to correlate structure and property, which alone by experiment does not look like a viable option.

5.5   MICROSCOPIC UNDERSTANDING

How can one design and predict the synthesis of a new type of zeolite according to the desired application? As mentioned before, there are various types of experimental works that try to deliver a lot over the microscopic understanding on the nucleation and the growth of the zeolite. However, there remain challenges to explore the exact mechanism behind the zeolite synthesis experimentally. The creative nature of the scientist thus introduced computer simulation to assure the correctness of the scenario as obtained through experiments. This is a great challenge for the theoretical chemist, to predict the best possible solutions and compare with experimental findings to confirm the correctness of the calculations.

Zeolite nucleation has been widely studied by computer simulation using several techniques. Today we have confidence that we can generate a virtual zeolite model and compute pore dimensions, framework density, diffraction patterns, or vibrational spectra. Molecular modeling techniques now are applied routinely to predict the structure, but still there are some aspects from the synthesis part of zeolite which remain difficult to answer using simulation; for example, the role of inorganic cations like Na^+, K^+, etc. What are the actual roles they play in nucleation? How does the pH or the concentration of hydroxide ions effect the crystallization? Moreover, the structural stability of zeolites, which is an important parameter, is a challenging problem in molecular modeling techniques. Depending on the external conditions, the stability of zeolites varies, though there are several reports on the thermal stability of zeolites where experimental data require a reliable model for a complete understanding of the process.

Considering mesoporous structures, which are now widely used in various field such as sensing, catalysis, sorption, etc., there is still not enough structural understanding. For example, in catalysis one key goal is to achieve high conversion and high selectivity and a tailor-made catalyst is necessary to save time and energy consumption. It is obvious that mesoporous materials having amorphous structure and

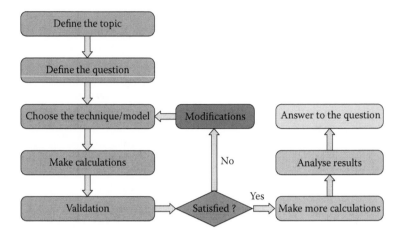

FIGURE 5.7   Flowchart of simulation to solve the problem related to experiment.

their characterization technique based on the experiment are also limited, which requires the attention of the theoretical chemist. Thus, proper characterization to gain insight on the synthesis mechanisms and the interior of nanoporous materials is a great challenge for the material scientist.

Overall, molecular modeling plays a significant role in explaining zeolite synthesis a part of nanoporous materials, to optimize their structures with varied compositions, some part of its stability factor, and most importantly the template–zeolite framework interactions.

Finally, it is time to create a recipe from the discussion so far and there lies the idea of this book. We are going to provide a procedure to approach the problem of zeolite synthesis from a simulation point of view to provide options for the bench chemist either to use simulation to rationalize the ambiguity or unknown in the experiment or to validate an experimental procedure. The procedure depends on the problem you are looking to solve. Figure 5.7 is a flowchart that describes the procedure for solving the problem by simulation. This is a very general overview of the procedure to shape up the problem.

You certainly have to ask yourself then, what is the main issue you wish to resolve by this? Is it the synthesis process mechanism? If that is the case, then you may start with a simple component of the system from the sources of the synthesis and try to see what is the closest experimental data you have in that domain to compare with experiments. You may have a synthesis condition like temperature and pressure where you mix SiO_4 and AlO_4 in ratio in the presence of water and a template. To reproduce the steps you have to go in a sequence, and you have to think with experimental numbers that how the silica and alumina tetrahedra binds, so you will first try to test their binding in gas phase, or in solvent phase, or in a hydrated phase where you add water molecules around the groups of interest, not immersing the molecule in a solvent. Then you will see the individual interaction of these groups with a structure-directing agent to find the affinity of these molecules to the groups you are working on. These calculations can be performed by quantum mechanics, preferably DFT, where one will be able to handle larger sizes and get semiquantitative information. Further, it is possible to do some kinetics on these systems by transition state calculations to see the rate order where you may use kinetic Monte Carlo to feed in that information comparable to experiment. The final stage is to take the information in terms of the charges; the localized geometry and interaction information to a force field–based calculation domain like GULP. This was generally used to incorporate the hydrothermal environment in the procedure.

Anther example is screening of the best material from the range of materials. Screening can be done by simulation where the localized interaction or affinity matters. It is not only about affinity but the geometry and architecture of the pores, and that is what zeolites are made with. A substitution of a metal by another changes the localized environment, so before making a new form of nanoporous material with metal incorporation or other substitution, one can screen the desired material though quantum calculation within DFT to know and then go for the synthesis. One can further test the component ratios, the best Si/Al ratio, or a metal percentage in a nanoporous material.

The most important aspect in simulation or modeling is the analysis of results. Synthesis of material in real, there are the possibilities of gaining or losing in terms of product formation or product yield as numerous parameters associated. In simulation we explain the results mostly in terms of energy values, either total energy or binding energy, substitution energy, interaction energy, etc. What we determine from simulation is how two components in the synthesis can combine, we will see the energy value to see whether their combination results in a lower energy value and then we will see the localized geometry, if the bonding is changed in terms of the bond order or the inter atomic distance is enhanced, like when we talk about a substitution of silica by aluminum or other heteroatoms, we have to find those feasibility in terms of localized geometry. The structure obtained is then validated by some characterization experimental tools, which we will elaborate on in the following chapters. But at this point the loading of a metal or a substitution composition can be validated by lattice energy and the structural stability. In experiments it is difficult to explain the reason for the affinity of the components to result in the final product. In simulation we deal with the charges of the individual atom, their interactions with the neighbors in terms of orbital contributions, or their potential behaviors with the constituents. This will explain the reason for a certain interaction within the components based on the chemistry in combination with the forces within the laws of physics. The exemplary references attached is an explanation to the facts one can handle within simulation, the quality of results justify the advancement in simulation where it is becoming an additional tool to experiment to add on to one's existing workbench.

REFERENCES

1. S. M. Auerbach, K. A. Carrado, P. K. Dutta (eds.), *Handbook of Zeolite Science and Technology*, Marcel Dekker, New York (2003).
2. Y. Wada, T. Okubo, M. Ryo, T. Nakazawa, Y. Hasegawa, S. Yanagida, *J. Am. Chem. Soc.*, 122 (2000) 8583.
3. S. I. Zones, M. E. Davis, *Curr. Opin. Solid State Mater. Sci.*, 1 (1996) 107.

4. M. E. Davis, R. F. Lobo, *Chem. Mater.*, 4 (1992) 756.
5. (a) S. M. Auerbach, M. H. Ford, P. A. Monson, *Curr. Opin. Colloid Interface Sci.*, 10 (2005) 220; (b) B. Slater, *Stud. Surf. Sci. Catal.*, 154 (2004) 1197.
6. P. A. Jacobs, E. G. Derouane, J. Weitkamp, *J. Chem. Soc.Chem. Comm.*, (1981) 591.
7. E. G. Derouane, S. Detremmerie, Z. Gabelica, N. Blom, *Appl. Catal.*, 1 (1981)101.
8. Z. Gabelica, E. G. Derouane, N. Blom, in *Catalytic Materials: Relationship between Structure and Reactivity*, T. E. Whyte, Jr., R. A. Dalla Betta, E. G. Derouane, R. T. K. Baker (eds.), American Chemical Society, Washington, DC, (1984).
9. D. P. Serrano, R. V. Grieken, *J. Mater. Chem.*, 11 (2001) 2391
10. S. M. Auerbach, M. H. Ford, P. A. Monson, *Curr. Opin. Colloid Interface Sci.*, 10 (2005) 220.
11. B. J. Schoeman, *Microporous and Mesoporous Materials*, 22 (1998) 9.
12. J. N. Watson, L. E. Iton, R. I. Keir, J. C. Thomas, T. L. Dowling, J. W. White, *J. Phys. Chem. B*, 101 (1997) 10094.
13. P. P. E. A. de Moor, T. P. M. Beelen, R. A. van Santen, *J. Phys. Chem. B*, 103 (1999) 1639.
14. R. Ravishankar, C. E. A. Kirschhock, P. Knops-Gerrits, E. J. P. Feijen, P. J. Grobet, P. Vanoppen, *J. Phys. Chem. B*, 103 (1999) 4960.
15. D. D. Kragten, J. M. Fedeyko, K. R. Sawant, J. D. Rimer, D. G. Vlachos, R. F. Lobo, *J. Phys. Chem. B*, 107 (2003) 10006.
16. S. Yang, A. Navrotsky, D. J. Wesolowski, J. A. Pople, *Chem. Mater.*, 16 (2004) 210.
17. S. Hamad, S. Cristol, C. R. A. Catlow, *J. Am. Chem. Soc.*, 127 (2005) 2580.
18. (a) J. C. G. Pereira, C. R. A. Catlow, G. D. Price, *J. Phys. Chem. A*, 103 (1999) 3252; (b) J. C. G. Pereira, C. R. A. Catlow, G. D. Price, *J. Phys. Chem. A*, 103 (1999) 3268; (c) M. J. Mora-Fonz, C. R. A. Catlow, D. W. Lewis, *Angew. Chem.*, 44 (2005) 3082; (d) T. T. Trinh, A. P. J. Jansen, R. A. van Santen, *J. Phys. Chem. B*, 110 (2007) 23099.
19. M. G. Wu, M. W. Deem, *J. Chem. Phys.*, 116 (2002) 2125.
20. (a) B. P. Feuston, S. H. Garofalini, *J. Phys. Chem.*, 94 (1990) 5351; (b) P. Vashishta, R. K. Kalia, J. P. Rino, I. Ebbsjö, *Phys. Rev. B*, 41 (1990) 12197; (c) N. Z. Rao, L. D. Gelb, *J. Phys. Chem. B*, 108 (2004) 12418.
21. A. Chatterjee, T. Iwasaki, *J. Phys. Chem. A*, 105 (2001) 6187.
22. A. Chatterjee, T. Iwasaki, *Stud. Surf. Sci. Catal.*, 135 (2001) 265.
23. (a) V. Nikolakis, E. Kokkoli, M. Tirrell, M. Tsapatsis, D. Vlachos, *Chem. Mater.*, 12 (2000) 845; (b) W. H. Dokter, H. F. Vangarderen, T. P. M. Beelen, R. A. van Santen, W. Bras, *Angew. Chem.*, 34 (1995) 73.
24. S. Cundy, P. A. Cox, *Microporous and Mesoporous Materials*, 82 (2005) 1.
25. M. Jorge, S. M. Auerbach, P. A. Monson, *J. Am. Chem. Soc.*, 127 (2005) 14388.
26. B. Slater, T. Ohsuna, Z. Liu, O. Terasaki, *Faraday Discuss.*, 136 (2007) 125.
27. S. A. Pelster, W. Schrader, F. Schuth, *J. Am. Chem. Soc.*, 128 (2006) 4310.
28. M. J. Mora-Fonz, C. R. A. Catlow, D. W. Lewis, *Angew. Chem. Int. Ed.*, 44 (2005) 3082.
29. N. J. Henson, A. K. Cheetham, J. D. Gale, *Chem. Mater.*, 6 (1994) 1647.
30. M. J. Sanders, M. Leslie, C. R. A. Catlow, *J. Chem. Soc. Chem. Comm.*, (1984) 1271.
31. J. D. Gale, A. L. Rohl, *Mol. Simulat.*, 29 (2003) 291.
32. M. J. Mora-Fonz, C. R. A. Catlow, D. W. Lewis, *J. Phys. Chem. C*, 111 (2007) 18155.
33. (a) B. Slater, C. R. A. Catlow, Z. Liu, T. Ohsuna, O. Terasaki, M. A. Camblor, *Angew. Chem.*, 114 (2002), 1283; (b) *Angew. Chem. Int. Ed.*, 41 (2002) 1235.
34. T. T. Trinh, A. P. J. Jansen, R. A. van Santen, *J. Phys. Chem. B*, 110 (2006) 23099.
35. J. Sefcik, A. V. McCormick, *AIChE J.*, 43 (1997) 2773.
36. Z. Nan, M. Wang, B. Yan, *J. Chem. Eng. Data*, 54 (2009) 83.
37. J. Zhang, Z. Luz, D. Goldfarb, *J. Phys. Chem. B*, *101* (1997) 7087.
38. A. Ramanan, M. S. Whittingham, *Cryst. Growth Des.*, 6 (2006) 2419.
39. J. C. G. Pereira, C. R. A. Catlow, G. D. Price, *J. Phys. Chem. A*, 103 (1999) 3252.

40. V. A. Ermoshin, K. S. Smimvov, D. Bougerad, *J. Mol. Struct.*, 393 (1997) 171.
41. G. F. Jiao, M. Pu, B. H. Chen, *Struct. Chem.*, 19 (2008) 481.
42. D. W. Lewis, C. M. Freeman, C. R. A. Catlow, *J. Phys. Chem.*, 99 (1995) 11194.
43. A. Chatterjee, *J. Mol. Catal.*, 120 (1997) 155.
44. A. E. Boyett, A. P. Stevens, M. G. Ford, P. A. Cox, *Zeolites*, 17 (1996) 508.
45. B. H. Toby, N. Khosrovani, C. B. Dartt, M. E. Davis, J. B. Parise, *Microporous and Mesoporous Materials*, *39* (2000) 77.
46. P. Sun, Q. Jin, L. Wang, B. Li, D. Ding, *J. Porous Mater.*, 10 (2003) 145.
47. A. Chatterjee, R Vetrivel, *Faraday Trans.*, 91 (1995) 4313.
48. (a) G. Sastre, S. Leiva, M. J. Sabater, I. Gimenez, F. Rey, S. Valencia, A. Corma, *J. Phys. Chem. B*, 107 (2003) 5432; (b) G. Sastre V. Forner A. Corma *J. Phys. Chem. B*, 106 (2002) 701.
49. (a) J. H. Yun, T. Duren, F. J. Keil, N. A. Seaton, *Langmuir*, 18 (2002) 2693; (b) Y. He, N. A. Seaton, *Langmuir*, 19 (2003) 10132.
50. M. W. Maddox, J. P. Olivier, K. E. Gubbins, *Langmuir*, 13 (1997) 1737.
51. B. P. Feuston, J. B. Higgins, *J. Phys. Chem.*, 98 (1994) 4459.
52. (a) G. E. Martin, S. H. Garofalini, *Journal of Non-Crystalline Solids*, 171 (1994) 68; (b) K. Yamahara, K. Okazaki, *Fluid Phase Equil.*, 144 (1998) 449; (c) N. Z. Rao, L. D. Gelb, *J. Phys. Chem. B*, 108 (2004) 12418.
53. M. G. Wu, M. W. Deem, *J. Chem. Phys.*, 116 (2002) 2125.
54. (a) J. C. G. Pereira, C. R. A. Catlow, G. D. Price, R. M. Almeida, *Journal of Sol-Gel Science and Technology*, 8 (1997) 55; (b) J. C. G. Pereira, C. R. A. Catlow, G. D. Price, *J. Phys. Chem. A*, 106 (2002) 130; (c) J. R. Hill, J. Sauer, *J. Phys. Chem.*, 98 (1994) 1238.
55. C. Schumacher, J. Gonzalez, P. A. Wright, N. A. Seaton, *J. Phys. Chem. B*, 110 (2006) 319.
56. A. Zheng, L. Chen, J. Yang, M. Zhang, Y. Su, Y. Yue, C. Ye, F. Deng, *J. Phys. Chem. B*, 109 (2005) 24273.
57. J. Limtrakul, J. Tantanak, *Chem. Phys. Lett.*, 208 (1996) 331.
58. A. K. Chandra, A. Goursot, F. Fujula, *J. Mol. Catal. A*, 119 (1997) 45.
59. J. G. Fripiat, F. Berger-Andre, J. M. Andre, E. G. Derouane, *Zeolites*, 3 (1983) 306.
60. E. G. Derouane, J. G. Fripiat, *Zeolites*, 5 (1985) 165.
61. E. G. Derouane, J. G. Fripiat, *Proceedings of the 6th International Zeolite Conference*, RENO USA, 1983 Reno, A. Olson, A. Bislo (eds.) Butterworths, London (1984).
62. P. J. O. Maley, J. Dwyer, *Zeolite*, 8 (1996) 317.
63. R. Vetrivel, *Zeolites*, 12 (1992) 424.
64. R. Vetrivel, S. Pal, S. Krishnan, *J. Mol. Catal.*, 66 (1991) 385.
65. A. E. Swasigood, M. K. Barr, P. J. Hay, A. Redondo, *J. Phys. Chem.*, 95 (1991) 10033.
66. K. P. Schroder, J. Sauer, M. Leslie, C. R. A. Catlow, *Zeolites*, 12 (1992) 20.
67. A. Chatterjee, R. Vetrivel, *Zeolites*, 14 (1994) 225.
68. N. F. Mott, M. J. Littleton, *Trans. Faraday Soc.*, 34 (1938) 485.
69. E. Terres, *Microporous and Mesoporous Materials*, 431 (1996) 111.
70. M. G. Wu, M. W. Deem, *J. Chem. Phys.*, 116 (2002) 2125.
71. W. Fan, B. M. Weckhuysen, *J. Nanosci. Nanotech.*, 3 (2003) 271.
72. A. J. M. De Man, W. P. J. H. Jacobs, J. P. Gilson, R. A. van Santen, *Zeolites*, 12 (1992) 826.
73. (a) M. H. Ford, S. M. Auerbach, P. A. Monson, *J. Chem. Phys.*, 126 (2007) 144701; (b) M. H. Ford, S. M. Auerbach, P. A. Monson, *J. Chem. Phys.*, 121 (2004) 8415.
74. M. D. Rutter, U. Takeyuki, R. A. Secco, Y. Huang, Y. Wang, *J. Phys. Chem. Solid.*, 62 (2001) 599.
75. Y. Huang, *J. Mater. Chem.*, 8 (1998) 1067.
76. Y. Lee, T. Vogt, J. A. Hriljac, J. B. Parise, J. C. Hanson, S. J. Kim, *Nature,* 420 (2002) 485.
77. C. L. I. M. White, A. Rabdel Ruiz-Salvador, D. W. Lewis, *Angew. Chem. Int. Ed.*, 43 (2004) 469.

78. J. D. Gale, *J. Chem. Soc. Faraday Trans.*, 93 (1997) 629.
79. R. A. Jackson, C. R. A. Catlow, *Mol. Simulat.*, 1 (1988) 207.
80. (a) N. H. de Leeuw, S. C. Parker, *Phys. Rev. B*, 58 (1998) 13901; (b) N. H. de Leeuw, S. C. Parker, *J. Am. Ceram. Soc.*, 82 (1999) 3209.
81. J. Gulín-González, G. B. Suffritti, *Microporous and Mesoporous Materials*, 69 (2004) 127.
82. B. Slater, J. O. Titiloy, F. M. Higgins, S. C. Parker, *Curr. Opin. Solid State Mater. Sci.*, 5 (2001) 417.
83. M. Eddaoudi, J. Kim, N. Rosi, D. Vodak, J. Wachter, M. O'Keeffe, O. M. Yaghi, *Science*, 295 (2002) 469.
84. O. M. Yaghi, M. O'Keeffe, N. W. Ockwig, H. K. Chae, M. Eddaoudi, J. Kim, *Nature*, 423 (2003) 705.
85. H. K. Chae, D. Y. Siberio-Perez, J. Kim, Y. B. Go, M. Eddaoudi, A. J. Matzger, M. O'Keeffe, O. M. Yaghi, *Nature*, 427 (2004) 523.
86. N. Ockwig, O. D. Friedrichs, M. O'Keeffe, O. M. Yaghi, *Accounts Chem. Res.*, 38 (2005) 176.
87. Y. Kang, Y. Yao, Y. Qin, J. Zhang, Y. Chen, Z. Li, Y. Wen, J. Cheng, R. Hu, *Chem. Comm.*, (2004) 1046.
88. J. L. C. Rowsell, E. C. Spencer, J. Eckert, J. A. K. Howard, O. M. Yaghi, *Science*, 309 (2005) 1350.
89. D. Dubbeldam, K. S. Walton, D. E. Ellis, R. Q. Snurr, *Angew. Chem. Int. Ed.*, 46 (2007) 4496.
90. S. S. Han, W. A. Goddard, III, *J. Phys. Chem. C*, 111 (2007) 15185.
91. W. Zhou, T. Yildirim, *Phys. Rev. B*, 74 (2006) 180301.
92. W. Zhou, H. Wu, T. Yildirim, J. R. Simpson, A. R. Walker, *Phys. Rev. B*, 78 (2008) 054314.
93. H. Wu, W. Zhou, T. Yildirim, *J. Am. Chem. Soc.*, 129 (2007) 5314.
94. M. C. Kemp, P. F. Taday, B. E. Cole, J. A. Cluff, A. J. Fitzgerald, W. R. Tribe, *Proc SPIE*, 44 (2003) 5070.
95. P. F. Taday, *Philos. Trans. R. Soc. London, Ser. A*, 351 (2004) 362.
96. K. Schröck, F. Schröder, M. Heyden, R. A. Fischer, M. Havenith, *Phys. Chem. Chem. Phys.*, 10 (2008) 4732.

6 Characterization of Nanoporous Materials

Characterization of nanoporous materials is important for the quantification and prediction of their physical, chemical, and mechanical properties. Successful applications of nanoporous materials require detailed characterization. As discussed in Chapter 5, nanoporous materials are mainly of two types, such as crystalline-like zeolites or amorphous mesoporous materials. Characterization of nanoporous materials is thus based on several techniques, which are applicable for crystalline as well as amorphous structures. Generally, the structure elucidation of nanoporous materials can be carried out by powder X-ray diffraction (XRD) techniques in combination with scanning electron microscopy (SEM) and high-resolution transmission electron microscopy (HRTEM). Other spectroscopic techniques like Fourier transformation infrared (FTIR) and ultraviolet-visible (UV-vis) spectroscopy are also to measure the hydroxyl bond strength of zeolites and mesoporous materials. In addition, the thermal stability, which is an important parameter for the application of nanoporous materials, particularly as catalysts, has also been studied by different calorimetric techniques like thermogravimetry–differential thermal analysis (TG-DTA). Nanoporous materials with simple geometries can be characterized by means of transport coefficients such as diffusion and viscosity of the fluid or the gas flow through the material. Due to the porous structure, adsorption–desorption is an important method to serve as a fingerprint of the internal pore structure of the material and allows one to evaluate specific surface area, pore size, pore volume, and wall thickness. All the methods mentioned above require a lengthy sample processing period, depending on the quality of the sample, having all the instruments to analyze the samples is demanding as for the routine analysis visualization is not there, either for the bulk or the surface. Compared with different surface and bulk characterization techniques like XRD, UV-vis, IR, etc., the calculation time for molecular simulation techniques is reasonable, requiring only software and multiple processor units, which consequently provide valid data to support or design new materials. One can visualize the bulk, the surface, location of impurity, and the reason for a specific site to be active. One has to consider that there are several methods to characterize nanoporous materials, but due to their complex structures and unavailability of proper methods, molecular-level understanding is still distant. It has already been mentioned that experiments have their own limitations; thus, the combination of experiment and computer simulation is necessary to obtain a desired material with specific functionalities for proper applications. This chapter covers structural characterization of the material mainly in terms of chemical composition, spectroscopic analysis, mechanical stability, and porosity to show the capability of simulation to justify and validate experimental observation. It also can provide lot of space to the experimentation where it is hard and analysis is ambiguous.

6.1   CHEMICAL COMPOSITION

The framework of crystalline zeolite material is a delicate balance between bonding forces in the tetrahedral skeleton and the perturbation caused by extraframework cations and sorbed molecules [1]. Thus, variation of chemical composition could be the key parameter for the structural variation of zeolite. Investigations related to the influence of synthesis factors on specific features such as zoning, crystal defects, morphology, sorption behavior, order of elements in framework T atoms, location, and orientation of cations and molecules hosted in the pores (especially template ions and molecules) are found in the works by Barrer [2], Lyman et al. [3], Milton [4], Erdem and Sand (1980) [5], and Ueda and Koizumi [6]. Figure 6.1a and Figure 6.1b show the results of an experimental study of the effect of gel composition on the structure depending on the temperature and crystallization time in conventional hydrothermal synthesis [7]. The influence of the composition of zeolites (Si/Al ratio and cation loading) on the framework structure is also evident from the combination of IR and *ab initio* calculations, which predict that the T–O bond strength increases as the electronegativity of the framework increases. Using a similar technique, the influence of the composition of zeolites on the interaction of the surface hydroxyls with adsorbed molecules has been predicted. It was concluded that several properties of the unperturbed OH vary linearly with the geometric average electronegativity. On the other hand, for hydrogen-bonded OH groups, the interaction strength increases with the average electronegativity of the framework, and the OH bond properties are much more sensitive for changes in the composition [8].

The aluminosilicate framework of the zeolites is characterized by pore systems consisting of channels and cages. The initial days of simulation when computation was expensive and information on zeolite architecture was experimentally not well validated, cluster models were the best choice to handle zeolitic structures. An example of the effect of aluminum substitution depends on the localized bond length and bond angle distribution was shown in our previous work [9]. Models during the earlier days of simulations vary from monomeric (with nine atoms) to pentameric (with thirty-three atoms) clusters. There are also crystal structure reports for the low-temperature monoclinic phase of ZSM-5 as well as for the high-temperature orthorhombic phase of ZSM-5 with or without templates. Due to the above variations, there is always an inconsistency among the reports in the literature regarding the preferred sitting of aluminum in ZSM-5. We determined that it is very important to maximize the neighboring effect and a pentamer cluster is a reasonable choice. The larger pentameric cluster models give consistent results for different crystal structures. From the calculations performed for clusters with regular geometry, it is predicted that the contributions of different geometric parameters for the relative energy and substitution energy are not uniform. Further, it is evident that the charge density on oxygen, which is an indication of acidity in zeolites, possesses a better correlation to geometric parameters such as Si–O bond distance and Si–O–Si bond angle rather than to the total energy and the substitution energy. The charge on four oxygen

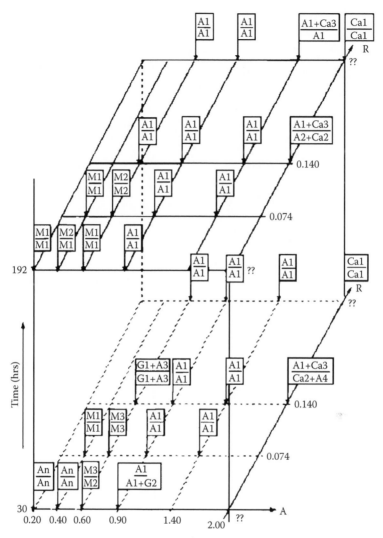

Structure abbreviations: Am = Amorphous; F = FAU; G = GIS; C = CHA; S = SOD;
Ca = CAN; M = MOR; A = ANA
Intensity scale: 4 = very low; 3 = low; 2 = medium; 1 = strong

FIGURE 6.1 (a) Effect of gel composition and crystallization time on the type of crystalline structure that is formed at crystallization temperature of 473 K. Flag register: upper/aged gels; lower/unaged gels. [7] With permission.

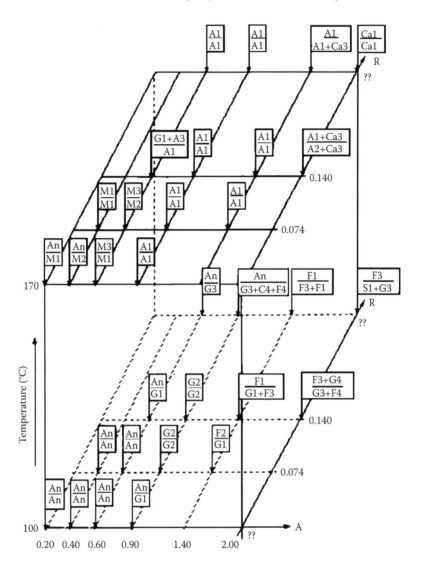

Structure abbreviations: Am = Amorphous; F = FAU; G = GIS; C = CHA; S = SOD;
Ca =CAN; M = MOR; A = ANA
Intensity scale: 4 = very low; 3 = low; 2 = medium; 1 = strong

FIGURE 6.1 (Continued). (b) Effect of gel composition and crystallization temperature on the type of crystalline structure that is formed from aged gel samples. Flag register: upper/crystallization time 30 h; lower/crystallization time 192 h. [7] With permission.

atoms attached to an Si site were averaged and correlated with Si–O bond distance and Si–O–Si bond angle. For dimeric cluster models, the Si–O distance and Si–O–Si angle values are again the average of the four Si–O distances and four Si–O–Si angles, whereas for pentameric clusters the distance and angle values are average values of four corresponding clusters. The charge density on oxygen calculated for

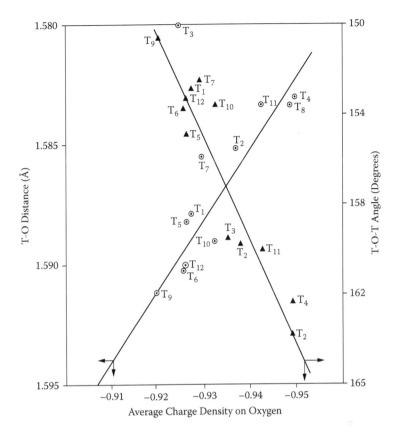

FIGURE 6.2 Place of silicon in the ZSM-5 lattice calculated according to Equation (6.1): $H_{12}Si_5O_{16} + AlO_4^- \rightarrow H_{12}Si_4AlO_{16}^- + SiO_4$.

dimeric cluster models is directly proportional to Si–O–Si angle and inversely proportional to the Si–O distance with few exceptions, as shown in Figure 6.2. A preferred site for aluminum substitution is therefore prescribed in terms of the bond and angular distribution or in other way the pore architecture matters. The anionic character of the lattice generate due to the substitution of silica by aluminum is neutralized by exchangeable cations located in very well-defined surface sites surrounded by oxygen atoms of the framework [11]. Knowledge of the partition of these extra-framework cations among the different sites is a crucial point because it influences the distribution of the electron density in the framework [12] and hence the interactions with the adsorbed molecules, leading to changes in the adsorption and catalytic properties of the zeolites. Maurin et al. [13] have performed a molecular simulation to rationalize the effect of hydration on the behavior of extraframework cations in zeolite Na$^+$-mordenite. Energy minimization techniques, combined with appropriate interatomic potentials to describe the potential energy surface of this complex system, have been used to determine the site selectivity of both cations and water molecules as a function of the hydration level. This modeling technique has been compared with experimental data obtained by dielectric relaxation spectroscopy and

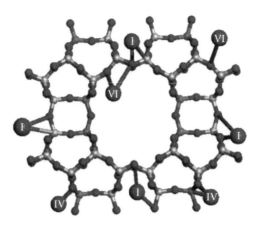

FIGURE 6.3 (For color version see accompanying CD) Typical (Si, Al) mordenite lattice used for the simulation. Distribution of the extraframework cations in the dehydrated structure among the three distinct crystallographic cation sites I, IV, and VI.

showed an exciting comparison. The cations in the dehydrated mordenite are located half in the main channels (sites IV and VI) and half in the small side channels (site I; Figure 6.3). The figure is a model designed with a similar nomenclature based on the work of Maurin et al. [13]. It is well-known fact that water plays a key role in many applications involving adsorption and, more particularly, in ion exchange carried out in aqueous solution. During ion exchange, water improves the efficiency of this process by coordinating the cations and hence increasing their mobility [14]. A first attempt was made by Mortier et al. to elucidate the distribution of the twenty-four H_2O among the thirty-six possible sites grouped in five different crystallographic sites named II(w), III(w), IV(w), V(w), and VI(w) (Figure 6.4) [10].

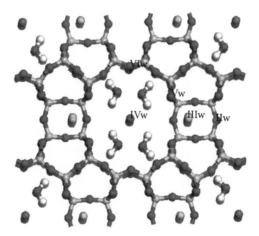

FIGURE 6.4 (For color version see accompanying CD) Unit cell of mordenite and description of the thirty-six crystallographic sites for water named IIw, IIIw, IVw, Vw, and VIw according to Mortier's classification.

A general procedure is described and can be used as a trick of the trade. The idea behind this procedure is to develop knowledge of simulation techniques, which are helpful with the visualization of the structure and make it more realistic in comparison to the experiment. In the first step, one has to select the relevant nanoporous lattice with a specific Si/Al ratio from experimental data or supported simulation to get the best structure to start with; such as (Si, Al) mordenite lattice, characterized by Si/Al = 5. In the next step, the location of the extraframework cations in the totally dehydrated state was determined by means of the energy minimization code GULP (General Utility Lattice Program) [15]. The starting positions for those extraframework cations can be determined by using a visualization algorithm available through the Material Studio Visualizer of Accelrys (Material Studio module Vizualizer, Version 4.4, Accelrys Inc., San Diego), which can readily change the Si/Al ratio and the loading according to the Lowenstein rule. The third step is the use of simple grand canonical Monte Carlo (GCMC) methodology to load extraframework cations to neutralize the charges resulting from the Si/Al ratio. It was also observed that the extraframework cations are always located close to the aluminum atoms. Therefore, one can introduce these cations in a user-defined way. Normally, simulation is performed using periodic boundary conditions and the crystallographic unit cell of a desired matrix such as in mordenite $Si_{40}Al_8O_{96}$-Na_8 with space group $P1$ to avoid any symmetry constraint. It is observed that for the current calculation, which uses a force field-based energy minimization technique, four cations are split equally between sites IV and VI in the main channels and four are located on sites I in the small side channels according to Figure 6.3 and Figure 6.4. Hence, the choice of aluminum configuration is particularly judicious because it leads to a cation distribution close to that observed experimentally by X-ray diffraction [16] and dielectric relaxation spectroscopy [17]. Because statistical treatment of the effect of hydration on a nanoporous lattice with mordenite containing 144 atoms per unit cell corresponding to various distributions of the aluminum atoms is highly demanding. However, this calculation is not feasible in terms of CPU if one considers normal PC hardware. This insertion was performed by using Accelrys (Cerius[2], Accelrys, Molecular Simulation Inc., San Diego), which allows putting the oxygen atoms in the sites reported by the X-ray diffraction data [18] and then generating the hydrogen positions manually. These thirty-six generated structures can then be minimized to select the most stable for the next adding step. This procedure can be repeated for each hydration rate until the system is fully hydrated. In simulation it is necessary to look into minimization of energy in terms of force and electrostatic and/or other long-range interactions depending on which type of algorithm is used. In the current case, due to the complexity of the potential energy surface, a stepwise minimization approach was adopted to reach a convergence to the lowest energy structures. Those steps consist of (1) minimizing only the water positions under constant volume, (2) optimization of the extraframework cations, and (3) optimization of the zeolite framework under constant volume. Finally, a minimization of the whole structure was performed under constant pressure. This computationally demanding methodology was only to reach a suitable convergence of the gradients [19]. There is another issue of hydration, where one need to pay attention to the multitude of local minima for each given hydration state, which occurs probably due to the variable orientations of the hydrogen atoms of water. To avoid this problem, several starting configurations corresponding to different orientations of the water hydrogen atoms for each hydration state needs to be chosen. This complex route will allow isolating the lowest energy structures corresponding to hydration levels ranging from one to twenty-four water molecules per unit cell, where it can categorically justify the positions of the constituent ions. The current simulations show the difference in behavior between two cations depending on their locations; the cations situated in the main channels are progressively extracted from their initial sites upon hydration, whereas those located in the small side channels remain trapped whatever the water content.

Molecular dynamics simulations, using classical potential models to represent the cation–framework interactions, were performed in order to predict the low-frequency region of vibrational spectra for mordenite zeolites. The position and the shape of the bands assigned to the cation vibrations have been studied as a function of the nature of the extraframework charge-balancing cations (alkali, alkaline earth) and of the Si/Al ratio characterizing the zeolite framework. The critical role of the force field is also demonstrated by computing the low-frequency spectra using two different force fields that include the flexibility of the host framework [20]. A follow-up study by the Maurin group [21] examined the role of the extraframework cations and their position variances at different Si/Al ratios. The idea comes from the fact that extraframework cations needed to balance the charges resulted from variable Si/Al ratio can be located at different crystallographic positions depending on the architecture. The cations can migrate and the energy needed to extract the cations from their sites, called *detrapping energies*, is measured by a thermally stimulated depolarization current (TSDC) for increasing Si/Al ratios. The detrapping energy T is interpreted as an activation barrier for some jumps of Na^+ responsible for polarization of the zeolite. Some variation of these activation energies for Si/Al ratios is observed from TSDC but the reasoning is not known, exactly here comes the different need for performing simulation. The modeling technique for cation location is similar to what we have discussed before. Just to add with it is the Monte Carlo calculation protocol, where for each Si/Al ratio it generates a large number of possible configurations (10,000) and the populations of five possible Si centers as identified by nuclear magnetic resonance (NMR) was tested. The average occupation of these five possible environments defines the average configuration for a given Si/Al ratio in the sample and the Si–0Al was the predominant population followed at higher Si/Al ratios and a similar population of Si–0Al and Si–1Al for the smaller ratios. Once the types of Si centers are identified, the next job is to find the equilibrium extraframework cation configurations for both a rigid and nonpolarizable framework. A possible dielectric relaxation mechanism corresponding to that motions of cations between different metastable configurations of Na^+ was calculated and a new site for the monovalent cation was identified.

The position of monovalent cations in a zeolite — e.g., mordenite — with increasing Si/Al ratio is possible to calculate using a combination of Monte Carlo techniques and ^{29}Si NMR spectroscopy. For each Si/Al ratio, a combination of ^{29}Si NMR and Monte Carlo simulations is preferred. This leads to the proposal of a realistic structural model, where Al atoms are distributed among the four possible crystallographic sites. Positions of the cations stabilizing the mordenite lattice were calculated by Monte Carlo simulated annealing. The described model predicted the populations of the sites occupied by the cations and their variations with the Si/Al ratio were in very good agreement with those measured by thermally stimulated current spectroscopy [22]. NVT Monte Carlo simulations were first used to describe the distribution of Na cations in faujasite for several Si/Al ratios. These calculations were performed by combining two different sets of potential parameters combined with both T atoms and explicit Si, Al models. GCMC simulations were then employed to investigate the influence of water adsorption on the distribution of cations in the case of a faujasite sample with fifty-six cations (NaY56) and are comparable with the available experimental data [23].

Another approach to address the effect of the aluminum content on the properties of acidic zeolites is the use of sophisticated quantum mechanics/molecular mechanics (QM/MM) technique. To account for both electrostatic and mechanical interactions between the QM cluster and its MM environment is fruitful technique to describe the embedded cluster models as present in the covalent variant of the elastic polarizable environment (covEPE) [24]. To apply the covEPE method, it was necessary to develop a new force field for Al-containing zeolites. Two types of zeolite materials, FAU and MFI, were chosen as examples because of their specific pore architectures. The advantage of QM/MM method is that it can handle bigger system with a compromise of accuracy but at the same time with an improvised ability to handle bigger system size so it would be easier to incorporate the neighboring effect. The neighboring effect is important for the situation when the there is a variation in the Si/Al loading, which is dependent on the residual charges. The residual charge is balanced by Brønsted protons and hence the proton sites to be identified as OH in the cluster. The neighboring Al and Si present in the network influences the relative properties as well the frequency. Such QM/MM schemes have the capability to model easily the point defects (including charged ones) from strategies based on traditional (isolated) cluster models but overcome their deficiencies connected with neglecting steric constraints and the electrostatic field of the environment through the MM methodology. However, for proper construction of QM/MM methods one has to account for long-range electrostatic effects of the environment and the mechanical coupling of the QM clusters with their immediate surroundings. In addition, it is highly desirable to preserve the variability of the method as well as to reduce and control the influence of the QM/MM over border region. In this case, the variation of the Al content both in the MM environment and in the QM cluster were performed to predict pertinent properties of bridging OH groups of the zeolite frameworks, OH vibrational frequencies, and deprotonation energies. Calculations are performed on both high and low Si/Al ratio to compare the Al siting, whose position changes with the variation of Si/Al ratio. An example can be considered for cases when the Brønsted site is located at O1(H) and O3(H) crystallographic positions of a faujasite lattice to that of at Al7–O17(H)–Si4 sites of zeolite HZSM-5 (with MFI structure). The embedded cluster model will give an idea of the neighborhood (Figure 6.5). The most important aspect of the embedded model is the parameterization of the connecting atom between the MM and QM regime. Shor et al. [25] performed that by tuning the size by doing a variable cluster size with QM and when the cluster size was optimized they built the MM zone beyond that cluster size for convergence. The MM parameters were chosen with charge parameters and bonding information

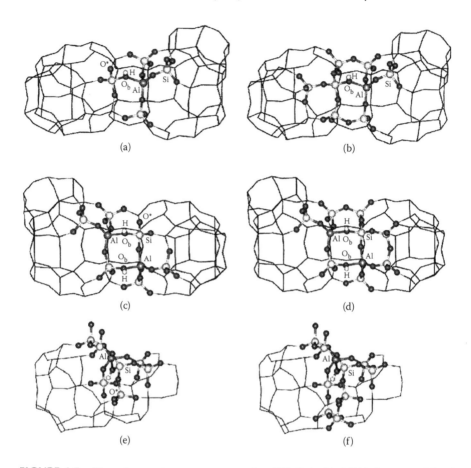

FIGURE 6.5 (For color version see accompanying CD) Embedded QM cluster models of various sites of faujasite: (a) 5T–1Al–O1, (b) 8T–1Al–O1, (c) 8T–2Al–O1, (d) 10T–2Al–O1, (e) 8T–1Al–O3, and (f) 10T–1Al–O3. [25] With permission.

using the GULP code [25]. The embedded model was constructed with various Si/Al ratios. The computational results suggest that the local structure and the location of the OH groups exert a stronger effect than the variation of the aluminum content of the framework. This embedding strategy was able to reproduce the trend of decreasing proton removal energies (by about 20–25 kJ/mol) along the series FAU > CHA > HZSM-5, in agreement with experiments. In addition, ordering and differences between simulated OH frequencies of the three bridging hydroxyl groups, O1(H) ($3,720$ cm^{-1}) and O3(H) ($3,635$ cm^{-1}) of faujasite and the OH group of HZSM-5 ($3,715$ cm^{-1}), fit the experimental data given as follows: $3,623$ cm^{-1}, $3,550$ cm^{-1}, and $3,610$–$3,617$ cm^{-1}, respectively [26].

   Another new methodology [27] came from Lewis et al. to probe the influence of framework ordering, cation sitting, and hydration of pores on the structure and its stability. The developed methodology allows the location of aluminum within the framework to be determined together with the position of extraframework cations, in a stepwise fashion, progressing from an anhydrous model, via a dielectric continuum model, to, finally, a fully atomistic model of the water within the zeolitic pore space. In parallel to the many developments in experimental structural methods, computational methods have been highly effective in describing the structure of zeolites and many other solids. Modeling situations with variable Si/Al ratios has its challenges, so it is important to make a realistic model both in terms of structure and composition. For example, consideration of the energetic of replacing only single silica by aluminum does not reproduce experimental T sites occupations in low-silica materials, because the critical interactions between other aluminum atoms and extraframework cations are ignored [28]. But certainly such calculations are valuable in determining the geometry of T sites in high-silica zeolitic matrices. Moreover, the results clearly demonstrate that aluminum in the lattice results in distortions of the structure up to the fourth coordination sphere when compared to a purely siliceous structure or to an average structure (where silica and aluminum are treated the same). Therefore, it is difficult to create a model in terms of Löwenstein's rule [29]. The method described by Salvador et al. [30] has been applied to anhydrous zeolite models, which proposed a contraction of the zeolite cavities and the shifting of the extraframework cation toward the framework walls (from those found in the hydrated state). Such displacements and contractions were also observed during the dehydration of many zeolites [30]. This procedure works well for low or medium Si/Al ratios, due to the increased attractive electrostatic interactions between the extraframework cation and the framework oxygen atoms in presence of water was removed. However, it should be mentioned that for low-silica zeolites, the electrostatic field arising from incorporation of Al in the framework is much more complex. It needs to incorporate a large number of cations within the pore to make the direct coordination of all cations with the entire aluminum-containing sites. This demands that the calculation must be performed in the presence of water, which is computationally very challenging. Almora-Barrios et al. [31] studied the geothermal properties of a low-silica natural zeolite, Goosecreekite, to show the amount of water present in these matrices. They expanded the method of Salvador et al. based on force fields involving polar interaction depending on core-shell-types of interactions using the GULP code and prescribed a stepwise procedure to rationalize the understanding of the geothermal process. The first step is to start with pure silica structure at the anhydrous mode, minimize the energy, and then gradually load aluminum through a high Si/Al ratio to a low one. Then the resulting structure will be further monitored in presence of water as continuum model or as a water cluster describing a localized hydration scenario.

Periodic *ab initio* and cluster model calculations using density functional theory (DFT) were applied [32] to study the structure of Li^+, Na^+, K^+, Mg^{2+}, Ca^{2+}, Ni^{2+}, and La^{3+} exchanged zeolite. Periodic calculations were performed to monitor the cation migration in the zeolite lattice. A correlation to monitor the strength of Lewis acidity of the cations was established by Fukui function–based reactivity descriptors. The activity order does not depend on the size of the cations but on the localized geometry of cations. The Lewis acidity order for different corresponding cationic sites was derived and compared with the experimental trend. In another simulation procedure [33], a combination of periodic first principle calculations and the localized reactivity index was used to address the cation exchange phenomenon inside the zeolite. The

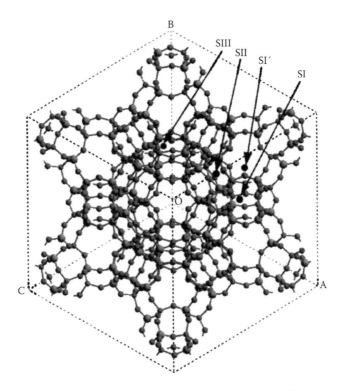

FIGURE 6.6 Zeolite faujasite structure showing idealized cation positions.

structure of Na–Y as determined from neutron diffraction [33] with typical cation positions is shown in Figure 6.6. Cations in the SI position have a more symmetric environment (D3d) in the center of double six-rings. The SII cation position in the faujasite is located in the supercage and is coordinated to three oxygen atoms of the six-ring window to the sodalite cage as shown in Figure 6.7a, whereas the SI′ position is in the sodalite cage. Quadruple coupling constants (QCC) determined for the sites SI, SI′, and SII were 4.8, 4.2, and approximately 0.1 MHz [34]. The SIII cations were located in the supercages at the entrance of zeolite cage with an eight-member ring and hinder the passage of incoming molecule and hence SIII cation was neglected for this calculation. According to NMR, SI and II are two of the best possible catalytic sites in faujasite. Auerbach et al. first performed the periodic calculation to monitor the structural behavior of the lattice after cation exchange specifically for SI and SII sites [35]. The calculation comprised localized cluster calculation, which was performed on two clusters representative of two sites and then their activity was compared through relative electrophilicity/nucleophilicity. The cluster models representing SI and SII are shown in Figure 6.7b and Figure 6.7c, respectively. Depending on the nature, cations are selective about their position; for example, trivalent cations prefer the SI site, rather than the SII site, whereas Na^+ was least stable when placed in an SI site and hence may have more mobility. On the other hand, Li^+ showed the most stability when present in an SII site. It was observed that those cations in SII sites are strongly bonded than

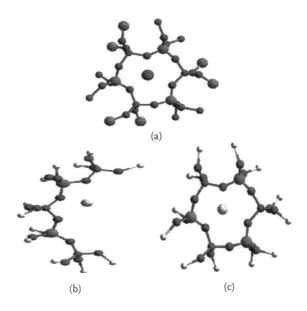

(a)

(b)   (c)

FIGURE 6.7 (For color version see accompanying CD) (a) cluster model representing the SII site of faujasite with the metal coordination with three oxygens, (b) cluster model representing the SI site of faujasite zeolite with formula $M3[Al_3Si_2O_{16}H_{12}]$, (c) cluster model representing the SII site of faujasite zeolite with formula $M3[Si_3Al_3O_{18}H_{12}]$. The color code is as follows: white, hydrogen; yellow, metal cation; red, oxygen; brownish yellow, silicon; and pink, aluminum.

the cations in an SI site. As a result, the cation in an SI site has ample room for relaxation, which makes Na^+ a better performer. However, no quantitative order can be predicted when the cations are in the fully relaxed condition. Cation migration is least feasible for Li^+. We have not optimized the full structure because complete relaxation of the cluster leads to a different structure, which does not resemble the experimental geometry as mentioned by Pal and Chandrakumar [36]. Reactivity index calculation is performed by removing one electron from the neutral system to minimize it and get the nucleophilic reactivity, whereas an addition of electron to a neutral system gives electrophilicity. We have carried out the ratio of electrophoilic and nucleophilic activity at the same site to get the relative activity of the cation. The point here is that from periodic symmetric point of view, there may be mobility of the cation and it is sensitive to the architecture of the pore. However, we may not be able to suggest about the activity of the atom center conclusively, because there exists a size and activity difference of exchanged cations when they changes from monovalent to trivalent. This is the first study to correlate the behavior of exchangeable cations with a variable oxidation state in faujasite-type zeolites. The stabilization energy trend shows that when fully relaxed, the cation migrates from an SII site to an SI site with some constraints depending on its stability. The reactivity descriptor results in a very interesting qualitative trend. From the reactivity index we have seen that the activity order is different from that in an SI site and an SII site, which is solely dependent on the particular architecture of the site. This trend matches with the trend observed from periodic calculation. The relative electrophilicity scale fits with the periodic trend for SI, whereas the relative nucleophilicity trend matches with that of SII. The trend thus produced is the semiquantitative scale to compare the Lewis acidity and basicity of the cations. The role of the zeolite architecture on the activity of the cations is thus established. The optimistic results force us to study this behavior with other zeolites and propose a prior rule for cation activity in zeolites.

We have followed a similar study with $AlPO_4$ materials [37]. The influence of both bivalent and trivalent metal substituents from a range of metal cations (Co, Mn, Mg, Fe, and Cr) on the acidic property (both Brønsted and Lewis) of metal-substituted aluminum phosphate MeAlPOs was monitored. The influence of the environment of the acid site was studied both by localized cluster and periodic calculations to propose that the acidity of AlPOs can be predictable with accuracy so that AlPO materials with desired acidity can be designed. A semiquantitative reactivity scale within the domain of the hard–soft acid–base (HSAB) principle was proposed in terms of the metal substitutions using DFT. It was observed that for the bivalent metal cations, Lewis acidity linearly increases with ionic size, whereas the Brønsted acidity is solely dependent on the nearest oxygen environment. Intramolecular and intermolecular interactions show that once the active site of the interacting species is identified, the influence of the environment can be prescribed. Mg(II)-doped AlPO-34 shows highest Brønsted acidity, whereas Cr(III)-doped species show lowest acidity. Fe(II)/Fe(III)-doped AlPO-34 shows the highest Lewis acidity, whereas Mn(III), Mg(II) show the lowest acidity. The combination of different methodologies like bulk structural optimization with the CASTEP code of Accelrys [38] and cluster calculations were performed with DMol^3 code of Accelrys [39]. As we have seen so far, these exchangeable cations were influenced by pore architecture and as well these are detrimental toward the acidic property of the matrix, independent of the pressure of either a Lewis acid or a Brønsted acid site. The periodic model is shown in Figure 6.8. Throughout the calculations one dopant per unit cell was used to see the effect of the size of the cations on the bulk geometry. It would be easier to build clusters depending on the localized behavior of the cations and then combine them to propose the activity of the cations in AlPOs. The advantage of this combination of methods with different accuracy is to go further details along with experiment to predict the new material within the domain studied. Figure 6.9a and Figure 6.9b mimic the dopant scenario of bivalent and trivalent metal ions, respectively. The periodic calculation results show that the activity of hydroxyl protons is much more dependent on the environment as the distortion in the tetrahedral geometry propagates beyond the doping region. This phenomenon is not observed in the case of nonprotonated structures resulting from trivalent dopants. These results show us the distortion propagation or, in other words, the deformation resulting in the structure due to inclusion of the dopant can affect its nearest- or next-to-nearest-neighbor atom. This will further influence the activity resulting from the substitution of the dopant and hence the acidity. For the bivalent cation, the replacement energy increases with increase in the M–O distance and the larger the distortion, the greater the substitution energy. In case of trivalent dopants, the trend is the same, but for Cr^{3+}, substitution energy remains very high compared to the rest of the dopant and hence may not be explained by the current method. The relative large values of the substitution

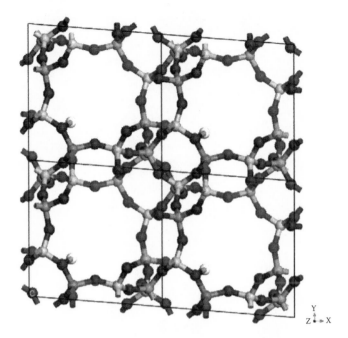

FIGURE 6.8 The super cell model to represent the bivalent metal AlPO-34 framework used for periodic calculation. This is an alternate network of Al and P connected through oxygen. One of the Al is replaced by a bivalent dopant. For trivalent dopant the structure will be without this proton.

energy support the phenomenon observed experimentally that AlPOs are unstable upon high metal doping. The substitution energy at this point allows predictions about the feasibility of dopant substitution, but it does not correlate with trends in the activity and therefore it is not possible to predict the activity based on the computed substitution energy. To visualize this effect, one needs to look more closely at the localized interaction at the close environment of the dopant atom. This can only be done through the localized reactivity index calculation. The relative nucleophilicity order for the active moiety in the AlPO cluster represents the trend nicely. We therefore can account for the acidity of AlPOs comprehensively. The trend for Brønsted acidity shows $Cr^{2+} < Mn^{2+} < Fe^{2+} < Co^{2+} < Mg^{2+}$, whereas the trend for Lewis acidity mainly for trivalent metal dopants is $Fe^{3+} > Co^{3+} > Cr^{3+} > Mn^{3+}$. The trend matches with experimental observations for Brønsted and Lewis acidity, respectively [40]. It is worth mentioning that all calculations for AlPO were carried out in the unhydrated rather than hydrated form because only calcined AlPO exhibits catalytic activity.

Ultrastable Y zeolites (USY) are the main components of cracking catalysts. They are normally produced from an NH_4Y derivative, upon treatment with steam at 773–973 K. Under these synthesis conditions, there is a partial loss of aluminum from the zeolite structure, which improves the thermal stability and catalytic properties of the zeolite. Nevertheless, the aluminum atoms released from the framework stay inside the cavities and channels as extraframework aluminum (EFAL) species. The nature

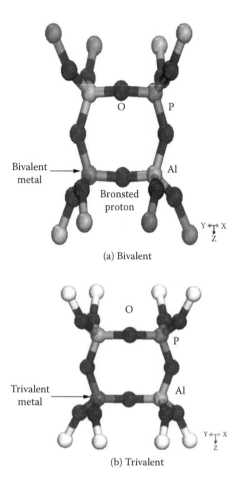

(a) Bivalent

(b) Trivalent

FIGURE 6.9 Two independent cluster with the formula (a) $M^{+2}AlP_2O_{12}H_9$ and (b) $M^{+3}AlP_2O_{12}H_8$ to represent the bivalent and trivalent dopant incorporated clusters. The first cluster represents the bivalent dopant and the second one a trivalent dopant. The cluster termination here is with hydrogen.

of the EFAL species is not completely known, but it is postulated that oxoaluminum cations, such as AlO^+, $Al(OH)^{2+}$, and some neutral compounds such as $AlOOH$ and hydrated Al_2O_3, could account for some of the EFAL species. The distribution of silica and aluminum cations over the framework of dealuminated faujasites has been analyzed by a Monte Carlo procedure, and the results have been compared with the ^{29}Si NMR data. Modeled materials are dealuminated according to different criteria in order to determine the main characteristics of the actual dealumination processes. The distribution of cations after partial aluminum removal depends on the dealumination method. In particular, this distribution is more homogeneous for specimens dealuminated with $SiCl_4$ than for those dealuminated by following a hydrothermal method. In the latter case, an appreciable gradient of atom concentrations is present, indicating that the dealuminating agent acts effectively only on a partial volume of the treated material. In all cases considered, dealumination yields cation distributions far from that of the state of lowest free energy [41].

One of the issues related to synthesis gel composition that remains ambiguous to experimentalists is the site preference after substitution in zeolites. Several questions have been raised to know the best site for aluminum substitution and how does it matter. ZSM-5 has the MFI framework topology, characterized by a three-dimensional pore system with straight and sinusoidal channels. The pore openings are defined by number of member rings that are wide enough to allow passage of molecules as large as benzene. Among the twelve crystallographically distinct T sites, T4 and T10 sites occur in the ten-member ring that forms the straight channel. Similarly, T8 and Tl sites never occur in the ten-member ring that forms the sinusoidal channel. Thus, there are eight T sites that are common to both straight and sinusoidal channels [42]. The difference in catalytic and separation properties resulting from framework substitution offers the potential to design zeolites for new applications. Boron-substituted zeolites find their application in the Assoreni process of methyl butyl ether conversion to methanol and isobutene. They have also been used in the xylene isomerization and ethylbenzene conversion process [43]. Fe-ZSM-5 has been shown to be an active catalyst for the production of methanol by the direct oxidation of methane [44]. The substitution site of Fe and B in MFI [45] can be calculated using a cluster model of [T(OH)], dimer [(OH>,–Si–0–T–(OH)] and pentamer T(OSiO,H$_3$)$_4$ cluster models where T = Fe or B and kept fixed throughout the calculations. The boundary oxygens are saturated by hydrogen atoms, occupying the position of the nearest T site. In the monomer cluster models the adjacent T sites are approximated by H atoms. The dimer cluster model consists of two TO groups bridged by commonly shared oxygen and the pentamer cluster model represents a TO group that shares a corner with four adjacent TO groups through the bridging oxygen atom. The boundary oxygens are saturated by hydrogen atoms, with hydrogen atoms occupying the position of the nearest T site. Hydrogen atoms, necessary to maintain the cluster neutrality for a substituted situation, were located at 1 A along the bond axes connecting with the bridging oxygen. The choice of cluster models involved in this calculation has been discussed elsewhere [46]. From Figure 6.10 it is observed that larger bond lengths (T–O) are easily correlated with smaller bond angles (T–O–T). The figure indicates that substitution is preferable at the T9 site, with T6 and T12 being the other choices. The internal coordinates of the all silicon and substituted ZSM-5 models show a dependence on the number of shells of atoms surrounding the central T–OH–Si bridge present in the protonated clusters. Figure 6.11 shows the variation of optimized O–H bond lengths in protonated models of increasing size from dimer to pentamer. The changes in bond length are largely due to the change in electronic environment of the optimized atoms of the central bridge; in the dimer model T atoms are linked with OH groups. Further replacement of terminal hydroxyls by OSi(OH) groups leads to a change in geometry, which leads to an increase in the O–H bond lengths in case of B and Fe and a decrease in case of aluminum. The T–O bond lengths behave in a more or less similar fashion as shown in Figure 6.12. The T–O bond lengths in the all silicon models are relatively constant. The substituent atoms appear to cause significant changes in

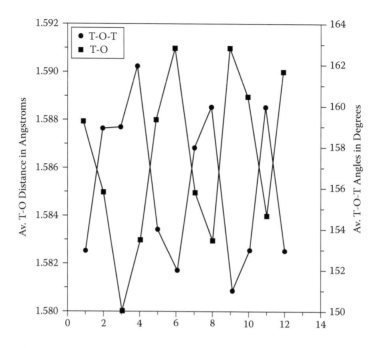

FIGURE 6.10   Average T–O bond lengths and T–O–T bond angles as obtained from crystallographic data of van Koningsveld et al. [10]

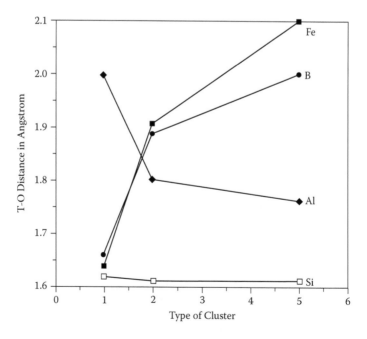

FIGURE 6.11   T–O bond length variation with increasing cluster size.

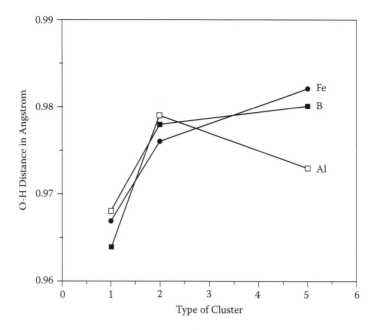

FIGURE 6.12   O–H bond length variation with increasing cluster size.

the zeolite geometry and are much more sensitive to model size. Figure 6.13 exhibits the variation of T–OH–S1 angles between the dimer and pentamer models. It is observed that for B and Fe the bond angles widen by ~20 degrees. The T–OH–T angles appear to converge slower with increasing model size. The results show that the preferable sites for Fe substitution are T3 and T11 and for B the preferred sites are T2 and T12 in the ZSM-5 zeolite framework. The calculation on the pentameric cluster model has demonstrated the importance of including distant oxygen shells, which again reinforces the role of calculations in attempting to determine the local structural effects of substitution in host materials. The relative acidity of substituted zeolites can be predicted from the calculated proton affinities of the zeolite model. Models with high proton affinity have a low Brønsted acidity, whereas those with low proton affinity will have more Brønsted acidity. The results of acidity have been compared with vibrational frequency calculation of experiment for the OH stretching as in Table 6.1. The vibrational frequency of the O–H bond decreases with increasing acidity. The relative acidity of the hydroxyl groups calculated in terms of E_{dp} matches very well with the experimental acidity trend [47], predicted from IR frequencies. The large difference between the E_{dp} values of B-ZSM-5 and other substituent correlates well with the difference in the IR frequencies of bridging hydroxyls in case of B with those of other substituted ZSM-type materials. The credibility of simulation regarding the composition variation of synthesis gel one can reproduce the trend of experiment, designed newer material within a short range of time and without any expensive and hazardous chemicals. Recently, Oumi et al. [48] used the combination of three techniques to differentiate the best place

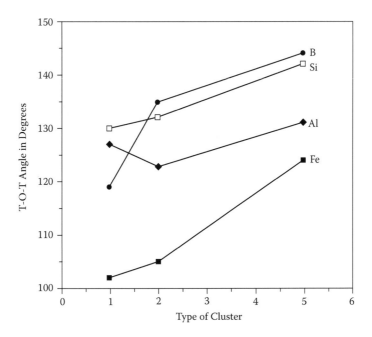

FIGURE 6.13   Si–OH–T bond angle variation with increasing cluster size.

for Al in mordenite and compared the results with experimental pressure temperature scenario. The bulk model is shown in Figure 6.14 and the adsorption pattern
in Figure 6.15.

There is a completely new scenario to this composition topic, with the work of
Simperler et al. [49], where they present the results of lattice energy optimizations
of hypothetical AlPOs and compare them to those of their iso-structural silica polymorphs. They used the Tilling theory as proposed by Foster et al. [50] to generate
model silica polymorphs and only focus on the uninodal structures; i.e., those with
only one topologically distinct T site. The network points were converted to chemical structures with oxygen atoms inserted between each pair of adjacent T sites.

TABLE 6.1

Experimental and Theoretical Comparison of Acidity

Property	Al	Fe	B
O–H[a]	3610	3630	3745
E_{dp}[b]	317.2	323.5	335.4
Charge on H[c]	0.25	0.24	0.19

Notes: [a] O–H stretching frequency in cm^{-1} (experimental).
[b] Deprotonation energy in kcal/mol.
[c] Partial charge on protonic hydrogen.

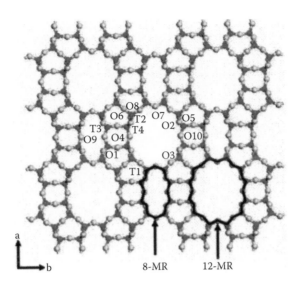

FIGURE 6.14 (For color version see accompanying CD) MOR zeolite framework viewed along the *c* axis. T1–T4 and O1–O10 indicate four and ten nonequivalent crystallographic sites of Al and O atoms, respectively. [48] With permission.

Lattice energy minimization was performed using the GULP code [14] using the ionic potential model of Sanders et al. [51]. Here the methodology was applied to AlPO structures, which were obtained using strictly alternating phosphorous and aluminum atoms at T sites. A total of 134 uninodal hypothetical silica polymorphs with suitable AlPO counterparts were optimized with GULP using the potential parameter as obtained by Gale and Henson [52]. For both compositions of zeo-type, the dependence of the relative stability on framework densities and on the number of T sites in the fourth coordination shell was examined [53]. To be useful as a catalyst or sorbent, a chemically feasible structure should comprise voids or channel systems that are accessible from outside the crystallite. Therefore, accessible volumes and internal surface areas were measured and examined to quantify the stability of the various structures as a function of these quantities. To calculate the accessible volume and surface area, the Free Volume module of the Cerius^2 package was used (Cerius^2 v. 4.2, Molecular Simulations Inc., San Diego). This applies the Connolly method [54] consisting of "rolling" a probe sphere with a radius of 1.4 Å over the van der Waals surface of the framework atoms (using the van der Waals radii of 1.32, 0.9, 1.05, and 0.85 Å for O, silica, aluminum, and phosphorous, respectively). The void volume was obtained first, and by forcing the probe molecule to enter the unit cell from the outside, the accessible volume was calculated. The Connolly surface is the surface enclosing the void volume. Average r(T–O) bond lengths, r(T...T) bond distances, and T–O–T angles have been calculated, as has the difference between the largest and smallest tetrahedral α(O–T–O) angles, to estimate the distortions within a TO4 unit. Table 6.2 lists the experimentally obtained enthalpies [55] for a selection of known AlPO and known ZEO structures and their calculated relative lattice

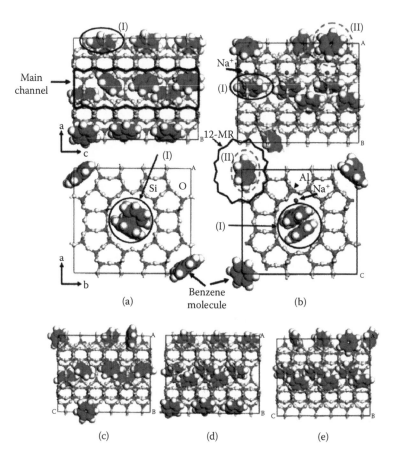

FIGURE 6.15 (For color version see accompanying CD) The equilibrium adsorption states of benzene molecules on MOR zeolite. (a) Pure silica MOR (Si/Al = 1), and Al-containing MOR zeolites (Si/Al = 47) substituted by Al atom for (b) T1, (c) T2, (d) T3, and (e) T4 sites. [48] With permission.

energies. Eight structures were selected with $\Delta E_{latt} < 20$ kJ/mol, which is considered as chemically feasible, and show features relevant for experimentalists, such as a continuous channel system and/or large voids: (a) AlPO_1_272, (b) AlPO_1_195, (c) AlPO_1_120, (d) AlPO_1_122, (e) AlPO_1_73, (f) AlPO_1_66, (g) AlPO_1_14, and (h) AlPO_1_11, listed in order of decreasing stability (Figure 6.16a–h). Table 6.2 lists the most important data for these hypothetical AlPOs as well as their isostructural hypothetical silica polymorphs. Note that the numbering AlPO_1_n refers to structures in the database of Simperler et al. [49] and not to the AlPO-n convention used to identify known AlPO structures. Excellent correlations between accessible volume and framework density prove the internal consistency of the calculation methodology, and correlations between surface area and framework density for both known and hypothetical structures were pretty reasonable.

We may therefore comment here that simulation can help one to guide experiments with reasonable accuracy. The challenge here is the model in relevance to

TABLE 6.2

Experimental Values of Enthalpies of Transition, ΔH_{trans}, 298 K, and Calculated Lattice Energies, ΔE_{latt}, of AlPOs and ZEOs per TO$_2$ Unit

	ΔH_{trans}, 298 K (AlPO)$_{exp}$ (kJ/mol)	ΔH_{trans}, 298 K (ZEO)$_{exp}$ (kJ/mol)	$\Delta E_{latt,barlinite}$ (kJ/mol)	$\Delta E_{latt,quartz}$ (kJ/mol)
R-cristobalite	3.05[a]	2.48[b]	−1.50	3.22
AEL	6.19[a]		4.03	10.99
AET	5.77[a]		7.32	14.44
AFI	7.01[a]	7.20[c]	5.35	11.68
AST		10.86[d]	11.38	18.14
BEA		9.29[d]		14.39
CFI		8.82[d]		13.00
CHA		11.43[d]	8.65	16.30
FAU		13.60[c]	13.20	19.91
FER		6.60[c]		11.78
IFR		10.04[d]		15.00
ISV		14.37[d]		16.44
ITE		10.08[d]		14.12
LTA	7.78[a]		11.66	19.26
MEI		13.90[e]		
MEL		8.19[d]		10.76
MFI		6.78[d]		9.96
MTW		8.70[c] 8.14		8.70[c] 8.14
MWW		10.42[d]		14.68
STT		9.19[d]		14.70
VFI	8.37[a]		10.95	21.12

[a] N. Hu, A. Navrotsky, Chen, C.-Y.; Davies, M. E. *Chem. Mater. 7*, (1995) 1816.

[b] Richet, P.; Bottinga, Y.; Denielou, L.; Petitet, J. P.; Tequi, C.*Geochim. Cosmochim. Acta 46*, (1982) 2639.

[c] Petrovic, I.; Navrotsky, A.; Davis, M. E.; Zones, S. I. *Chem. Mater. 5*, (1993) 1805.

[d] Piccione, P. M.; Laberty, C.; Yang, S.; Camblor, M. A.; Navrotsky, A.; Davis, M. E. *J. Phys. Chem. B 104*, (2000) 10001.

[e] Navrotsky, A.; Petrovic, I.; Hu, Y.; Chen, C.-Y.; Davis, M. E. *Micropor. Mater. 4*, (1995) 95.

the question to be answered, simulation is the best way to screen without wasting chemicals and time in the bench. Among different methods of simulations, QM is the safest one though it has size limitation, can guide the position of substitution in nanoporous material for composition variation using bigger model with atomistic calculation. In simulation, main points are structural atomic distribution and the stability of the structures, which can be calculated in terms of total energy, binding energy, or stabilization energy — and the general mathematics in interpreting the results; more negative stands for more stability and sometimes more stability may as well say lesser activity, which depends on the binding energy of different intermolecular interactions.

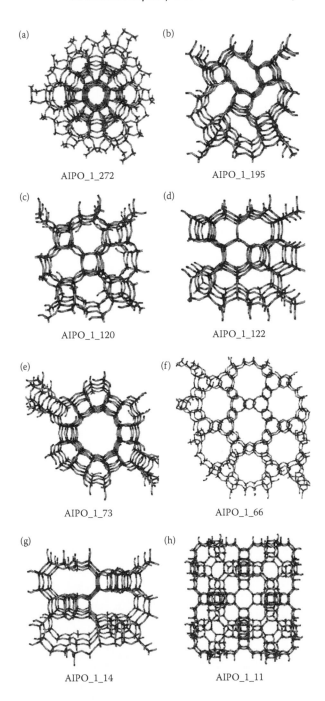

FIGURE 6.16 Eight chemically feasible hyp-AlPO structures: (a) AlPO_1_272, (b) AlPO_1_195, (c) AlPO_1_120, (d) AlPO_1_122, (e) AlPO_1_73, (f) AlPO_1_66, (g) AlPO_1_114, and (h) AlPO_1_11. [49] With permission.

6.2 SPECTROSCOPIC ANALYSIS

Spectroscopic analysis is the backbone of any characterization for experiments. Techniques mainly used are NMR and IR for the specific positions of the constituent atoms for the bulk, whereas XPS is a tool for surface analysis.

Harmonic vibrational frequencies may be obtained from the matrix of Cartesian second derivatives, also known as the *Hessian matrix*, of a molecular or periodic system [56]. For the case of molecules (or any finite system) the elements of the Hessian, $H_{i,j}$, are given by the second derivative of the total energy E:

$$H_{i,j} = \frac{\partial^2 E}{\partial_{qi} \partial_{qj}}$$

where q_i is a Cartesian coordinate of a system with N atoms, so that $1 < i < 3N$.

For the periodic case, there is an infinite number of atoms, but due to the periodicity ($H_{i,j} = H_{i+Tj+T}$), the infinite Hessian matrix can be Fourier transformed into an infinite set of $3N \times 3N$ matrices, where N now stands for number of atoms in the unit cell.

$$H_{i,j}^k = \frac{1}{V} \sum_T H_{i,j+Te^{-kT}}$$

Here, **T** is the lattice translations vector and **k** represents the vector in the first Brillouin zone.

For many phenomena, such as infrared spectra, for example, the most important factor is the Hessian matrix for the case where **k** = 0; i.e., at the Γ-point. Thus, the Hessian for periodic systems will hereafter be referred to as the Hessian at the Γ-point.

The mass-weighted Hessian is obtained by dividing Hessian elements by the square roots of the atomic masses:

$$F_{i,j} = \frac{H_{i,j}}{\sqrt{m_i m_j}}$$

According to the harmonic approximation, the vibrational frequencies are the square roots of the eigenvalues of F and the normal modes are the eigenvectors of F.

The infrared intensities are obtained from the atomic polar tensors (*A*), which are conventionally called *Born effective charges* in solid-state calculations, of all atoms in the system. *A* is a second derivative of the total energy with respect to the Cartesian coordinates and dipole moments.

$$A_{i,j} = \frac{\partial E}{\partial_{qi} \partial \mu_i}$$

where q_i is a Cartesian coordinate of a system with N atoms, so that $1 < i < 3N$; and μ_i is the dipole moment.

The intensity of a given mode can be evaluated as a square of all transition moments of this mode and expressed in terms of the **A** matrix and eigenvectors of the mass-weighted Hessian, F:

$$l_i = \left(\sum_{j,k} F'_{i,j} A_{j,k} \right)^2$$

Here, F' are eigenvectors of the normal mode, i.

A molecule composed of N atoms has $3N$ degrees of freedom, of which six are translations and rotations of the molecule itself. This leaves $3N - 6$ degrees of vibrational freedom (or $3N - 5$ if the molecule is linear). The frequency at which a given vibration occurs is determined by the strengths of the bonds and the masses of the atoms involved. The number of observed absorption peaks is usually different from what would be predicted by the formula above, because their number may be increased by additive and subtractive interactions, leading to combination tones and overtones of the fundamental vibrations. Furthermore, the number of observed absorption peaks may be decreased by molecular symmetry and spectroscopic selection rules.

Negative eigenvalues correspond to modes in which the energy is not a local minimum. Generally, these indicate that the system is in a transition state. The normal mode corresponding to this eigenvalue is the mode that moves the system in the direction of a local energy minimum.

The frequencies obtained from negative eigenvalues are obviously imaginary numbers. However, by convention, such frequencies are given as real negative numbers.

$3N$ frequencies are generated by diagonalizing the mass-weighted Hessian. Of these, $3N - 6$ correspond to the normal modes of vibration and the remainder to translations and rotations of the entire system. (For a linear molecule, there are $3N - 5$ normal modes; for a periodic system, there are $3N - 3$.) If the geometry of the system has been optimized so that the atomic forces are small ($\sim 1.0 \times 10^{-4}$ Hartree $Bohr^{-1}$), then the frequencies corresponding to translations and rotations will only be a few wave numbers in magnitude. If, however, the geometry of the system has not been optimized, then the frequencies of the translations and rotations can be rather large, of the order of several hundred wave numbers.

It is possible to remove the translations and rotations from the Hessian matrix before computing the frequencies. This will result in values for the translational and rotational modes that are rigorously zero. The procedure involves constructing a matrix representation of the translations and rotations, P, which is used to project the translations and rotations from the Hessian. The remaining normal modes and frequencies are slightly altered from the unprojected values [57].

NMR can be calculated mainly by two commercial software packages: one is Gaussian and other is CASTEP from Accelrys. In Gaussian, NMR shielding tensors and magnetic susceptibilities are calculated using the Hartree-Fock method, all DFT methods, and the Mollar Plasset 2 method. NMR shielding tensors may be computed with the continuous set of gauge transformations (CSGT) method [58–60] and the gauge-independent atomic orbital (GIAO) method [61]. Magnetic susceptibilities may also be computed with both GIAOs and CGST. Gaussian also supports the IGAIM method [62] (a slight variation on the CSGT method) and the single origin method, for both shielding tensor and magnetic susceptibilities. The calculations are very expensive and it is still unable to handle the crystal structures well, the accuracy is limited. Therefore, the plane wave code of Accelrys called CASTEP [63] can be used to solve the problem. NMR CASTEP has been implemented in its framework.

A uniform external magnetic field, **B**, applied to a sample induces an electric current. In an insulating nonmagnetic material, only the orbital motion of the electrons contributes to this current. In addition, for the field strengths typically used in NMR experiments, the induced electric current is proportional to the external field, **B**. This first-order induced current, $j^{(1)}(\mathbf{r})$, produces a nonuniform magnetic field:

$$B_{in}^{(1)}(r) = \frac{1}{c}\int d^3r'\, j^{(1)}(r') x \frac{r-r'}{|r-r'|^3}$$

The shielding tensor, $\sigma(\mathbf{r})$, connects the induced magnetic field to the applied magnetic field:

$$B_{in}^{(1)}(r) = \overline{\sigma(r)}B$$

and the isotropic shielding is given by:

$$\sigma(r) = \frac{Tr[\overline{\sigma(r)}]}{3}$$

Information about $\sigma(\mathbf{r})$ at nuclear positions can be obtained from NMR experiments. In CASTEP, the first-order induced current, $j^{(1)}(\mathbf{r})$, is calculated and then used to evaluate the induced magnetic field, $\mathbf{B}^{(1)}$, via the equation mentioned. The isotropic chemical shift $\sigma(r)$ for a nucleus in the position r is defined as

$$\partial(r) = -[\sigma(r) - \sigma^{ref}]$$

where σ^{ref} is the isotropic shielding of the same nucleus in a reference system. The references for 17O and 29Si are spherical water and tetramethylsilane, respectively. The 17O quadrupolar coupling constant C_q and the asymmetry parameter η are obtained from the EFG (electric field gradient) tensor $\overline{G(r)}$:

$$G_{\alpha\beta}(r) = \frac{\partial E_\alpha(r)}{\partial r_\beta} - \frac{1}{3}\partial_{\alpha\beta}\sum_\gamma \frac{\partial E_\gamma(r)}{\partial r_\gamma}$$

where α, β, γ denote the Cartesian coordinates x, y, z and $E\alpha(r)$ is the local electric field at the position e, which can be calculated from the charge density $n(r)$:

$$E_{\alpha}(r) = \int d^3r' \frac{n(r')}{|r-r'|^3}(r_{\alpha}-r_{\alpha}')$$

The EFG tensor is then equal to:

$$G_{\alpha\beta}(r) = \int d^3r' \frac{n(r')}{|r-r'|^3}\left[\partial_{\alpha\beta} - 3\frac{(r_{\alpha}-r_{\alpha}')(r_{\beta}-r_{\beta}')}{|r-r'|^2}\right]$$

If one then levels the eigenvalues of the EFG tensor V_{xx}, V_{xy}, V_{zz} so that $|V_{zz}| > |V_{yy}| > |V_{xx}|$, then

$$C_q = \frac{eQV_{zz}}{h}$$

where e is the absolute electronic charge, Q is the nuclear quadrupolar moment, and h is the Planck constant.

Calculations are performed with the gauge-including projector augmented-wave method (GIPAW) developed by Pickard and Mauri [64]. The projector augmented-wave formalism originally introduced by Blöchl [65] makes it possible to obtain expected values of all-electron operators in terms of pseudo-wave functions coming from a pseudo-potential calculation and the GIPAW formalism reconciles the requirement of translational invariance of the crystal in a uniform magnetic field with the localized nature of Blöchl's projectors. The theory was initially applied with Troullier-Martins norm-conserving pseudo potentials [64] and has recently been extended to the case of ultrasoft pseudo-potentials [66]. The calculation of the induced current involves evaluation of expected values of the position operator, whose definition for extended systems is not straightforward. Two approaches have been developed [64] and implemented in CASTEP. For translationally invariant systems with well-defined Blöch vectors, k, the application of the position operator can be reduced to calculating the gradient in k-space. Alternatively, for finite systems, which are represented by supercell structures (a molecule in a box) in CASTEP, one can apply a real-space representation of the position operator. The artificial periodicity imposed on the system causes the position operator to acquire a sawtooth shape. Thus, in the vicinity of the box boundaries, the position operator is distorted. Because of this, a sufficiently large supercell must be constructed so that the electronic density does not reach the regions where the position operator behaves unphysically. This condition is easily satisfied by the usual physical requirement that the molecules in neighboring boxes do not interact. Finally, the induced magnetic field, $B^{(1)}$, is calculated from the induced current and its value at the nuclear position determines the NMR shielding tensor. The calculated tensor is rather sensitive to numerical errors in both the unperturbed wave function, $\Psi^{(0)}$, and its first-order perturbation, $\Psi^{(1)}$, so the convergence requirements for NMR calculations are more stringent than, for example, routine band structure calculations. To obtain reliable results, the convergence of the wave function with respect to the number of basis functions (plane waves) must be ensured and a well-converged conjugate gradient procedure put in place to determine the coefficients of the plane waves. Apart from that, in calculating solid-state NMR properties (a small unit cell as opposed to a molecule in a box), the convergence of the k-space sum over the Brillouin zone must be achieved. The NMR observables are not particularly sensitive to uncertainties in the ground-state density distribution in the system, which determines the unperturbed Hamiltonian. Here, the usual convergence criteria for self-consistent DFT calculations are sufficient. NMR properties are very sensitive to atomic positions, which make NMR such a useful experimental tool for structure analysis. This implies that it is highly recommended to perform a geometry optimization run prior to the NMR calculation.

^{29}Si solid-state NMR spectroscopy plays an important role in the characterization of silicate materials. For instance, NMR remains the most valuable tool in the determination of the connectivity of the SiO_4 tetrahedral. To know if they are in monomeric Q1, dimeric Q2, trimeric Q3, and tetrameric Q4 form, their ratio gives a better understanding of the structure of nanoporous material. There is a way to correlate chemical shift to the Si–O–Si angles in silicate structures [67]. ^{17}O is a nucleus center with a spin of 5/2, which makes it possible to measure both the chemical shift and the quadrupolar magnetic moments. In zeolite-type microporous materials, oxygen is at the acid centers, which remain the site for catalytic activity, and an understanding of the NMR shift for different oxygens will allow one to rationalize the structure property relationship. Due to the low magnetic moment and natural abundance of ^{17}O isotope, solid-state NMR spectroscopy remains challenging. In a recent paper, Profeta et al. [68] apply the GIPAW method to the calculation of the ^{17}O chemical shifts in several SiO_2 polymorphs and two zeolites. The structures studied were α-cristobalite, α-quartz, coesite, and the all-silica zeolites ferrierite and faujasite. The unit cells of these structures contain 12, 9, 24, 108, and 144 atoms, respectively. The calculations were performed for ^{17}O NMR parameters and the ^{29}Si chemical shifts. Absolute shielding tensors were generated. To fix the ^{29}Si scale, Profeta et al. have chosen $\sigma_{ref} = 337.3$ ppm, in such a way that the experimental and theoretical $\delta(r)$ of quartz coincide. Given the larger experimental uncertainty in the ^{17}O chemical shifts, for the ^{17}O scale the σ_{ref} for ^{17}O is 262.6 ppm. The NMR calculations were performed with the experimental geometries determined by X-ray or neutron diffraction. Finally, in Table 6.3 and Table 6.4, the theoretical and experimental ^{29}Si chemical shifts and ^{17}O chemical shifts were compared. For ^{17}O chemical shift calculation, δ the chemical shift is chosen relative to water in ppm. The NMR parameters are computed at the experimental geometries in all systems but ferrierite, for which the theoretically relaxed structure for both ^{17}O and ^{29}Si were used. For ^{29}Si the chemical shift δ is chosen relative to tetramethylsilane in ppm. The experimental data are taken from Lewis et al. [69a] and Smith and Blackwell [69b]. This justifies the fact that one cannot only assign peaks and justify the NMR behavior to rationalize the structure; if cleverly used it can use this to design new material of interest.

The point of this book is to show how simulation can aid experiments and in some cases may guide the experiment to design selective materials. NMR is a useful but difficult tool with its interpretation is a critical issue as we had discussed before and

TABLE 6.3

Comparative results for experimental and theoretical ^{29}Si chemical shift

Structure	Site	Experimental δ	Theoretical δ
Coesrite	Si1	−113.9	−114.69
	Si2	−108.1	−108.41
Cristobalite		−108.5	−109.05
Quartz		−107.1	−107.1
Ferririte	Si1	−116.5	−117.67
	Si2	−112.3	−113.66
	Si3	−111.9	−112.23
	Si4	−117.2	−119.45
	Si5	−116.2	−116.29
Faujasite		−107.8	−106.69

Source: [68] With permission.

TABLE 6.4

Comparative Results for Experimental and Theoretical ^{17}O Chemical Shift

Structure	Site	Experimental δ	Theoretical δ
Coesite	O1	29	25.8
	O2	41	39.2
	O3	57	56.0
	O4	53	52.4
	O5	58	57.8
Cristobalite		37.2	39.3
Quartz		40.8	44.3
Ferririte	O34	43.1	42.1
	O23	41.6	41.3
	O12	40.7	40.0
	O22	39.6	39.3
	O24	39.0	38.1
	O35	37.0	36.9
	O43	37.0	37.1
	O15	35.9	35.8
	O45	34.8	32.7
	O55	28.0	28.2
Faujasite	O3	47.3	48.6
	O1	42.3	44.3
	O2	37.2	38.3
	O4	34.8	36.8

along with the regular issue of sample preparation assignment of the peak and else. One of the uses of simulation will certainly be to assign the peaks to a particular visualization and therefore resolve a particular issue for a system. NMR-CASTEP has been used to address the problem of the assignment of the high-resolution ^{27}Al and ^{31}P NMR spectra in AlPOs [70]. AlPO-14 has been widely investigated by high-resolution magic-angle spinning NMR spectroscopy and the ^{27}Al and ^{31}P NMR parameters are well documented in the literature. [71–76]. It is only once the NMR peaks have been assigned to the known crystallographic aluminum and phosphorous sites that one will be able to compare site-specific dynamic information derived from NMR experiments with, for example, the results of molecular dynamics calculations. The reason behind the idea of MD simulation is that by experiment such behavior is absent in the calcined phase [76] as observed by ^{27}Al multiple-quantum MAS (MQMAS) and satellite transition MAS (STMAS) NMR experiments [77, 78]. Three forms of AlPO-14 as a model system have been chosen: (1) a fully calcined–dehydrated form; (2) an as-synthesized form, prepared using isopropylamine ($CH_3CH(NH_2)CH_3$); and (3) a second as-synthesized form, prepared using piperidine ($C_5H_{10}NH$) as the template molecule. Ashbrook et al. [78] perform first principles calculations of NMR parameters using a gauge-including projector augmented-wave (GIPAW) formalism [64], implemented within CASTEP [79], a plane wave, pseudo-potential code that exploits the periodic nature of crystalline solids. Calculated values for the ^{27}Al and ^{31}P isotropic chemical shifts and the quadrupolar parameters of the spin I = 5/2 ^{27}Al nucleus can be compared with the measured NMR parameters. Valence orbitals considered for the study were 3s, 3p (^{27}Al), 2s, 2p (^{17}O), 3s, 3p (^{31}P), with core radii of 2.0, 1.3, and 1.81 Å, respectively. Integrals over the Brillouin zone were performed using a Monkhorst-Pack grid with a k-point spacing of 0.04 Å_1. Wave functions were expanded in plane waves with a kinetic energy 700 eV. All calculations were converged with respect to both k-point spacing and cutoff energy. Of course, this method is not possible for as-synthesized AlPO-14, for which no crystal structure appears in the literature. These calculations still remain memory intensive and for this typical case Ashbrook et al. used 136 AMD Opteron processing cores partly connected by Infinipath high-speed interconnects. Typical NMR calculation times were 48 (AlPO-14) and 96 h (as-synthesized AlPO-14) using twelve cores. It was observed that the NMR parameters calculated from the published crystal structures appear to be different from those seen experimentally. This is particularly true for the calcined material, where large forces are observed on many of the atoms in the CASTEP calculation. However, once the local geometry of the structures was optimized prior to the calculation of NMR parameters, a much better agreement was observed, supporting the spectral assignments suggested by the two-dimensional NMR correlation experiments. These modifications to the structure are reasonably small, involving small changes in the positions of certain atoms; for the cases where the unit cell size remains fixed, only minor intensity differences were observed between simulated diffraction patterns and those reported from experiments. However, the differences observed in the NMR parameters are considerable (in particular for the ^{27}Al quadrupolar coupling), demonstrating the sensitivity of NMR to only small changes in the local environment. When the unit cell size was also allowed to change within the structural optimization, an expansion of the cell was observed (a typical observation when the GGA functional is used). Although this results in poor agreement with the cell parameters obtained by powder XRD, the resulting agreement with the NMR experiments was better than when the cell size was fixed. In this context, it can be noted that just as a range of experimental techniques (NMR, FTIR, XRD, etc.) often yield slightly different but equally valid values for bond lengths, owing to their differing sensitivities to time and ensemble averaging, so too a computational approach can be expected to yield slightly different values for the cell parameters. Though spectral assignment is achieved for both as-sysnthesized AlPO-14 and the calcined–dehydrated material, no crystal structure has yet been published for the second as-synthesized AlPO-14 and so a similar approach cannot be employed in this case. However, the MQ-J-HETCOR spectrum of as-synthesized AlPO-14 (II) demonstrates that the framework Al–O–P connectivity in this material is the same as that found for as-synthesized AlPO-14 and calcined–dehydrated AlPO-14, and that the framework hydroxide ions are connected at the same aluminum centers. We can therefore assign the ^{27}Al and ^{31}P resonances by their resulting positions in the structure after calcinations. Overall, the results indicate that a combination of NMR spectroscopy (using high-resolution approaches and through-bond correlation information) and first principles calculation of NMR parameters may soon be considered a generally useful step in the refinement of the structures of microporous materials derived from powder diffraction data.

Let us take an example where simulation is looking at distribution of aluminium sites which can be observed and confirmed through NMR, and depending on the Si/Al ratio can be confirmed as described earlier in this chapter. A work by Himei et al. [80] described the capability of molecular dynamics (MD) and NMR in combination with molecular modeling to investigate the sites and distribution of framework aluminum atoms in faujasite-type microporous zeolite structures, which cannot be derived only by experimental techniques. MD calculations can successfully predict the optimum aluminum distribution in Na–Y zeolite (Si/Al = 2.43), because they reproduce the sites and occupancies of Na$^+$ cations that were reported by neutron diffraction techniques. The validity of the aluminum distribution model in Na–Y was strongly confirmed by the simulated ^{29}Si MAS NMR spectra in agreement with the experimental chemical shifts and intensities. Although the sites and occupancies of the exchanged cations in faujasites can be determined by analytical experimental techniques, their atomistic detailed distribution cannot be derived. A methodology is proposed to predict the detailed distribution of the Na$^+$ cations in Na–Y by using neutron diffraction spectrum simulation. This is another aspect of simulation where experiment has its own limitations and needs simulation to prove the experimental findings. Several studies have been done to compare spectroscopic analysis for characterization of the material along with simulation, in which the main target is to obtain the synthetic spectra that can reproduce most of the features of the experimental spectra satisfactorily. The zero-wave vector vibrational modes of an infinite sodalite framework with atomic positions corresponding to the sodalites $M_8[Al_6Si_6O_{24}]Cl_2$ (M = Li, Na, K) and silica sodalite are calculated by the Wilson GF matrix method. The result shows that the synthetic spectra reproduce most of the features of the experimental spectra very satisfactorily, and for the aluminosilicate sodalites the wave numbers of the bands are simulated to within 30 cm^{-1}.

Moreover, the changes in the spectra with the substitution of M = Li, Na, K are also well reproduced and enable an extensive analysis of the spectra to be made, including the assignment of the symmetries of the modes. The analysis in terms of the contributions from the characteristic vibrations of the TO4 and four-ring or six-ring structural subunits of the aluminosilicate framework has identified new relationships between both the intensities and the frequencies in the spectra. Moreover, structural features between the intensity of an infrared band near 750 cm^{-1} and the Si–O–Al angle and between the intensity of a Raman band near 1,000 cm^{-1} and the degree of ordering of silica and aluminum atoms over the tetrahedral sites of the framework was also obtained [81]. A new molecular sieve topology of SSZ-77 consisting of alternating layers has been determined from a multistep Monte Carlo simulation procedure using the program ZEFSAII. The material, SSZ-77, consists of alternating layers present in the RUT and AST topologies, and intergrowths may be possible. The product first arose from a synthesis where the degradation of the quaternary ammonium acts as a structure-directing agent (SDA) to produce the viable organo-guest molecule in the structure formation. NMR investigations show that the larger molecules are occluded within the cages of the SSZ-77 structure, and the primary occluded species is trimethylamine or tetramethylammonium [82].

Translational motion is an important form of transport in chemical and biochemical systems. Pulsed-field gradient nuclear magnetic resonance (PFG NMR) provides a convenient and noninvasive means for measuring translational motion. PFG NMR allows studying the molecular transport on various displacements for the rate of molecular exchange between catalyst particles and their surroundings and consequently the overall rate and selectivity of different reaction like FCC. Mesoscopic kinetic Monte Carlo simulations and PFG NMR measurements were compared in order to investigate the transport of ethane in a bed of NaX crystals. A novel molecular mechanics particle-based reconstruction method was employed for the digital representation of the bed, enabling for the first time a parallel study of the real system and of a computer model tailored to reproduce the void fraction, particle shape, and average size of the real system. Simulation of the long-range diffusion of ethane in the bed over the Knudsen transient and molecular diffusion regimes is consistent with the PFG NMR measurements in yielding tortuosity factors that depend upon the regime of diffusion; more specifically, tortuosity factors defined in the conventional way are higher in the Knudsen than in the molecular diffusion regime. Detailed statistical analysis of the computed molecular trajectories reveals that this difference arises in a nonexponential distribution of the lengths and in a correlation between the directions of path segments traversed between collisions with the solid in the Knudsen regime. When the Knudsen tortuosity is corrected to account for these features, a single, regime-independent value is obtained within the error of the calculations [83].

Chiolite, a tetragonal mineral of formula $Na_5Al_3F_{14}$, was used to develop a general approach for solving inorganic structures from powders by combining NMR, modeling, and X-ray diffraction. The different steps of the strategy were successfully performed, building the candidate integrant units using NMR, simulating candidate crystal structures using the computational Automated Assembly of Secondary Building Units (AASBU) method, and checking the consistency of the candidate structures against the diffraction data analyzed with the FOX program [84]. Titanium-substituted ultrastable Y zeolite was subjected to multinuclear solid-state NMR spectroscopy (^{29}Si MAS, ^{27}Al MAS/3Q-MAS) in combination with computer simulation. ^{27}Al MAS and ^{27}Al MAS 3Q-MAS revealed the presence of aluminum in four, five, and six coordination and the multiplicity within Al-4 and Al-6, respectively. The emergence of signal with higher intensity at −101 ppm in the ^{29}Si MAS spectrum of Ti–USY samples indicates the possible occurrence of Q(4)(3Si,1Ti)-type silicon environments due to Ti substitution in the faujasite framework. Moreover, the chemical shielding and electric field gradient tensors for the titanium environment in the zeolite have been determined by computer simulation of the quadrupolar broadened static Ti-47, Ti-49 NMR spectra [85]. The use of bare cluster models to understand the nature of zeolite-substrate interactions may be improved to take account of the environment of the Brønsted acid site. Two models for introducing the electrostatic effects of the zeolite lattice were considered. The first method involves generating a specialized correction potential by fitting a nonperiodic array of ca. sixty point charges to the difference between the bare cluster and periodic potentials. The second part of the method starts by fitting a periodic array of atomic charges to the potential of the infinite lattice and then builds up a classical cluster of ca. 2,000 atoms into which the QM cluster is embedded. Such embedded cluster calculations, employing a T3 cluster, with electron correlation at the density functional theory level, are described to model the interaction of water at a Brønsted acid site. The calculated structures of the water-zeolite complex–associated vibrational frequencies and ^{1}H NMR shift were successfully compared with experimental data [86]. A combination of NMR spectroscopy and computer simulation is also helpful to characterize the location of the cations in dehydrated zeolite. ^{23}Na MAS and DOR NMR spectroscopy were applied to describe the location of Na$^+$ ion in dehydrated NaX (Si/Al = 1.23) zeolite by decomposing the spectra recorded at three different magnetic fields using computer simulation and attributed to the crystallographically distinct cation sites. The assignments of the lines follow from electric field gradient calculations at the ^{23}Na nuclei applying a simple point charge model based on crystal structure data. A weak Gaussian line at low field (δ_{iso} = −6 ppm) is assigned to Na$^+$ at site I, two broad quadrupole patterns at the high field side of the spectra are attributed to site I′ (δ_{iso} = −19 ppm, QCC = 5.2 MHz, η = 0) and site II cations (δ_{iso} = −15 ppm, QCC = 4.6 MHz, η = 0), and two quadrupolar lines dominating the central region of the spectra originate from Na$^+$ at two different III′ sites (δ_{iso} = −13 and −29 ppm, QCC = 2.6 and 1.6 MHz, η = 0.7 and 0.9, respectively) [87].

It should be mentioned that solid-state NMR spectroscopy is one of the most important experimental techniques for characterizing zeolitic materials. It provides information on the type of (NMR active) nuclei and on their abundance in the framework. The position of the chemical shift describes the chemical environment of a specific nucleus. Because protons are an excellent nucleus for NMR study, Brønsted acidic zeolites have been the subject of many investigations using this technique. ^{1}H MAS NMR spectroscopy can clearly distinguish between nonacidic silanol groups (1.3–2.2 ppm), which are located at the surface or at defect sites, and acidic bridging ("structural") hydroxyl groups (3.8–4.3 ppm) [88]. Low field shifts of about 4.8 ppm result from protons located in small cavities or small rings. Relaxation and line-width studies have been employed to investigate proton mobility at bare Brønsted sites [89] and spinning sideband analysis yields the distance between the Brønsted proton and Al atom [90]. In attempting to understand the catalytic function of acid zeolites, it is necessary to consider not only the intrinsic acidity of a particular site (related to its proton affinity) but also its accessibility. By this term, one refers broadly to the way in which the interaction between the acidic proton and an adsorbed species may be mediated by the topology of the zeolite framework. Calculations [91, 92] have been carried out of ^1H, ^{15}N, and ^{13}C NMR chemical shift parameters used to characterize the acid strength and accessibility of Brønsted acid sites in seven high-silica zeolites of structure types CHA, FAU, FER, MFI, MOR, MTW, and TON. The acid sites had previously been selected by a systematic minimization procedure, and an acetonitrile molecule was inserted using a Monte Carlo method. The main interactions between acetonitrile and the zeolite are an N...H–O-type hydrogen bond to the Brønsted proton and van der Waals interactions with the framework. Calculations using Gauge Including Atomic Orbital (GIAO) (with B3LYP/TZV) were performed on optimized clusters, which included the acid site and sufficient atoms to represent the surrounding pore topology. The isotropic shielding values were obtained, as well as the principal components of the shielding tensors and the chemical shift parameters. Both the isotropic shifts of the acid proton, $\delta_{Iso}(^1H)$, and $\delta_{Iso}(^{15}N)$ correlate well with the hydroxy bond length, $r(OH)$, and hence to acid strength. In contrast, $\delta_{Iso}(^{13}C)$ is strongly dependent on the orientation of the acetonitrile molecule, as influenced by the local zeolite geometry. The principal tensor components, δ_{11}, δ_{22}, and δ_{33}, exhibit recognizable trends only for the nitrogen atoms. Figure 6.17a to Figure 6.17g depict the seven clusters that were selected to represent the Brønsted site and surrounding pore system of the zeolite. For clarity, only the connections of T sites are shown, except in the immediate vicinity of the active site. Terminating OH groups are also omitted. The first stage involved a systematic series of lattice energy minimization calculations, using GULP [15] with the Schröder-Sauer potentials [93], in which the lowest energy Al/OH configuration was determined for each structure type. There was an additional condition that the selected Brønsted site must be accessible from the principal pore system. The composition of the zeolites was otherwise siliceous, and because there was one Al/Brønsted site per unit cell (as usually defined for each structure), the Si/Al ratios were 11 for CHA, 191 for FAU, 35 for FER, 95 for MFI, 95 for MOR, 55 for MTW, and 23 for TON. Into each of these optimized periodic zeolite models an acetonitrile molecule was inserted using the Monte Carlo docking procedure of Freeman et al. [94] with the cff91_czeo force field [95]. Low-energy docked configurations served as starting positions for the subsequent DFT cluster calculations. The clusters excised from the periodic structures were of a reasonable size (between 116 and 137 atoms) to include both the active site and enough framework atoms to represent the diameter of the zeolite channels or cages. During the optimization, only acetonitrile (in case of Z–A) and the first coordination shell around the Brønsted site, TO3Si–(OH)–AlO3T (T = SiO_4 tetrahedral, which remain fixed), was allowed to relax. All of the other atoms were kept fixed at their positions, as obtained from the lattice energy minimizations. Dangling bonds were saturated with hydroxy groups in such a way that the cluster remained neutral and could not interact either with one another or with acetonitrile. The clusters Z and Z–A, respectively, were optimized with the $DMol^3$ program of Accelrys Inc. as discussed before

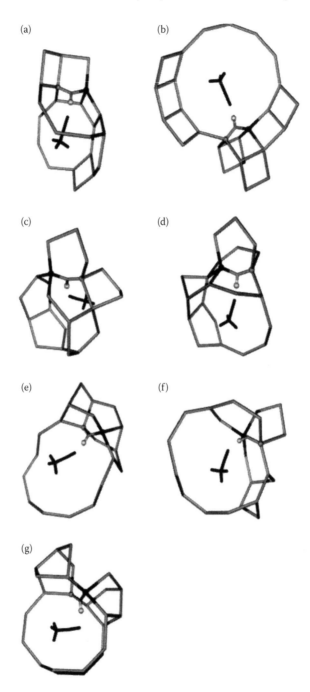

FIGURE 6.17 Seven cluster models to represent the Brønsted site and surrounding pore system of the zeolite. Only the connectivity between T sites is shown and the terminating hydroxyl groups are also omitted: (a) CHA, (b) FAU, (c) FER, (d) MFI, (e) MOR, (f) MTW, and (g) TON. [91] With permission.

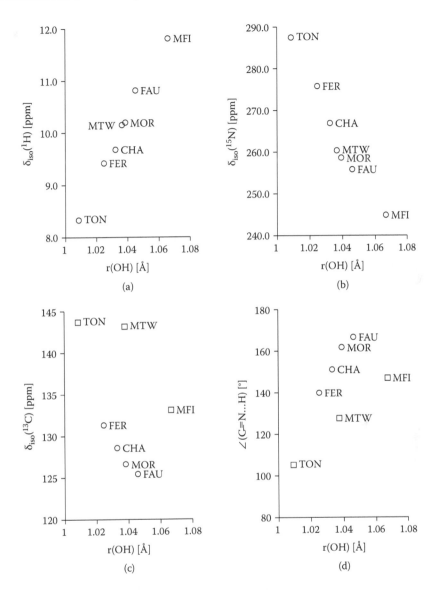

FIGURE 6.18 Average bond lengths (a) r(Al–O), (b) r(P–O), and (e) r(Si–O); bond distances (c) r(Al...P) and (f) r(Si...Si), and bond angles (d) α(Al–O–P) and (g) α(Si–O–Si) vs. relative lattice energy, $\Delta E_{latt,berlinite/quartz}$, for known AlPOs and known ZEOs. [91] With permission.

employing the PW91 [96] density functional and the DNP basis set (i.e., a double numerical basis function together with polarization functions), comparable to the Gaussian basis set 6-31G(d,p) [97]. Subsequently, ^{1}H, ^{15}N, and ^{13}C NMR chemical shift parameters were calculated using GIAO33, 34(B3LYP35/TZV36) as implemented in Gaussian 98 [98]. Figure 6.18 exhibits the plots of these isotropic shifts versus the r(O–H) distance to measure of the acid strength of the Brønsted sites.

An excellent correlation between $\delta_{iso}(1H)$ and the bond length, $r(OH)$ was found (Figure 6.18a): the stronger the hydrogen bond (i.e., the longer the $r(OH)$ bond length), the less shielded the proton is. Obviously, the nitrogen atom becomes more shielded, as a stronger hydrogen bond must result in electron density being shifted toward the nitrogen atom; indeed the plot in Figure 6.18b shows this correlation. A strong hydrogen bond interaction should also increase the electron density within the triple bond, so one might also expect an effect on the isotropic shielding of the nitrile carbon atom, and, therefore, the same trend for the carbon shifts as for the nitrogen shifts. Figure 6.18c, though, fails to demonstrate a clear trend. Moreover, the series appears to fall into two groups: FER < CHA < MOR < FAU and TON < MTW < MFI, in order of increasing shielding. Although H-bond strength is still an important factor, it is clear that, being further removed from the acid site, other factors influence the carbon shift. The most obvious of these is the orientation of the molecule, as constrained by the zeolite framework. Because acetonitrile is not conformationally flexible, the angle $\angle(C \equiv N\&H)$ effectively measures the angle of the principal axis of the molecule with respect to acid proton, as well as the proximity of the nitrile carbon to the zeolite framework. The graph in Figure 6.18d shows a plot of the angle $\angle(C \equiv N\&H)$ vs. the $r(OH)$ bond length, and again the data split into two groups, FER < CHA < MOR < FAU and TON < MTW < MFI, in order of increasing angle. In fact, to a first approximation, the graph is a mirror image of Figure 6.18c, confirming the strong influence of $\angle(C \equiv N...H)$ on nitrile. Framework curvature thus has an influence on the hydrogen bond strength, as investigated by Onida et al. [99] to detect the interactions between the surroundings of the acidic hydroxyl species and the probe molecule. The higher framework curvature and the relatively less approachable position of the proton in the channel walls in TON constrain the acetonitrile to approach at a more acute angle and thus the hydrogen bond is less optimally established. In MTW, despite being a large-pore zeolite, the molecular orientation is constrained by the proton location, although in this case $\angle(N...H-O)$ is still 161.6 degrees and hence the acidity is comparatively high. In MFI also, the end-member of the series, acetonitrile adopts a hydrogen bond angle of 177.8 degrees and encounters the strongest proton interaction, though the probe molecule orientation is otherwise not particularly favorable, as evidenced by the low $\angle(C \equiv N...H)$ and the relatively low heat of adsorption [100]. The span Ω describes the width of the powder pattern, and these data are summarized in Table 6.5. For the Brønsted proton in clusters Z, Ω (1H) covers a range between 19.25 and 25.10 ppm, whereas for the Z–A clusters, it ranges between 27.83 and 43.64 ppm. For acetonitrile in the Z–A clusters, the K (15N) and K (13C) values are all above 0.88 (see Table 6.5), which reflects the axially symmetric nature of the probe molecule. As may be readily seen from Table 6.5, δ_{11} and δ_{22} of these nuclei are of a similar magnitude. The calculated isotropic chemical shifts, $\delta_{iso}(1H)$, of the proton in clusters Z (see Table 6.5) range between 3.61 and 5.03 ppm and do not show any significant trend attributable to intrinsic acidity. Experimental studies on bare acid sites in zeolites find shifts of around 4–5 ppm: a review article by Hunger [88a] reports the following ranges of experimental ^1H signals from bridging OH groups (3.8–4.3 ppm) in large cavities or channels of zeolites and 4.6–5.2 ppm if they are situated in small channels or cages within a zeolite. It has been

TABLE 6.5

1H Parameters of the Brønsted Proton for Acetonotrile and the Cluster Z and Cluster Z–A; 13C and 15N Parameters of C≡N for Acetonitrileand Cluster Z–A; Isotropic Chemical Shifts, δiso, Principle Components δ11, δ22, δ33, Span Ω, Skew k, Anisotropy δaniso, Asymmetry η

	δ_{iso} (ppm)	δ_{11} (ppm)	δ_{22} (ppm)	δ_{33} (ppm)	Ω (ppm)	k	δ_{aniso} (ppm)	η
^{1}H								
MFI	3.95	12.70	5.71	−6.55	19.25	0.27	−10.50	0.67
FAU	4.33	16.43	4.39	−7.83	24.26	0.01	−12.16	0.99
MOR	3.97	12.69	6.50	−7.30	19.99	0.38	−11.27	0.55
MTW	3.64	11.71	8.00	−8.78	20.49	0.64	−12.42	0.30
CHA	4.01	15.36	3.88	−7.20	22.56	−0.02	11.35	0.98
FER	5.03	16.95	6.27	−8.15	25.10	0.15	−13.18	0.81
TON	3.61	10.87	8.33	−8.39	19.26	0.74	−12.00	0.21
^{1}H								
MFI–A	11.79	28.67	17.98	−11.30	39.97	0.46	−23.09	0.46
FAU–A	10.80	29.31	16.72	−14.33	43.64	0.41	−25.13	0.50
MOR–A	10.18	25.53	16.87	−11.87	37.40	0.54	−22.05	0.39
MTW–A	10.14	20.87	17.42	−7.88	28.75	0.76	−18.02	0.19
CHA–A	9.66	24.48	15.09	−10.61	35.09	0.46	−20.27	0.46
FER–A	9.41	25.26	12.33	−9.37	34.63	0.25	−18.78	0.69
TON–A	8.32	18.72	15.35	−9.11	27.83	0.76	−17.43	0.19
^{15}N								
Acetonitrile	307.44	485.06	485.05	−47.80	532.86	1.00	−355.24	0.00
MFI–A	244.73	403.02	386.04	−54.89	457.91	0.93	−299.62	0.06
FAU–A	255.65	418.91	404.15	−56.10	475.01	0.94	−311.75	0.05
MOR–A	258.53	418.21	410.14	−52.75	470.96	0.97	−311.28	0.03
MTW–A	260.17	420.70	410.76	−50.94	471.64	0.96	−311.11	0.03
CHA–A	266.48	429.41	419.21	−49.18	478.59	0.96	−315.66	0.03
FER–A	275.67	444.37	430.25	−47.60	491.97	0.94	−323.27	0.04
TON–A	287.28	457.92	451.45	−47.52	505.44	0.97	−334.80	0.02
^{13}C								
Acetonitrile	127.66	243.63	243.63	−104.30	347.94	1.00	−231.96	0.00
MFI–A	133.03	250.46	242.41	−93.80	344.26	0.95	−226.83	0.04
FAU–A	125.37	242.20	236.73	−102.80	345.00	0.96	−228.17	0.02
MOR–A	126.49	242.40	236.31	−99.25	341.65	0.96	−225.74	0.03
MTW–A	143.08	261.25	254.03	−86.04	347.29	0.96	−229.12	0.03
CHA–A	128.51	240.48	238.85	−93.81	334.29	0.99	−222.32	0.01
FER–A	131.26	246.65	238.33	−91.19	337.84	0.95	−222.45	0.04
TON–A	143.66	269.85	249.15	−88.04	357.89	0.88	−231.70	0.09

Source: [92] With permission.

clearly pointed out by Gorte [100] that ^{13}C NMR isotropic shifts of simple probe molecules such as acetonitrile provide no information on acid strength whatsoever. However, the obtained results additionally pointed out that a study of ^{13}C NMR can elucidate the topology of the region surrounding the active site and, thus, probe the molecule–cavity interaction. Both of these statements are in agreement with the current simulated results, which suggest that the orientation of the molecule, as influenced by the framework topology, is the main influence on ^{13}C shift. In general, the closer the nitrile carbon is to the framework, the more deshielded it becomes due to the additional interaction. The first group (FER, CHA, MOR, and FAU) has δiso(^{13}C) shifts closer to the gas phase value, as can be seen in Table 6.5.

Among the most widely used techniques to characterize nanoporous materials, FTIR is an important method to yield information not only on short-range bond order and characteristics but also long-range order caused by lattice coupling and electrostatic effects and serves as a rapid and useful tool. Therefore, the structural characterization of zeolites like crystalline materials obtained from FTIR would be more informative when compared with the simulated spectra. The aluminum distribution in high-silica mordenite (MOR) zeolites with various Si/Al ratios was investigated by FTIR spectroscopy in the presence of CD_3CN probe molecules and benzene adsorption. Two adsorption bands assigned to CN stretching vibration were observed at 2,280–2,295 cm^{-1} and approximate to 2,315 cm^{-1}, which are due to interaction of CN with acidic hydroxyl groups in the main channels and the side pockets of H–MOR, respectively. The relative intensity of the peak at 2,315 cm^{-1} increased with an increase in the Si/Al ratio, indicating that the proportion of aluminum atoms in the main channels relatively decreased with the Si/Al ratio. This was confirmed from the linear relationship between the number of benzene molecules adsorbed in a unit cell and the number of aluminum atoms in the main channels. Furthermore, computer simulation goes deeper to characterize the position of aluminum atoms and suggests that the preferential sitting is in the T-3 site [101]. The catalytic potential of Co-containing molecular sieves has stimulated the study of spectroscopy to rationalize the sitting of Co^{2+} ions. In a review by Verberckmoes et al. [102], a critical overview is given of the spectroscopic tools to (1) decide about the cation sites of Co^{2+} and their occupancy, (2) determine the spectroscopic signatures of framework and extraframework cobalt, (3) determine the amount of Co incorporated into the framework of molecular sieves, and (4) discuss local distortions of framework cobalt. This can be used as a basic guideline to address with metal impurity.

In addition, this was also suggested from the computer simulation result that aluminum atoms are preferentially sitting in the T3 site. One therefore will be able to predict the location of the metal substitution probability as well as the active site and hence the prediction of the reaction mechanism becomes easier and more relevant.

IR simulation is very well explored to predict the Brønsted acidity of nanoporous materials. Within the harmonic approximation the normal modes for the species studied can be routinely computed by most of the available *ab initio* quantum chemical calculations [104]. Spectroscopic data for adsorbates interacting with protons are widely available but interpretation requires skill. In this situation, simulation can help and guide that process by visualization. For zeolite-type materials, DFT is a useful method to calculate the protonation and deprotonation process and normally produces the ground state interaction energy for protonation within ~5 kcal/mol. The simulation of infrared spectra is not straightforward, because it is necessary to consider the anharmonicity, which is expensive. Secondly, it appears that weakening of the vibrational bonds by H-bonding is significantly overestimated. The procedure still remains based on the deprotonation energy calculation for the adsorbate at the proton site and consequently the IR shift induced by the adsorbate to have a qualitative trend. Indeed, this process is difficult to proceed for experimentalists but may still give some reasonable information. Now, DFT works well when a dedicated adsorbate is used for a dedicated proton. However, the absorption scenario change when there are multiple interactions like two ammonia approaching the proton or two water molecules approaching the proton. Then a network of H-bonding occurs and *ab initio* plane wave–type molecular dynamics methodology is an obvious choice to probe the interaction. A DFT method was used to study the Brønsted acidity in supercages of MCM-22 with double aluminum atom substitution at T1 and T4 sites considering the effect of neighbors. The neighboring groups were varied from one closest to at least three nonbonded neighbor groups [105]. The calculation of the deprotonation energy, atomic charge on proton, bridging hydroxyl stretching frequencies, and the adsorption energy of NH_3 to predict the influences of neighboring groups or otherwise the effect of pore architecture on Brønsted acidity was already performed. For this purpose, the whole lattice of the zeolite is not optimized and hence consideration for the neighbors is not included. Furthermore, MCM-22 crystal structure is difficult to model, so it is better to consider a cluster model to replicate the scenario. The calculations were performed with DFT. The calculated results revealed that the larger the distance between two aluminum atoms, the stronger the acidity of the zeolite. The acidity calculated results with more neighbors is stronger than clusters with fewer neighbors. The results are strongly dependent on the models selected, especially the O–H stretching vibrational frequency and deprotonation energy. However, the adsorption energy of NH_3 interacting with protonic zeolites shows minor dependency on the model structures. That means that NH_3 adsorption energy can give a reasonable measure on Brønsted acid strength. In addition, v_{OH} is not only correlated with Al–OH–Si angle but also depends on the surrounding lattice structure. Therefore, it is strongly affected by the model size and configuration. It is generally accepted that the stretching frequency of O–H bond can be taken as an indicator of Brønsted acidity; i.e., the lower the frequency, the greater the acidity. However, by combining theoretical and experimental studies [106, 107], it was indicated that acidity (which is principally a measure of the ease at which an O–H bond dissociates heterolytically) is not well characterized by the stretching frequency of the bridging O–H bond. The calculation further confirmed this finding. The calculated O–H stretching frequencies provided a statistic tendency of acid strength.

In continuation to the acidity, let us look into a recent study, which is a combination of experiment and theory handling a complicated matrix of mesoporous ZSM-5 and compared the acidity in those matrices. Normally, mesoporous structures are neutral and to improve the acidity the system demands incorporation of heteroatoms like aluminum, gallium, etc. The number and the strength of Brønsted acid sites were determined quantitatively by infrared–mass spectroscopy/temperature-programmed desorption (IRMS-TPD) of NH_3. The Brønsted and Lewis acid sites were measured quantitatively for HZSM-5, Al-MCM-41, and silica–alumina and compared. The number and strength of the Brønsted acid sites are used as parameters to judge the presence of the ZSM-5 structure in the materials. DFT calculations were used to confirm the experimental findings, because they supported the experimental measure of ΔH (enthalpy change of NH_3 adsorption) [108]. It is also possible to calculate the adsorption energy of NH_3 with the variation of Al–OH and Si–OH localized geometry. The results are in very good agreement with experiments.

6.3   MECHANICAL STABILITY

The evaluation of mechanical stability of nanoporous materials, especially microporous materials, is important for the machinery and building industries. Most construction materials except for some alloys or ceramics are microporous and their mechanical stability is important for practical applications. The mechanical property generally depends on the relationship between its preparation condition, structure, and properties. It is expected that computer simulation is the best choice to reduce the costs of experimental studies. A simple and computationally efficient classical atomistic model of silica was used to simulate silicon and oxygen as hard spheres with four and two association sites, respectively. To study the mechanical and phase behavior, isobaric–isothermal Monte Carlo simulations were used over the model corresponding to quartz, cristobalite, and coesite, as well as some zeolite structures, which are mechanically stable and highly incompressible. Ratios of zero-pressure bulk moduli and thermal expansion coefficients for quartz, cristobalite, and coesite are in quite good agreement with experimental values. The pressure–temperature phase diagram was constructed and shows three solid phases corresponding to cristobalite, quartz, and coesite, as well as a fluid- or glass-phase behavior qualitatively similar for silica [109].

Unlike crystalline zeolite material, mesoporous materials are highly dispersed amorphous powder and frequently used as adsorbent after being compacted at high pressure into pellets. Hence, it is important to study the mechanical stability of mesoporous materials. Since the first published data of Gusev et al. [110], several studies [111] have been performed on MCM-41, MCM-48, FSM-16, SBA-15, and KIT. However, due to the difficulties in elucidation of the amorphous structure of mesoporous materials, computer simulation studies are limited. Chytil et al. [112] calculated the mechanical stability of SBA-15 material. They exposed the calcined solids to a unilateral external pressure in the range 16–191 MPa in order to monitor the impact of the mechanical pressure on the properties of SBA-15. From their results it appears that the elevated pressure has no influence on the hexagonal cell parameter. Through the N_2 sorption measurements the fraction of the preserved mesoporous structure was estimated to be 60% when the highest pressure was used and the remaining part becomes irreversibly disintegrated into small particles, the pressed sample is considered to be heterogeneous. However, the preserved fraction is slightly modified, showing the pressure of a smaller pore width and plugs located within the mesopores. The plugs most likely originate from a disintegrated fraction of the SBA-15. UV-Raman spectroscopy shows that the relative intensity of the band associated with the siliceous network (001) decreased on the pressed samples resulting in a less ordered material, possessing an enhanced population of silanols as compared to parent SBA-15 [112]. The activity of SBA-1 was also investigated using n-heptane and cyclohexane adsorption in addition to nitrogen adsorption. The mechanical stability of SBA-1 is high. When SBA-15 is subjected to compression of about 217 MPa, the specific pore volume calculated from nitrogen adsorption decreased by 19.7%, whereas the pore volume calculated from the n-heptane and cyclohexane adsorption decreased by 12.5 and 7.5%, respectively. These results together with the large difference between the pore volumes obtained by nitrogen and organics adsorption indicate that the presence of microporosity in the SBA-1 pore walls has to be considered. SBA-1 is mechanically more stable as compared with hexagonal materials such as MCM-41 and SBA-15 but exhibits similar mechanical stability compared to the cubic MCM-48 [113]. SBA-15 is a novel porous material with uniformly sized mesopores arranged in a regular pattern. The adjacent mesopores are connected to each other by microporous walls. The major disadvantages of these materials are amorphous wall structure and low thermal, hydrothermal, and mechanical stability. There have been a few attempts to either coat the walls of SBA-15 by microporous crystalline zeolites or to fabricate SBA-15 using CMK-3 in such a way that the walls are made up of ZSM-5. The present work provides a first-ever study of RMM (replicated mesoporous materials), which are ordered like SBA-15 and the walls are made up of ZSM-5 using molecular modeling. A random orientation of the unit cells and the distribution of sizes of the supercells located at nucleation sites would be ideal to model the RMM. However, such a study would introduce more uncertainties with regard to voids between the individual supercells, noncrystalline silica, and the location of active sites where the nucleation occurs. In a simpler model studied in the present work, the walls of SBA-15 were made up of regularly arranged ZSM-5 having the same orientation. The structure was characterized by estimating the nitrogen accessible area/volume by Connolly surfaces, small-angle and wide-angle X-ray diffraction patterns, methane adsorption, and ice as a probe to study the pore structure. It was found that RMMs have significantly higher methane adsorption capacity compared to SBA-15 and the majority of methane is adsorbed in the microporous walls of RMM. Sonwane and Li [114] performed the first study of ordered mesoporous material SBA-15 coated with microporous zeolites ZSM-5 using molecular simulations. Several model structures with different variables like periodic arrangement of mesopores, randomly arranged micropores, surface hydroxyls, and bulk deformations of SBA-15 were used. A hundred faces of H-ZSM-5 unit cell were then placed on the surface of SBA-15 with the coordinates obtained from the well-established ZSM-5 lattice from the crystal structure. The entire structure including ZSM-5 and

SBA-15 was equilibrated to obtain a final configuration. The resulting structure was then characterized using simulated small angle and wide angle X-ray diffraction methodologies to compare with experimental observations. This is a tool to see how the model matches with experiments and whether there remains any ambiguity. Thus, it is a very important technique for the interface. Connolly surface area (to compare BET area), accessible pore volume for nitrogen molecules (to compare with t-plot volume of micro- and mesopores), and methane adsorption at 303 K were calculated using simulation where one can tune the pore radius to access the smallest possible domain and design the probe accordingly. It was observed that the orientation of ZSM-5 on the SBA-15 had no effect on the surface area, pore volume, or adsorption capacity. In order to determine whether the addition of microporous ZSM-5 should increase the total methane adsorption capacity due to addition of micropores, adsorption on bare and coated SBA-15 by using GCMC-type simulation methodology was performed. Total adsorption capacity was found to decrease, whereas the number of methane molecules adsorbed per unit cell of the SBA-15 structure increased. This phenomenon tells one about the typical pore architecture of the matrix and selective role of ZSM-5 and SBA-15. This study needs further extension to see the mechanical stability of the matrix with increasing wall thickness.

Molecular modeling is an effective technique to determine the mechanical stability of the nanoporous material but to replicate experimental findings it remains a challenge. The advantage of this property is immense because zeolites now find their application in composite types of materials. Molecular modeling can be applied to know the stability of a matrix in presence of external pressure and to predict a stress–strain relationship. A recent study explored the mechanical stability for known zeolite matrices like chlorosodallite [115] and calculated the elasticity constant using different computational methodologies [116–118]. One of the most significant outcomes of those studies [117, 118] is if one considers polymorphs of SiO_2, then many zeolite may possess frameworks with negative Poisson's ratio, which implies that a counterintuitive lateral widening of longitudinal stress can be applied in certain directions. Li et al. [119] have worked on the specific zeolite like chlorosodalite, $Na_8(Al_6Si_6O_{24})Cl_2$, as derived experimentally [120]. The methodology used for simulation was a force field–based methodology because it is simple and certainly able to qualitatively reproduce the experimental behavior. The energy has been calculated using CVFF force field [120, 121] and the nonbonded terms were summed using the traditional Ewald summation technique. The energy was minimized by a conjugant gradient method [122]. Minimization is an iterative procedure in which the coordinates of the atoms and possibly the cell parameters are adjusted so that the total energy of the structure is reduced to a minimum (on the potential energy surface) and results in a structural model that closely resembles the experimentally observed structure. No constraint was applied on the cell. The 6 × 6 stiffness matrix and its inverse, the compliance matrix, was calculated from the second derivative of the potential energy function

$$C_{ij} = \frac{I}{V} \frac{\partial^2 E}{\partial \varepsilon_i \partial \varepsilon_j} \quad i,j = 1,2, \dots, 6$$

TABLE 6.6

Mechanical Propoerties for Chlorosodalite

Index	Experimental Value in GPa	Modeling Results
C11	100.05	144.9
C12	24.90	38.58
C44	27.16	39.27

Source: [119] With permission.

where C_{ij} is a component of stiffness matrix, E is the energy expression, V is the volume of the unit cell, and ε_i and ε_j are strain components. The results for the experimental and theoretical values in comparison with Li et al. [119] are shown in Table 6.6. This gives an idea that the simulation reproduces a trend close to experiment; a qualitative trend is enough to design a new matrix, which alternately time consuming. With the understanding of the capability of modeling, it is possible to propose the tailor-made material of interest for specific application.

As mentioned before, the mechanical property is important for application with systems when one is looking at composite matrices, zeolitic films, or membranes. It is necessary to monitor the tensile strength, mechanical stability, and bulk compressibility. This can be handled by force field methodology for larger systems and one can use GULP-type force fields [15] for inorganic matrices.

There is an application of zeolitic material in low dielectric constant materials to replace dense silica. Amorphous porous silica such as sol-gel silica and organic-templated mesoporous silica have been used in some length as possible low-k materials [123–126]. However, the introduction of porosity into amorphous silica results in a dramatic reduction in mechanical strength (commonly manifested in lower values of the elastic modulus, E), which makes their survivability during chemical mechanical processing (CMP) and packaging difficult. A recent publication [127] proposes that pure silica zeolites (PSZs) offer several potential advantages over amorphous silicas, including crystalline structure, intrinsically uniform and small pore size, and hydrophobicity. The authors have computationally studied the limiting values of dielectric constants and elastic constants for PSZ single crystals for a range of zeolites with different crystalline symmetries and hence different pore architectures like MFI, FER, CHA, and BEA all are representative type of zeolite, there is no rule for the choice. It is to find out the effect of porosity on the behavior of the dielectric constant so this material can be used as substitute to the regular matrices. The elastic constants for each of the zeolites were determined with the MPDyn simulation code [128] and the BKS force field [129]. The bulk modulus, shear modulus, elastic modulus, and Poisson's ratio were determined by the Voigt-Reuss-Hill method [130]. Dielectric constants were calculated from energy-minimized structures using the GULP simulation package [131]. An ion-pair shell-model force field was used to represent the interactions between atoms by Jackson and Catlow [132]. The calculated E and k (infinite frequency) values for MFI, FER, BEA, and CHA are presented in Table 6.7. For comparison, experimental data for single-crystal MFI (E), CHA (E),

TABLE 6.7

Calculated and Measured Dielectric Constant and Elastic Constant for a Range of Zeolitic Material

Property	MFI x	MFI y	MFI z	FER x	FER y	FER z	CHA x	CHA y	CHA z	BEA x	BEA y	BEA z
Dielectric constant calc	1.71	1.71	1.70	1.69	1.75	1.75	1.59	1.59	1.59	1.58	1.58	1.67
Dielectric constant expt	[a]	2.7[b]	3.1[c]	[a]	1.78	[a]	[a]	[a]	[a]	2.3[e]		
Elastic constant calc	50.6 ± 0.7			62.4 ± 0.7			43.3 ± 0.8			47.1 ± 0.6		
Elastic constant expt	53.9 ± 0.5			49.4 ± 0.7			48.9 ± 1.2			[a]		

[a] Values not measurable due to experimental limitations.
[b] Measured on *in situ* films.
[c] Measured on seeded growth films.
[d] Measured on randomly oriented films.
[e] Measured on single crystals assuming a Poisson ratio of 0.25.
Source: [127] With permission.

FER (E and k), and BEA (k) are included in Table 6.7. Because of the geometries (non-thin-plate), it was not possible to measure the k values of single-crystal MFI, BEA, and CHA. The fact that the polycrystalline films have a much higher k value than calculated values for single crystals (2.70–3.10 vs. 1.70 for MFI, and 2.30 vs. 1.6 for BEA) suggests that grain boundaries and crystal defects play a major role in controlling the k value, which may be a topic of further research. The point of this example is to show the applicability of simulation to propose the efficiency of nanoporous materials in specific applications.

6.4 BULK POROSITY — COMPARISON WITH ADSORPTION ISOTHERMS

This section aims to explain the importance of simulation to elucidate the adsorption isotherm as observed in experiments to predict the porosity of nanoporous materials. The experimental adsorption isotherm can propose the porosity, which then can be visualized by TEM, supported by XRD data. In simulation, once you know the material or if you can visualize the material with a valid structure, you will then be able to tune the porosity and see how that varies with different probe adsorbents and temperature-dependent correlation between porosity with activity can be prescribed. There are several examples of the adsorption study of nanoporous materials. Among them, a very interesting topic deals with natural gas adsorption behavior. Natural gas is a mixture of hydrocarbons, water, and acidic gases like CO_2 and H_2S. Therefore, it is essential to get rid of these gases, which cause a decrease in the energy content and are responsible for corrosion of the transport passage. Depending on the pore structure of the nanoporous material, it would be easier to separate two species with very different kinetic diameters (the kinetic diameter of a molecule being the minimum diameter of a cylindrical pore in which the molecule can fit) and the zeolite with cylindrical straight channel, pores of intermediate diameter (between the diameters of two species) will be highly selective. Let us take the example of CO_2 and CH_4. Due to their low and similar kinetic diameters (0.33 nm for CO_2 and 0.38 nm for CH_4), only a few small-pore zeolites are possible candidates for the separation of these two molecules. CO_2/CH_4 separation has been reported in two types of small-pore zeolites: SAPO-34 (0.38 nm pore diameter) [133–137] and DDR (0.36–0.44 nm) [138]. These studies showed interesting selectivity for both the zeolites. Adsorption and permeation of CO_2 and/or CH_4 have also been investigated experimentally in large-pore-size zeolites like FAU-X [139] and FAU-Y [140] and medium-pore-size MFI [141] zeolites. In all these cases, the pore diameters are larger than all kinetic diameters of the sorbates and selectivity results from a combination of competitive adsorption, MCM-22 is a recently synthesized low Al/Si ratio aluminosilicate zeolite possessing an unusual framework structure made of two independent large-pore systems, offering a variety of possible sorption sites [142]. One of the pore systems is composed of two-dimensional sinusoidal ten-member ring (MR) sinusoidal channels (SC) with a pore diameter of 0.4–0.55 nm and the other one of large twelve-MR cavities (LC) of dimensions 1.7 ± 0.8 nm, connected by ten-MR windows. In view of its large-pore systems, MCM-22 has been mainly studied in the context of hydrocarbon adsorption for catalytic purpose [143, 144]. The high void space density, together with its unique crystal structure, also makes it interesting in the context of small gas adsorption and separation. The recently proposed ITQ-1 zeolite [145], the fully siliceous variant of MCM-22, is of particular interest for natural gas separation, because this material is hydrophobic. Molecular simulation techniques are now widely used to investigate sorption and transport properties of sorbates in zeolites as mentioned in other chapters.

A recent publication of Leyssale et al. has looked into the thermodynamics of the CO_2/CH_4 mixture in ITQ-1 [146]. The calculations were performed by the following procedure: all the structures of zeolite and the gases were equilibrated and the potential parameters were tested. The GCMC calculation was performed with single and multiple mixtures. Total quantities predicted from the simulation were partitioned into contributions coming from the LC and the SC of the zeolite. SC thus presents a very favorable environment for both sorbates. On the contrary, LC with very large side pockets communicating via large twelveMR openings in the z direction and narrow channels in the xy plane. Consequently, only a few locations inside the LC are able to form strong interactions with the sorbates. This point is highlighted in Figure 6.19, where high-density probability isosurfaces (in other words, highly favorable sorption sites) are shown for these two pore systems. This is an important observation obtained from simulation and is very difficult from the experiment. The adsorption isotherms are presented in Figure 6.20 for the pure sorbent case and two types of potential were used: the potential from Goodbody et al. [147] (referred to as GWMWQ) and from Dubbeldam et al. [148] (noted as DCVKMS). Methane molecules are described by these two united atom interaction potentials that have been invoked previously to model the CH_4-silicalite systems.

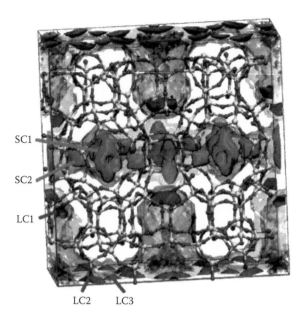

FIGURE 6.19 (For color version see accompanying CD) Isodensity surfaces showing the highest probability sorption sites of carbon dioxide obtained in a GCMC simulation in the Henry regime at 250 K. Sorption sites in SC are shown in green and those in LC in blue. The gray surfaces outline the accessible space in both pore systems. [146] With permission.

This potential fitting is a challenge for complex systems with multiple components to get reasonable results. From the results it is observed that the heat of adsorption, sorbate/sorbent interactions are always stronger in SC than in LC. Moreover, their evolution with pressure presents quite different behavior in LC and in SC. In the case of CO_2, the CO_2/sorbent energy in LC increases slightly with pressure. This comes from the fact that out of the Henry regime, molecules start occupying less favorable sites in LC. On the contrary, the CO_2/sorbent energy decreases with pressure in SC, showing an ordering transition in this pore system as pressure increases. The resulting total sorbate/sorbent interaction energies, which are constant at low pressure, show well-pronounced enhancement before reaching a higher plateau value at high pressures and correspond to the point at which the SC are fully filled and where the less favorable LC start to be filled. Hence, we can see that it is reasonable to use computer simulation to visualize the actual pore filling scenario and gain an understanding from the isotherm and the porosity behavior. Now, we consider the sorption of CO_2/CH_4 mixtures in ITQ-1. Sorption isotherms obtained at $T = 300$, 275, and 250 K for CO_2 mole fraction in the gas phase ranging from 0.1 to 0.9 are shown in Figure 6.21. The corresponding LC and SC components are shown in Figure 6.22 and Figure 6.23, respectively. In each figure, isotherms a, b, and c correspond to CO_2/DCVKMS CH_4, and isotherms d, e, and f correspond to CO_2/GWMWQ CH_4 mixtures at 300, 275, and 250 K. The CO_2 mole fraction yCO_2 (or partial pressure) in the ideal gas mixture evolves from 0.1 (circles) to 0.9 (stars) in steps of 0.1. The first

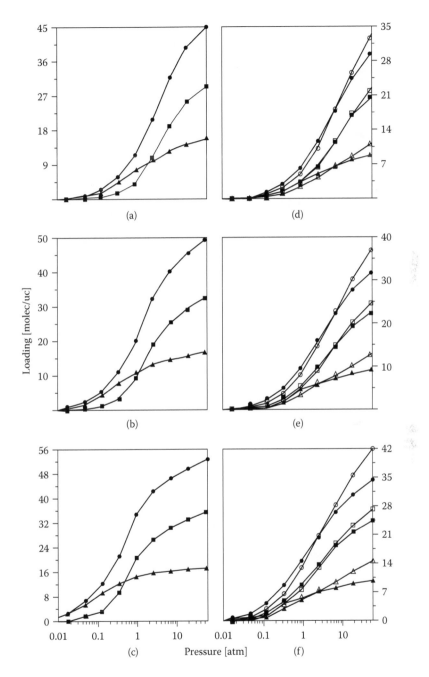

FIGURE 6.20 Sorption isotherms of carbon dioxide at 300 (a), 275 (b), and 250 K (c) and of methane (filled symbols: DCVKMS; open symbols: GWMWQ) at 300 (d), 275 (e), and 250 K (f). Circles: total loading; squares: LC loading; triangles: SC loading. [146] With permission.

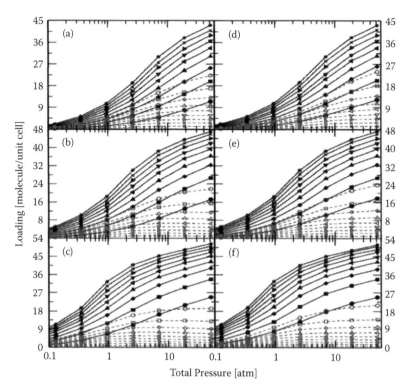

FIGURE 6.21 (For color version see accompanying CD) Full zeolite sorption isotherms of carbon dioxide/methane mixtures: (a) 300 K CO_2/DCVKMS; (b) 275 K CO_2/DCVKMS; (c) 250 K CO_2/DCVKMS; (d) 300 K CO_2/GWMWQ; (e) 275 K CO_2/GWMWQ; and (f) 250 K $CO2$/GWMWQ. Filled symbols and solid lines: CO_2 isotherms; open symbols and dashed lines: CH_4 isotherms. Carbon dioxide mole fractions in the gas phase $y_{CO2} = 0.1$ (circles), 0.2 (squares), 0.3 (diamonds), 0.4 (up triangles), 0.5 (left triangles), 0.6 (down triangles), 0.7 (right triangles), 0.8 (crosses), and 0.9 (stars). [146] With permission.

point is to note that CO_2 isotherms obtained with the two methane models are almost identical. Some slight differences can be observed in the methane isotherms obtained with the two models as for pure methane sorption isotherms. These isotherms show a strong selectivity for CO_2 adsorption over CH_4. For instance, looking at the global isotherms (Figure 6.21) we see that, for all the pressure studied, the sorbed amount of CO_2 is higher than that of CH_4 for $yCO_2 \geq 0.2$ at 250 K, $y_{CO2} \geq 0.2 - 0.3$ at 275 K and $y_{CO2} \geq 0.3$ at 300 K. Figure 6.22 and Figure 6.23 show that LC is weakly selective for CO_2, whereas SC has strong affinity toward CO_2. In fact, from Figure 6.23 it is clear that the sorbed amounts of CO_2 in SC are higher than CH_4 for almost all pressures and compositions at $T = 250$ K. Finally, Figure 6.24 shows a snapshot of a configuration obtained during a GCMC simulation of a CO_2/DCVKMS CH_4 mixture with $yCO_2 = 0.1$ at $T = 250$ K and $P = 1$ atm. This state is identified by the author because it corresponds to a state at which the zeolite is filled but not saturated and the total loadings of CO_2 and CH_4 have similar values. At this state point, CO_2 and CH_4 loadings are respectively 16.3 (6.2 in LC and 10.1 in SC) and 18.2 (15.0 in LC and 3.2

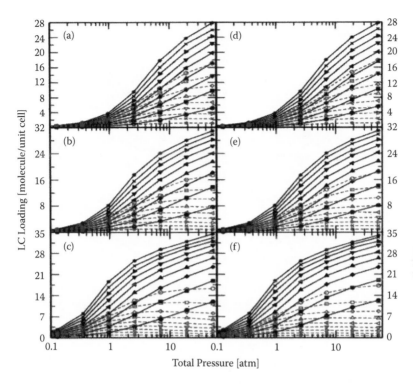

FIGURE 6.22   (For color version see accompanying CD) Identical to Figure 6.21 for LC sorption. [146] With permission.

in SC) molecules per unit cell. An isodensity surface plot showing the areas of high sorbate density is superposed on the bottom right-hand side of the figure. It shows that two sorbates tend to occupy different sorption sites. Whereas methane is mainly sorbed in LC, in the top of the supercage pockets (LC1), close to the oxygen atoms of the twelve-MR at the base of the LC (LC2) and in the intercavity space (LC3), CO_2 is mainly sorbed in SC and in the intermediate area between the top and base of LC.

In conclusion, we can say that the challenge is to establish the interaction of these sorbate molecules with zeolite domain; the potential parameter. There are many tools available, but forcefield is one of the greatest methodologies which helps to understand the process, and as well with the capability of visualization of pore filing based on the kinetic diameter of the ingredients and the zeolite matrix one can rationalize the sorption phenomenon. The simultaneous adsorption isotherm at different temperatures will allow one to look into the porosity of the system as the sorbate gets filled. This is a systematic study to justify experiments and design new materials with specific applications.

Transportation of molecules through porous solid is a never-ending research topic. The process is complicated and can occur through different mechanisms. It could depend on the size, connectivity of the void space, mobility, mean free path of molecules in the gas phase and on surfaces, and the arrangement of the surfaces in providing a connected macroscopic path within the porous particles. There are several transport mechanisms, ranging from pressure-driven flow at long length

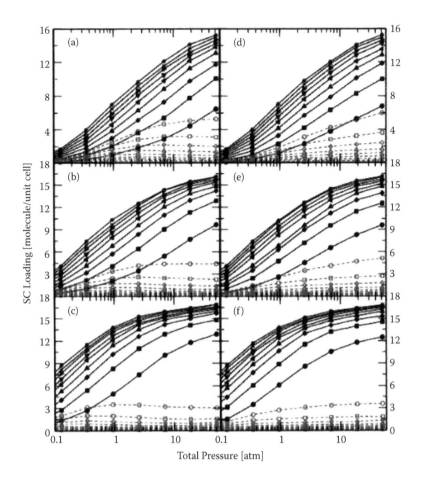

FIGURE 6.23 (For color version see accompanying CD) Identical to Figure 6.21 for SC sorption. [146] With permission.

and time scales to surface diffusion at small scales and short times. In this range of length and time scales, the operative processes evolve from the purely physical motion of noninteracting molecules to chemical reactions involving, in the case of surface diffusion, the sequential making and breaking of bonds between surfaces and adsorbed molecules.

In most situations, molecule–surface collisions are typically assumed to occur without significant energy redistribution and with instantaneous desorption in a random direction. In reality, desorption can occur, but if adsorbed species are immobile, the surface provides only a reservoir for molecules, without any contributions to steady-state diffusion rates. The pool of adsorbed molecules, however, can contribute to the apparent capacity of the system in tracer diffusivity measurements, such as those involving step or pulse chromatographic measurements or frequency response techniques. When adsorbed molecules can move, surfaces contribute a path for diffusion that adds to that provided by the void space. Application of classical techniques

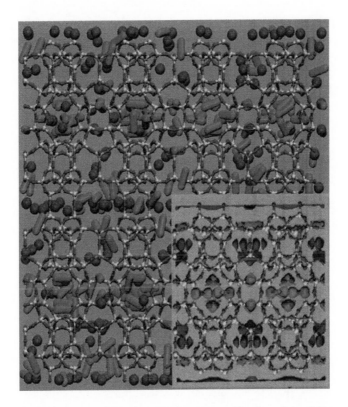

FIGURE 6.24 (For color version see accompanying CD) Snapshot of a configuration obtained in a GCMC simulation of a CO_2/DCVKMS CH_4 mixture at $T = 250$ K, $P = 7$ atm, and a gas-phase carbon dioxide mole fraction of 0.1. CO_2 molecules are displayed as light green sticks and methane as dark blue spheres. The inset (bottom right) shows high-density probability sites for carbon dioxide in green and methane in blue. [146] With permission.

where gaseous diffusion alone is considered to be contributing to the total diffusion leads to incorrect estimates of gas diffusivity. A recent study by Sonwane and Li [148] described the deterministic models along with molecular dynamics techniques with the variable charge transfer potential were used to fabricate alumina nanotubes and inverted porous. These nanostructures resemble recently invented MCM-41/ SBA-15 and other ordered nanoporous materials in terms of the structural aspect, mainly the nano dimension. The structure and thermal stability of these materials were investigated, and it was found that the nanoporous hollow spheres have greater thermal stability compared to the porous nanotubes. Also, nanotubes with a thick pore wall have greater thermal stability. A new method of the unbiased random walk of a tracer atom on the surface of a porous structure was successfully applied to study the surface tortuosity. The gas as well as surface tortuosity or in other way the surface roughness values of the coarsened solids were higher than that for densified solids. The increase in the breadth of distribution has an insignificant effect on the surface tortuosity (~2%) and gas-phase tortuosity (~5%) when changing from a small size to a broad distribution with variance of the porosity. With the boundary layer thickness of $\delta = 5\lambda$ the hybrid discrete-continuum simulations were able to give accurate estimates of both surface and gas-phase tortuosities in comparison to the discrete simulations. The First Passage Time Distribution (FPT) approximation was efficient over the conventional averaging method in terms of computational time and the statistics of the results. This is a novel methodology to handle the amorphous type of porous material that can be better handled by this type of mathematical model to explore the situation due to the absence of crystalline geometry of the matrix. Probably with the advent of modeling to describe the mesoporous structure one may be able to do a realistic atomistic modeling with real structures in future.

In this chapter we dealt with the characterization of nanoporous materials. There are many analytical techniques available and used for this characterization, which mainly consist of chemical composition analysis, followed by spectroscopic analysis apart from the X-Ray diffraction (XRD) study, which we kept away from the scope of this book. Modeling simulation can be a handy tool to solve the new structure of complicated matrix like nanoporous materials. Chemical composition is vital for the structural property and activity and one may wish to design a system with very low loading of metal or other active substituents for a specific application. This process requires a lot of trial and error and it is very difficult to identify the amount of a substance present at a very low level. The hardest part is to visualize its position, especially for nanoporous materials where the added substance can locate itself at the framework or in the extraframework. Spectroscopic analysis, which is the backbone of the characterization mainly focused on FTIR and solid-state NMR as the key methods. These characterization tools are expensive and sampling is critical, and in most cases *in situ* IR is preferable, but it is sometimes impossible to afford due to hard experimentation. Simulation plays a key role as support of the experimentation and it is possible to design a new material simply from the Q4, Q3 ratio obtained from NMR-sensitive species. It is also possible to design an amorphous porous architecture with novel hybrid zeolites and other nanoporous materials. Things again become very challenging with lower loading of metals. According to the IR measurement it is possible to predict the location of the metal from the shift of the bands resulted and the influence of the environment of the metal ion and mainly used to locate the metal ion existed as framework or extraframework cations. The advantage of simulation is to replicate the behavior and visualize the vibration to identify the translation or rotational behavior. Once the idea of the structure is in mind, then the mechanical stability of the matrix should be checked. The feature with porosity is very demanding, because it is important to decide the porous architecture of the system synthesized to tune its properties. Adsorption is one of the most informative experimental techniques for structural characterization of nanoporous materials. Because simulation has the potential to replicate the adsorption isotherm, it will be easier to predict the porous architecture along with the nature of pore filling, which is a tedious job and a combination with experiment therefore rationalizes the scenario.

REFERENCES

1. J. Datka, P. Geerlings, W. Mortier, P. Jacobs, *J. Phys. Chem.*, 89 (1985) 3483.
2. R. M. Barrer, *Hydrothermal Chemistry of Zeolites*, Academic Press, London (1982).

3. C. E. Lyman, P.W. Betteridge, E. F. Moran, *ACS Symposium Series*, American Chemical Society: Washington, DC (1983).

4. R. M. Milton, *ACS Symposium Series, Zeolite Synthesis*, M. L. Occelli, H. E. Robson (eds.), American Chemical Society, Washington, DC (1989).

5. A. Erdem, L. B. Sand, *Proceedings of the 5th International Conference on Zeolites*, L. V. C. Rees (ed.). Heyden, London (1980).

6. S. Ueda, M. Koizumi, *Am. Mineral.*, 64 (1979) 172.

7. S. Yang, A. C. Vlessidis, N. P. Evmiridis, *Ind. Eng. Chem. Res.*, 36 (1997) 1622.

8. J. Datka, P. Geerlings, W. Mortier, P. Jacobs, *J. Phys. Chem.*, 89 (1985) 3483; Part 2.

9. (a) A. Chatterjee, R. Vetrivel, *Zeolites*, 14 (1994) 225; (b) *Microporous Mater.*, 3 (1994) 211.

10. H. Van Konigsveld, J. C. Jansen, H Van Beckkum, *Zeolite*, 10 (1990) 235.

11. W. J. Mortier, J. J. Pluth, J. V. Smith, *Mater. Res. Bull.*, 11 (1976) 15.

12. E. Bosch, S. Huber, J. Weitkamp, H. Knozinger, *Phys. Chem. Chem. Phys.*, 1 (1999) 579.

13. G. Maurin, R. G. Bell, S. Devautour, F. Henn, J. C. Giuntini, *J. Phys. Chem. B*, 108 (2004) 3739.

14. (a) P. Pissis, D. Daoukaki-Diamanti, *J. Phys. Chem. Solid.*, 54 (1993) 701; (b) G. Artioli, J. V. Smith, A. Kvick, J. J. Pluth, K. Stahl, in *Zeolites*, B. Dram, S. Hocevar, S. Paovnik (eds.), Elsevier, Amsterdam (1985).

15. J. D. Gale, *J. Chem. Soc. Faraday. Trans.*, 93 (1997) 629.

16. B. Coughlan, W. M. Carrol, A. McCann, *J. Chem. Soc. Faraday Trans.*, 73 (1977) 1612.

17. (a) M. Pamba, G. Maurin, S. Devautour, J. Vanderschueren, J. C. Giuntini, F. Di Renzo, F. Hamidi, *Phys. Chem. Chem. Phys.*, 113 (2000) 4498; (b) S. Devautour, J. Vanderschueren, J. C. Giuntini, F. Henn, J. V. Zanchetta, J. L. Ginoux, *J. Phys. Chem. B*, 102 (1998) 3749.

18. W. M. Meier, *Z. Kristallogr.*, 115 (1961) 439.

19. G. Maurin, R. G. Bell, S. Devautour, F. Henn, J. C. Giuntini, *J. Phys. Chem. B*, 108 (2004) 3739.

20. G. Maurin, R. G. Bell, S. Devautour, F. Hennb, J. C. Giuntini, *Phys. Chem. Chem. Phys.*, 6 (2004) 182.

21. G. Maurin, P. Senet, S. Devautour, F. Henn, G. C. Giuntini, V. E. VanDoren, *Comput. Mater. Sci.*, 22 (2001) 106.

22. G. Maurin, P. Senet, S. Devautour, P. Gaveau, F. Henn, V. E. Van Doren, J. C. Giuntini, *J. Phys. Chem. B*, 105 (2001) 9157.

23. C. Abrioux, B. Coasne, G. Maurin, F. Henn, A. Boutin, A. Di Lella, C. Nieto-Draghi, A. H. Fuchs, *Adsorption*, 14 (2008) 743.

24. V. A. Nasluzov, E. A. Ivanova, A. M. Shor, G. N. Vayssilov, U. Birkenheuer, N. Rösch, *J. Phys. Chem. B*, 107 (2003) 2228.

25. E. A. Ivanova Shor, A. M. Shor, V. A. Nasluzov, G. N. Vayssilov, N. Rösch *J. Chem. Theory Comput.*, 1 (2005) 459.

26. E. A. Ivanova, A. M. Shor, V. A. Nasluzov, G. N. Vayssilov, N. Rösch, *J. Chem. Theor. Comput.*, 1 (2005) 459.

27. A. R. Ruiz-Salvador, N. Almora-Barrios, A. Gómezy, D. W. Lewis, *Phys. Chem. Chem. Phys.*, 9 (2007) 521.

28. (a) A. R. Ruiz-Salvador, D. W. Lewis, J. Rubayo-Soneira, G. Rodríguez-Fuentes, L. R. Sierra, C. R. A. Catlow, *J. Phys.Chem. B*, 102 (1998) 8417; (b) Y. M. Channon, C. R. A. Catlow, R. A. Jackson, S. L. Owens, *Microporous and Mesoporous Materials*, 24 (1998) 153.

29. W. Löwenstein, *Am. Mineral.*, 39 (1954) 92.

30. A. R. Ruiz-Salvador, A. Gómez, D. W. Lewis, G. Rodríguez-Fuentes, L. Montero, *Phys. Chem. Chem. Phys.*, 1 (1999) 1679.

31. N. Almora-Barrios, A. Gómez, A. R. Ruiz-Salvador, M. Mistry, D. W. Lewis, *Chem. Comm.*, (2001) 531.

32. A. Chatterjee, F. Mizukami, *Chem. Phys. Lett.*, 385 (2004) 20.

33. A. N. Fitch, H. Jobic, A. Renouprez, *J. Phys. Chem.*, 90 (1986) 1311.
34. G. Engelhardt, M. Hunger, H. Koller, J. Weitkemp, *Stud. Surf. Sci. Catal.*, 84 (1994) 421.
35. S. M. Auerbach, L. M. Bull, N. J. Henson, H. I. Metiu, A. K. Cheetham, *J. Phys. Chem.*, 100 (1996) 5923.
36. S. Pal, K. R. S. Chandrakumar, *J. Am. Chem. Soc.*, 122 (2000) 4145.
37. A. Chatterjee, *J. Mol. Graph. Model.*, 24 (2006) 262.
38. (a) M. P. Teter, M. C. Payne, D. C. Allen, *Phys. Rev. B*, 40 (1989) 12255; (b) M. C. Payne, M. P. Teter, D. C. Allen, T. A. Arias, J. D. Johannopoulos, *Rev. Mod. Phys.*, 64 (1992) 1045.
39. (a) B. Delley, *J. Chem. Phys.*, 92 (1990) 508–517; (b) B. Delley, *J. Chem. Phys.*, 94 (1991) 7245; (c) B. Delley, *J. Chem. Phys.*, 113 (2000) 7756.
40. (a) F. Corà, G. Sankar, C. R. A. Catlow, J. M. Thomas, *Chem. Comm.*, (2002) 734; (b) A. Tuel, S. Caldarelli, A. Meden, L. B. McCusker, C. Baerlocher, A. Ristic, N. Rajic, G. Mali, V. Kaucic, *J. Phys. Chem. B*, 104 (2000) 5697.
41. C. P. Herrero, *J. Chem. Soc. Faraday Trans.*, 87 (1991) 2837.
42. (a) A. Chatterjee, R. Vetrivel, *J. Chem. Soc. Faraday Trans.*, 91 (1995) 4313; (b) A. Chatterjee, R. Vetrivel, *J. Mol. Catal. A*, 106 (1996) 75.
43. G. Coudurier, J. C. Vedrine, *Pure Appl. Chem.*, 58 (1986) 1389.
44. V. A. Durrant, D. A. Walker, S. N. Gussou, J. E. Lyons, U.S. patent 4918249 (1972).
45. A. Chatterjee, A. K. Chandra, *J. Mol. Catal. Chem.*, 119 (1997) 5l.
46. A. Chatterjee, R. Vetrivel, *Microporous Mater.*, 3 (1994) 211.
47. C. T. W. Chu, C. D. Chang, *J. Phys. Chem.*, 89 (1985) 1569.
48. Y. Oumi, T. Kanai, B. W. Lu, T. Sano, *Microporous and Mesoporous Materials*, 101 (2007) 127.
49. A. Simperler, M. D. Foster, R. G. Bell, J. Klinowski, *J. Phys. Chem. B*, 108 (2004) 869.
50. (a) M. D. Foster, O. D. Friedrichs, R. G. Bell, F. A. Almeida Paz, J. Klinowski, *Angew. Chem. Int. Ed.*, 42 (2003) 3896; (b) M. D. Foster, R. G. Bell, J. Klinowski, *Stud. Surf. Sci. Catal.*, 135 (2001) P13.
51. M. J. Sanders, M. Leslie, C. R. A. Catlow, *Chem.Comm.*, 19 (1984) 1271.
52. J. D. Gale, N. J. Henson, *J. Chem. Soc. Faraday Trans.*, 90 (1994) 3175.
53. D. E. Akporiaye, G. D. Price, *Zeolites*, 9 (1989) 321.
54. M. L. Connolly, *J. Am. Chem. Soc.*, 107 (1985) 1118.
55. (a) P. M. Piccione, C. Laberty, S. Yang, M. A. Camblor, A. Navrotsky, M. E. Davis, *J. Phys. Chem. B*, 104 (2000) 10001; (b) N. Hu, A. Navrotsky, C. Y. Chen, M. E. Davies, *Chem. Mater.*, 7 (1995) 1816; (c) P. Richet, Y. Bottinga, L. Denielou, J. P. Petitet, C. Tequi, *Geochim. Cosmochim. Acta*, 46 (1982) 2639; (d) I. Petrovic, A. Navrotsky, M. E. Davis, S. I. Zones, *Chem. Mater.*, 5 (1993) 1805; (e) A. Navrotsky, I. Petrovic, Y. Hu, C. Y. Chen, M. E. Davis, *Microporous Mater.*, 4 (1995) 95.
56. E. B. Wilson, J. C. Decius, P. C. Cross, *Molecular Vibrations*, Dover, New York (1955).
57. Accelrys Material Studio Manual version MS 4.4.
58. T. A. Keith, R. F. W. Bader, *Chem. Phys. Lett.*, 194 (1992) 1
59. T. A. Keith, R. F. W. Bader, *Chem. Phys. Lett.*, 210 (1993) 223.
60. J. R. Cheeseman, M. J. Frisch, G. W. Trucks, T. A. Keith, *J. Chem. Phys.*, 104 (1996) 5497.
61. K. Wolinski, J. F. Hilton, P. Pulay, *J. Am. Chem. Soc.*, 112 (1990) 8251, and references therein.
62. K. Ruud, T. Helgaker, K. L. Bak, P. Jørgensen, H. J. A. Jensen, *J. Chem. Phys.*, 99 (1993) 3847.
63. V. Milman, B. Winkler, J. A. White, C. J. Pickard, M. C. Payne, E. V. Akhmatskaya, R. H. Nobes, *Int. J. Quant. Chem.*, 77 (2000) 895.
64. C. J. Pickard, F. Mauri, *Phys. Rev. B*, 63 (2001) 245101.
65. P. E. Blöchl, *Phys. Rev. B*, 50 (1994) 17953.

66. J. Yates, *First Principles Calculation of Nuclear Magnetic Resonance Parameters*, Ph.D. Thesis, Cambridge University (2003).
67. F. Mauri, A. Pasquarello, B. G. Pfrommer, Y. G. Yoon, S. G. Louie, *Phys. Rev. B*, 62 (2000) R4786.
68. M. Profeta, F. Mauri, C. J. Pickard, *J. Am. Chem. Soc.*, 125 (2003) 541.
69. (a) J. E. Lewis, C. C. Freyhardt, M. E. Davis, *J. Phys. Chem.*, 100 (1996) 5039–5049; (b) J. V. Smith, C. S. Blackwell, *Nature*, 303, (1983) 223.
70. S. E. Ashbrook, M. Cutajar, C. J. Pickard, R. I. Waltond, S. Wimperis, *Phys. Chem. Chem. Phys.*, 10 (2008) 5754–5764.
71. J. P. Amoureux, C. Huguenard, F. Engelke, F. Taulelle, *Chem. Phys. Lett.*, 356, (2002) 497.
72. L. Delevoye, C. Fernandez, C. M. Morais, J. P. Amoureux, V. Montouillout, J. Rocha, *Solid State Nucl. Magn. Reson.*, 22 (2002) 501.
73. J. W. Wiench, M. Pruski, *Solid State Nucl. Magn. Reson.*, 26, (2004) 51.
74. J. W. Wiench, G. Tricot, L. Delevoye, J. Trebosc, J. Frye, L. Montagne, J. P. Amoureux, M. Pruski, *Phys. Chem. Chem. Phys.*, 8 (2006) 144.
75. J. P. Amoureux, J. Trebosc, J. Wiench, M. Pruski, *J. Magn. Reson.*, (2007) 184.
76. S. Antonijevic, S. E. Ashbrook, S. Biedasek, R. I. Walton, S. Wimperis, H. Yang, *J. Am. Chem. Soc.*, 128 (2006) 8054.
77. (a) Z. Gan, *J. Am. Chem. Soc.*, 122 (2000) 3242; (b) S. E. Ashbrook, S. Wimperis, *J. Magn. Reson.*, 156 (2002) 269.
78. S. E. Ashbrook, S. Wimperis, *Prog. Nucl. Magn. Reson. Spectros.*, 45 (2004) 53.
79. M. D. Segall, P. J. D. Lindan, M. J. Probert, C. J. Pickard, P. J. Hasnip, S. J. Clark, M. C. Payne, *J. Phys. Condens. Matter*, 14, (2002) 2717.
80. H. Himei, M. Yamadaya, Y. Oumi, M. Kubo, A. Stirling, R. Vetrivel, E. Broclawik, A. Miyamoto, *Microporous Mater.*, 7 (1996) 235.
81. J. A. Creighton, H. W. Deckman, J. M. Newsam, *J. Phys. Chem.*, 98 (1994) 448.
82. D. J. Earl, A. W. Burton, T. Rea, K. Ong, M. W. Deem, S. J. Hwang, S. I. Zones, *J. Phys. Chem. C*, 112 (2008) 9099.
83. G. K. Papadopoulos, D. N. Theodorou, S. Vasenkov, J. Karger, *J. Chem. Phys.*, 126 (2007) 094702.
84. B. W. Lu, T. Kanai, Y. Oumi, T. Sano, *J. Porous Mater.*, 14 (2007) 89.
85. S. Ganapathy, K. U. Gore, R. Kumar, J. P. Amoureux, *Solid State Nucl. Mag. Reson.*, 24 (2003) 184.
86. P. Sherwood, A. H. de Vries, S. J. Collins, S. P. Greatbanks, N. A. Burton, M. A. Vincent, I. H. Hillier, *Faraday Discuss.*, 106 (1997) 79.
87. M. Feuerstein, M. Hunger, G. Engelhardt, J. P. Amoureux, *Solid State Nucl. Magn. Reson.*, 7 (1996) 95.
88. (a) M. Hunger, *Solid State Nucl. Magn. Reson.*, 6 (1996) 1; (b) D. Freude, H. Ernst, I. Wolf, *Solid State Nucl. Magn. Reson.*, 3 (1994) 271.
89. (a) T. Baba, N. Komatsu, Y. Ono, *J. Phys. Chem. B*, 102 (1998) 804; (b) P. Sarv, T. Tuherm, E. Lippmaa, K. Keskinen, A. Root, *J. Phys. Chem.*, 99 (1995) 13763.
90. D. Freude, J. Klinowski, H. Hamdan, *Chem. Phys. Lett.*, 149 (1988) 355.
91. A. Simperler, M. D. Foster, R. G. Bell, J. Klinowski, *J. Phys. Chem. B*, 108 (2004) 7142.
92. A. Simperler, R. G. Bell, M. D. Foster, A. E. Gray, D. W. Lewis, M. W. Anderson, *J. Phys. Chem. B*, 108 (2004) 7152.
93. K. P. Schröder, J. Sauer, M. Leslie, C. R. A. Catlow, J. M. Thomas, *Chem. Phys. Lett.*, 188 (1992) 320.
94. C. M. Freeman, C. R. A. Catlow, J. M. Thomas, *Chem. Phys. Lett.*, 186 (1991) 137.
95. J. R. Hill, J. Sauer, *J. Phys. Chem.*, 98 (1994) 1238.
96. J. P. Perdew, Y. Wang, *Phys. Rev. B*, 33 (1986) 8822.
97. W. J. Hehre, J. A. Ditchfield, J. A. Pople, *J. Chem. Phys.*, 56 (1972) 2257.

98. M. J. Frisch, G. W. Trucks, H. B. Schlegel, G. E. Scuseria, M. A. Robb, J. R. Cheeseman, V. G. Zakrzewski, J. A. Montgomery, R. E. Stratmann, J. C. Burant, S. Dapprich, J. M. Millam, A. D. Daniels, K. N. Kudin, M. C. Strain, O. Farkas, J. Tomasi, V. Barone, M. Cossi, R. Cammi, B. Mennucci, C. Pomelli, C. Adamo, S. Clifford, J. Ochterski, G. A. Petersson, P. Y. Ayala, Q. Cui, K. Morokuma, D. K. Malick, A. D. Rabuck, K. Raghavachari, J. B. Foresman, J. Cioslowski, J. V. Ortiz, B. B. Stefanov, G. Liu, A. Liashenko, P. Piskorz, I. Komaromi, R. Gomperts, R. L. Martin, D. J. Fox, T. Keith, M. A. Al-Laham, C. Y. Peng, A. Nanayakkara, C. Gonzalez, M. Challacombe, P. M. W. Gill, B. G. Johnson, W. Chen, M. W. Wong, J. L. Andres, M. Head-Gordon, E. S. Replogle, J. A. Pople, *Gaussian 98*, rev. A.7; Gaussian, Inc., Pittsburgh, PA (1998).
99. B. Onida, B. Bonelli, L. Borello, S. Fiorilli, F. Geobaldo, E. Garrone, *J. Phys. Chem. B*, 106 (2002) 10518.
100. R. J. Gorte, *Catal. Lett.*, 62 (1999) 1.
101. S. Bordiga, C. Lamberti, F. Geobaldo, A. Zecchina, G. Turnes Palomino, C. Otero Arean, *Langmuir*, 11 (1995) 527.
102. A. A.Verberckmoes, B. M. Weckhuysen, R. A. Schoonheydt, *Microporous and Mesoporous Materials*, 22 (1998) 165.
103. L. Baowang, K. Takahide, O. Yasunor, S. Tsuneji, *J. Porous Mater.*, 14 (2007) 89.
104. R. A. Van Santen, *Catal. Today*, 38 (1997) 377.
105. D. Zhou, N. He, Y. Wang, G. Yang, X. Liu, X. Bao, *J. Mol. Struct.*, 756 (2005) 39.
106. M. Sierka, U. Eichler, J. Datka, J. Sauer, *J. Phys. Chem.*, 102 (1998) 6397.
107. B. Onida, F. Geobaldo, F. Testa, R. Aiello, E. Garrone, *J. Phys. Chem. B*, 106 (2002) 1684.
108. K. Suzuki, Y. Aoyagi, N. Katada, M. Choi, R. Ryoo, M. Niwa, *Catal. Today*, 132 (2008) 38.
109. M. H. Ford, S. M. Aurbach, P. A. Monson, *J. Chem. Phys.*, 121 (2004) 8415.
110. V. Y. Gusev, X. Feng, Z. Bu, G. L. Haller, J. A. O'Brien, *J. Phys. Chem.*, 100 (1996) 1985.
111. (a) A. Galarneau, D. Desplantier, F. DiRenzo, F. Fajula, *Catal. Today*, 68 (2001) 191; (b) M. Broyer, S. Valange, J. P. Bellat, O. Bertrand, G. Weber, Z. Gabelica, *Langmuir*, 18 (2002) 5083; (c) K. Cassiers, T. Linssen, M. Mathieu, M. Benjelloun, K. Schrijnemakers, P. Van Der Voort, P. Cool, E. F. Vansant, *Chem. Mater.*, 14 (2002) 2317; (d) M. A. Springuel-Huet, J. L. Bonardet, A. Gédéon, Y. Yue, V. N. Romannikov, J. Fraissard, *Microporous and Mesoporous Materials*, 44–45 (2001) 745.
112. S. Chytil, L. Haugland, E. A. Blekkan, *Microporous and Mesoporous Materials*, 111 (2008) 134.
113. A. Vinu, V. Murugesan, M. Hartmann, *Chem. Mater.*, 15 (2003) 1385.
114. C. G. Sonwane, Q. Li, *J. Phys. Chem. B*, 109 (2005) 17993.
115. J. Jennifer, K. E. Evans, I. Walton Richard, *Appl. Phys. Lett.*, 88 (2006) 021914.
116. J. N. Grima, R. Jackson, A. Alderson, K. E. Evans, *Adv. Mater.*, 12 (2000) 1912.
117. R. Astala, S. M. Auerbach, P. A. Monson, *J. Phys. Chem. B*, 108 (2004) 9208.
118. R. Astala, S. M. Auerbach, P. A. Monson, *Phys. Rev. B*, 71 (2005) 014112.
119. Z. Li, M. N. Nevitt, S. Ghose, *Appl. Phys. Lett.*, 55 (1989) 1730.
120. P. P. Knops-Gerrits, W. A. Goddard, *J. Mol. Catal. Chem.*, 166 (2001) 135.
121. A. Hagler, E. Huler, S. Lifson, *J. Am. Ceram. Soc.*, 96 (1974) 5319.
122. R. Fletcher, C. M. Reeves, *Comput. J.*, 7 (1964) 149.
123. S. Seraji, Y. Wu, M. Forbess, S. J. Limmer, T. Chou, G. Z. Cao, *Adv. Mater.*, 12 (2000) 1695.
124. D. A. Doshi, N. K. Huesing, M. C. Lu, H. Y. Fan, Y. F. Lu, K. Simmons-Potter, B. G. Potter, A. J. Hurd, C. J. Brinker, *Science*, 290 (2000) 107.
125. K. Landskron, B. D. Hatton, D. D. Perovic, G. A. Ozin, *Science*, 302 (2003) 266.
126. R. A. Pai, R. Humayun, M. T. Schulberg, A. Sengupta, J. N. Sun, J. J. Watkins, *Science*, 303 (2004) 507.
127. Z. Li, M. C. Johnson, M. Sun, E. T. Ryan, D. J. Earl, W. Maichen, J. I. Martin, S. Li, C. M. Lew, J. Wang, M. W. Deem, M. E. Davis, Y. Yan, *Angew. Chem.*, 118 (2006) 6477.

128. MPDyn is a freely available simulation package written by W. Shinoda and M. Shiga. See http://staff.aist.go.jp/w.shinoda/MPDyn.

129. B.W. H. Van Beest, G. J. Kramer, R. A. Van Santen, *Phys. Rev. Lett.*, 64 (1990) 1955.

130. R. F. S. Hearmon, *Physics of the Solid State*, Academic, New York (1969).

131. J. D. Gale, A. L. Rohl, *Mol. Simulat.*, 29 (2003) 291.

132. R. A. Jackson, C. R. A. Catlow, *Mol. Simulat.*, 1 (1988) 207.

133. S. Li, G. Alvarado; R. D. Noble, J. L. Falconer, *J. Membr. Sci.*, 251 (2005) 59.

134. S. Li, J. G. Martinek, J. L. Falconer, R. D. Noble, T. Q. Gardner, *Ind. Eng. Chem. Res.*, 44 (2005) 3220.

135. S. Li, J. L. Falconer, R. D. Noble, *J. Membr. Sci.*, 241 (2004) 121.

136. J. C. Poshusta, R. D. Noble, J. L. Falconer, *J. Membr. Sci.*, 186 (2001) 25.

137. J. C. Poshusta, V. A. Tuan, E. A. Pape, R. D. Noble, J. L. Falconer, *AIChE J.*, 46 (2000) 779.

138. T. Tomita, K. Nakayama, H. Sakai, *Microporous and Mesoporous Materials*, 68 (2004) 71.

139. Y. Hasegawa, T. Tanaka, K. Watanabe, B. H. Jeong, K. Kusakabe, S. Morooka, *Kor. J. Chem. Eng.*, 19 (2002) 309.

140. K. Kusakabe, T. Kuroda, A. Murata, S. Morooka, *Ind. Eng. Chem. Res.*, 36 (1997) 649.

141. W. Zhu, P. Hrabanek, L. Gora, F. Kapeteijn, J. A Moulijn, *Ind. Eng. Chem. Res.*, 45 (2006) 767.

142. M. E. Leonowicz, J. A. Lawton, S. L. Lawton, M. K. Rubin, *Science*, 264 (1994) 1910.

143. H. Du, M. Kalyanaraman, M. A. Camblor, D. H. Olson, *Microporous and Mesoporous Materials*, 40 (2000) 305.

144. M. K. Rubin, P. Chu, U.S. patent 4954325 (1990).

145. M. A. Camblor, A. Corma, M. J., C. Días-Cabañas, Baerlocher *J. Phys. Chem. B*, 102 (1998) 44.

146. J.-M. Leyssale, G. K. Papadopoulos, D. N. Theodorou, *J. Phys. Chem. B*, 110 (2006) 22742.

147. S. J. Goodbody, K. Watanabe, D. MacGovan, J. P. R. B. Walton, N. Quirke, *J. Chem. Soc. Faraday Trans.*, 87 (1991) 1951.

148. D. Dubbeldam, S. Calero, T. J. H. Vlugt, R. Krishna, T. L. M. Maesen, B. Smit, *J. Phys. Chem. B*, 108 (2004) 12301.

149. C. G. Sonwane, Q. Li, *J. Phys. Chem. B*, 109 (2005) 5691.

7 Surface Activity Measurement

The main challenge in the research of nanoporous materials includes the understanding of the structure–property relation, which is mainly based on the surface structure of the material. As the nanomaterials get smaller, their properties increasingly diverge from their bulk counterparts. The most relevant property of nanoporous materials is increased surface area per unit volume. This increased pore surface can interact with a range of adsorbents and make it a good candidate for catalyst. It is also possible to control the morphology of pores as well as the pore size distribution, controlling pore shrinkage and degradation, hence maintaining the material performance over time.

A typical example of nanoporous materials is zeolite, with crystalline, microporous structure comprising channels and cages with a strictly regular dimension. For instance, the MFI structure is described as a combination of two interconnected channel systems. The silicate framework forms sinusoidal ten-member-ring (MR) channels along the direction of the a axis, interconnected with a ten-MR straight channel that runs down the b axis. A tortuous pore path is present along the c axis. This is the reason behind the use of zeolites as shape-selective catalysts both for the reactant and the product when a chemical reaction is concerned. Zeolites have a large internal surface area, void volumes, and extremely narrow pore size distribution. Two widely used techniques to study the surface structure of zeolite are atomic force microscopy (AFM) and high-resolution transmission electron microscopy (HRTEM). However, the drawback is that the exact termination of the structure is not always revealed. Thus, molecular simulation has been introduced to understand the structure of zeolite.

The motivation of this chapter lies in the fact that one has to design new nanoporous materials in terms of their activity toward specific applications, like surface adsorption phenomena, sensors, or as catalysts for various organic syntheses. Nanoporous materials are widely used as catalysts because of their large internal surface areas and consequently the presence of controllable large voids. Microporous materials like zeolites are mainly used as heterogenous redox catalysts in the petroleum industry, for various shape-selective reactions, and in separation. When discussing the catalytic activity of microporous materials, it must be mentioned that the transition metal-substituted microporous materials with aluminosilicate or aluminophosphate frameworks cover a large part of catalysis. They mainly take part in the various oxidative transformations in the presence of mild oxidizing agents like hydrogen peroxide or oxygen. A number of applications in waste treatment processes, including removal of heavy metals and radioactive species, as well as ammonia, different phosphates, and toxic gases from water, soil, and air are due to the unique structural and surface physicochemical properties of microporous materials, such as excellent absorption and ion-exchange capacities. Details of these applications will be described in the Chapter 8. Each of the applications demands a specific surface property such as pore architecture, pore size, surface area, and acidity or basicity of the matrix. Considering the surface properties, generally adsorption is used to characterize their surface structures. However, it is a complicated process and it is difficult to derive the molecular-level scenario from the experimental basis. Thus, simulation can play a significant role in mimicking the actual situation occurring at the molecular level This has been done by comparing the binding energy or the attachment energy for different surfaces and produce a result for example the probability of physisorption and chemisorption, to rationalize the surface activity of nanoporous materials.

7.1   SURFACE AREA

One of the most important applications of zeolites is their application as heterogeneous catalysts. Several factors dictate the catalytic properties of zeolites. Among them the large internal surface area (300–700 m^2 g^{-1}) provides a high concentration of active sites, usually the Brønsted acid site, found in protonated zeolite and located as bridging hydroxyl group. To determine the internal surface area of zeolites, a so-called probe atom model has been used [1]. Moloy et al. [2] demonstrated that an increase in size of the probe atom decreased the internal surface area, based on their study of a series of crystalline zeolites. The internal surface area is calculated as

$$internal\ surface\ area = \left[X\frac{\text{Å}^2}{uc}\right]\left[y\frac{\text{mol}}{uc}\right]^{-1}\left[\frac{m}{10^{10}\text{Å}}\right]^2$$

where X is the magnitude of internal surface area in one unit cell and Y is the number of silica molecules contained within the unit cell (from the structural formula). The equation has then been modified as

$$internal\ surface\ area = \frac{1}{M}\left(\sum_{i=1}^{N} 4\pi[R_{coord}(i)+r_{probe}]^2\frac{pi}{p}\right)$$

where M refers to the amount of SiO_2 present in the structure considered and $R_{coord}(i)$ refers to the atom/ion radius. The ratio pi/p then provides the fraction of the accessible surface contributed by center i [3]. Because the internal surface area is much larger than the external surface area of zeolites, the internal surface area is thought to dominate the chemical behavior. Many studies have been carried out on the application of zeolites based on the internal surface area, such as rate of diffusion, relation between the geometry and energetics, etc. [4]. In contrast, there are relatively few experimental and simulation studies on the external surface area. The external surface is fully accessible to all molecules and behaves catalytically in a non-shape-selective manner; it is therefore of great interest to study the effects of passivating or inerting external acid sites in order to promote the shape-selective reactions. Experimentally, it has been found that the external surface area plays a crucial role in isomerization of long-chain hydrocarbons over zeolite MCM-22 [5]. Similarly, ITQ-2, whose internal surface area is comparable with the external surface area, also suggests the important aspects of the external surface area [6]. As we mentioned in the beginning of this chapter, it is difficult to predict the termination of a zeolite structure simply by spectroscopic techniques; thus, computer simulation has been introduced to rationalize the understanding. To reveal the external surface of zeolites, Whitmore et al. studied the fundamental importance of the external surface through the adsorption of benzene on faujasite. Their study suggested a significant impact of external surface on physisorption, which permits the surface to subsurface diffusion more readily than bulk and internal surface [7]. Generally, physisorption is mostly related to the external surface and an effect of water environment along with the stability of zeolite structure has been conducted using atomistic simulation of zeolite surface. Three terminations are predicted for [100] surface and the stabilization of the terminal hydroxyl group is found to relate with the extraframework cation [8]. Zeolite A has a cubic structure with space group Fm3c and unit cell $a = b = c$ of 23.69 Å [9]. The structure consists of sodalite units. An octahedral array of sodalite units joined together by double four-rings (D4R) via oxygen bridges to construct the zeolite A structure. Hence, three different surfaces are constructed to account for the orientational effect of the D4R rings. The first surface as in Figure 7.1 shows an example of D4Rs oriented perpendicular to the surface such that two edges of the D4R bisect the [100] plane. There exists as well a different termination that is related to surface four-rings; this exposes a different domain for molecular transport. This orientation depends on the architecture of the pore.

Considering the surface area of nanoporous materials, mesoporous molecular sieves have attracted significant attention due to their large surface area and unique pore topologies. A milestone in the assessment of the pore size is the introduction

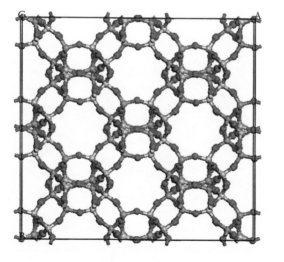

FIGURE 7.1   (For color version see accompanying CD) Relaxed hydroxylated pure Si LTA surface [100] exposing the DR4 termination.

of the Brunauer-Emmett-Teller (BET) [10] method, which has become a popular tool for the analysis of specific surface area. As far as the pore size analysis is concerned, a variety of methods have been proposed or implemented. Many are based on the Kelvin equation [11], which relates the pore size with the pressure of the capillary condensation or evaporation. Experimentally, the surface area of mesoporous materials is still estimated mainly by BET analysis of nitrogen adsorption at 77 K, but sometimes it is significantly less than actual. An alternate model based on molecular simulation in comparison to experimental values has been proposed. The model incorporates bulk heterogeneity of the material, surface hydroxyls, and, most importantly, physical deformations or indentations of the pore surface. The simulation results are consistent with the hypothesis that the interstitial space in MCM-41 is relatively amorphous despite the regular arrangement of the mesopores. The surface roughness associated with the amorphous structure increases the surface area beyond the nominal value produced by assuming smooth cylindrical pores [12]. Figure 7.2 shows a 3 × 3 unit cell of a structure that has been simulated here. To represent micelles, a tube or cylinder filled with randomly placed dummy atoms is used to create modeled micelles. One hexagonal unit cell model consists of one cylindrical modeled micelle (as shown in Figure 7.2) at the center of the cell and four quarters of a modeled micelle at the corner of the unit cell. The dummy atoms representing the modeled micelle were frozen during the entire simulation.

An easier way of calculating surface area is provided by the Accelrys Inc. Visualizer module. The model of calculation is shown in Figure 7.3, where a volume field for a simple diatomic molecule was created as if the molecule in consideration were within a solvent field. The surface calculation was performed with a probe to determine different surfaces in terms of van der Waals or solvent. A slice has been added, showing the location of a van der Waals (vdW) surface and two solvent surfaces. The blue and red circles illustrate the solvent probe position and radius. The van der Waals surface is the surface that intersects with the vdW radii of the atoms in the structure. This is

Dummy micelle   Silica

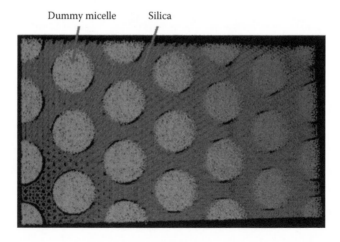

FIGURE 7.2   (For color version see accompanying CD) 3 × 3 × 1 structure of MCM-41 simulated in the present work showing the dummy atoms and the silicate network. [12] With permission.

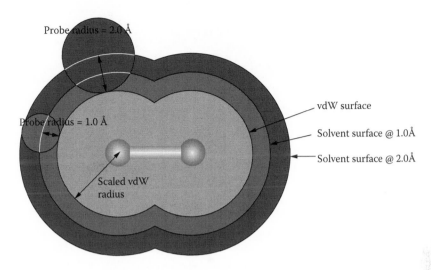

FIGURE 7.3 (For color version see accompanying CD) Visualizer model from Accelrys Inc. to calculate surface area of diatomic molecule.

equivalent to a solvent surface with a solvent probe radius of zero and to a Connolly surface with a Connolly probe radius of zero (the van der Waals surface is shown in cyan). The solvent surface is the locus of the probe center as the probe rolls over the scaled vdW surface. This surface describes a space that could, in principle, be occupied by a probe of given radius, ignoring the accessibility of such points (the solvent surface is shown in Figure 7.4b. As an example, the zeolite MFI framework was chosen as shown in Figure 7.4a. The vdW surface was calculated using a vdW scale factor of one, which specifies a factor to uniformly modify all vdW radii, representing a hard shell into which a probe may not pass. The results show that in terms of vdW surface, the occupied volume is 3,119.99 Å, the free volume is 2,212.03 Å, and the resulting surface area is 2,353.86 Å. This can be compared with the experimental BET surface area. As mentioned before, the most widely used tool to determine the surface area is the BET method represents less value than actual. Hence, it is expected that actual surface area can be predicted from the simulation of the nanoporous material through the surface prolee methodology mentioned earlier.

Let us consider the surface of the substrate be an array with N_S identical adsorption sites. No more than one atom can occupy one site and the atoms do not interact with each other. In the grand canonical ensemble, each site's grand partition function is

$$\Xi s = 1 + z \exp(\beta(\mu - E_0)) \tag{7.1}$$

E_0 is the surface binding energy, z is the partition function associated with possible internal degrees of freedom at every site (sometimes may be taken as unity), $\beta = \frac{1}{k_B T}$ is the inversed temperature, and μ is the chemical potential of the film. The grand canonical free energy is

$$\Omega = -\frac{1}{\beta} \ln \Xi = -\frac{Ns}{\beta} \ln \Xi s \tag{7.2}$$

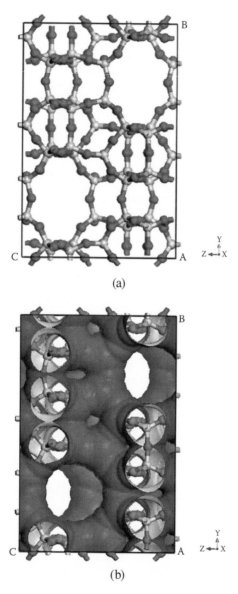

(a)

(b)

FIGURE 7.4   (For color version see accompanying CD) (a) vdW surface area of the zeolite pore. The blue spherical zone is occupied in terms of the van der Waals radii of the constituent atoms for MFI; (b) zeolite MFI structure model with a view from the *YZ* plane.

Because the mean number of particles in the ensemble satisfies:

$$N = -\left(\frac{\partial \Omega}{\partial \mu}\right)_\beta \tag{7.3}$$

the fractional occupation is

$$\theta = N/Ns = p/(p + p_L) \tag{7.4}$$

The characteristic scale of pressure is

$$p_L = \frac{g}{z\beta\lambda^3} \exp(\beta E_0) \tag{7.5}$$

Here $\lambda = \sqrt{2\pi\beta\hbar^2/m}$ is the de Broglie thermal wavelength and g is the spin degeneracy of the atom; Equation (7.4) is the Langmuir isotherm. It shows that the coverage grows linearly at low p according to Henry's law and saturates at $p \gg p_L$. Brunauer et al. [10] extended this lattice gas model to the case of multilayer films. Their model allows the particles to occupy a three-dimensional array of sites above the surface. The interactions between the sites are neglected, but the sites closest to the substrate experience additional attraction V_1. The relative probability, exactly N sites above a given surface to be occupied, is proportional to the corresponding term in the grand partition function for this site:

$$\Xi s = 1 + c \sum_{N-1}^{\omega} z^N \exp(N\beta\mu) \tag{7.6}$$

Here, $c = \exp(-\beta V_1)$ and z is the internal partition function per site of the bulk adsorbate. Thus, analogously to the Langmuir isotherm, one obtains the BET isotherm:

$$\frac{p}{p_L}\frac{N_s}{\left(1 - \frac{p}{p_L}\right)N} = \frac{1 + \frac{p}{p_L}(c-1)}{c} \tag{7.7}$$

According to the literature, the surface area values obtained from other methods are always less than the BET method. The observed difference is attributed to different reasons, such as multilayer capillary condensation, heterogeneity of the wall structure, and the error in the BET analysis. At first, Feutson and Higgins [13] proposed a model of MCM-41 incorporating the wall heterogeneity, surface roughness, and the surface hydroxyl group to account for the exact surface area that was very close to experimental values. Their model was modified [12] and in addition to the surface roughness, bulk heterogeneity, and surface hydroxyl groups, physical deformation or indentions of the pore surface was added. The surface area calculation based on this model structure of MCM-41 matches well with experimental values (experimental value of surface area = 960 m²/g, pore volume = 0.79 cm³/g; calculated value of surface area = 910 m²/g, pore volume = 0.51 cm³/g). The simulation results are consistent with the hypothesis that the interstitial space in MCM-41 is relatively amorphous despite the regular arrangement of the mesopores. The surface roughness associated with the amorphous structure increases the surface area. Furthermore, the grand canonical Monte Carlo (GCMC) simulation of Ar adsorption on SBA-15 for the calculation of surface area and pore size distribution is in good agreement with experimental values. The representative model of the material was prepared by mimicking the synthesis process using Monte Carlo simulations. The main feature of this model was the incorporation of micropores along with the mesoporous structure [14].

7.2 SURFACE CHEMISTRY

According to the definition of *surface chemistry*, it is the study of surface-related properties, mainly catalytic reactions, and is related to surface engineering. The surface structure of nanoporous materials has primary importance, based on which several applications can be decided. The surface structure can easily be modified by changing the chemical composition and incorporation of the selected elements or functional groups to get the desired effects or improvements in the surface properties. Microporous and mesoporous materials are widely used as adsorbents, for purification of water, and also in chromatographic columns. A realistic way to mimic the surface of the zeolite structure is a challenging task because there are different factors associated. Regardless of composition, the structures of such zeolite materials are based on three-dimensional, four-connected nets built of TO_4 units [15]. Surface chemistry is the key factor in determining the efficiency of these processes and has an impact on the thermodynamic parameter. It is known that the surface of most of the zeolite is covered by the silanol groups and it is necessary to understand the surface chemical reaction for its application as host to the different guest molecules. Several spectroscopic techniques deal with the surface chemistry of zeolites. FTIR studies on H-FER [16] and H-MFI [17] exhibits that the terminal silanols and Lewis acid sites exist at the external surface of zeolite and the stronger bridging Brønsted acid sites are located at the internal surface. Hydrogen forms of zeolites are active catalysts in numerous acid-catalyzed reactions. Hydroxyl groups are responsible for the Brønsted acidity of amorphous silica–alumina and zeolites. Generally, the acid strength of OH groups in H forms of zeolites depends on the Si/Al ratio, the degree of cation exchange, and the degree of dehydroxylation. The effects of the geometry of Si–O and Al–O bond lengths and Si–O–H as well as Si–O–Al angles of zeolites on the vibrational frequencies of their OH groups were modeled by $H_3SiOHAlH_3$ 1.1, whereas to understand the influence of chemical composition, models with formula of $H_3SiOSi(OH)_3$ 1.2 and $H_2AlOSi(OH)_3$ 1.3 have been studied. It was observed that structural characteristics influence the vibrational frequency more in comparison to chemical composition. A case study with H–Y zeolite has been performed for models with formula $(HO)_3SiOHAl(OH)_3$ 2.1 and $[(HO)_3SiOAl(OH)_3]^{-1}$ 2.2. It was observed that structural characteristics influence the vibrational frequency more in comparison to chemical composition. Now, starting from the equilibrium generally we varied Si–O, Al–O distance and Si–O–H, Si–O–Al bond angles at a regular interval. The difference between the initial and final bond distance and bond angles are noted as ΔR and $\Delta\phi$, respectively. Table 7.1 shows total energy and the relative energy

TABLE 7.1

Geometric Parameters as a Function of Bond Distance ΔR and Bond Angle Δφ, Total Energy, Relative Energy, and Vibrational Frequency of Bridging OH (μ_{OH}) Calculated for 1.1 Cluster

Geometric Parameters	Equilibrium Geometries	Deformations Introduced into this Cluster			
		ΔR = +0.02 Å Δφ = +10°	ΔR = +0.02 Å Δφ = −10°	ΔR = −0.02 Å Δφ = +10°	ΔR = −0.02 Å Δφ = −10°
R SiO (Å)	1.687	1.696	1.702	1.678	1.683
R AlO (Å)	1.829	1.844	1.851	1.816	1.827
φ Si–O–H (°)	114.2	112.6	111.9	117.3	116.8
φ Al–O–Si (°)	128.9	135.5	138.3	120.2	119.9
Total energy (kcal/mol)	−663,014.5	−663,013.22	−663,013.81	663,013.86	663,014.05
Relative energy (kcal/mol)	0.0	1.28	0.69	0.64	0.45
μ_{OH} (cm^{-1})	4186	4218	4202	4222	4214

after the optimization of geometry of the studied molecule. The total energy values listed in Table 7.1 reveal a maximum relative instability of ~1.2 kcal/mol. The interesting phenomenon to be noted here is that the O–H stretching vibrational frequency changes by 40 cm^{-1}. This result is attributed to an increase in the valence bond angle on the O atom and the angle (Si–O–H) results in an increase in the percentage of s character of the hybrid atomic orbitals, which forms Si–O and O–H bonds. All the results were compared with experimental IR frequencies. Comparison of the results of calculations on cluster models 1.2 and 1.3 reveals that the substitution of silica by aluminum in the third coordination sphere has little influence on the O–H stretching vibrational frequency. The observed vibrational frequency of 4,159^{-1} and 4,162 cm^{-1} for the respective clusters justifies the fact that the structural factor plays a decisive role in the determination of vibrational frequencies of the bridging hydroxyl group rather than the variation in chemical composition. The terminal hydroxyls are not affected by structural or compositional variation. To study the structural effect in a real situation, the electronic properties of the H–Y zeolite cluster models 2.1 and 2.2 were calculated using LDF on both types of oxygen to compare the shift in vibrational frequency for the respective oxygen. The LDF calculation results are given in Table 7.2. This shows that the net charge on oxygen is the same, but the total energy for the individual oxygen shows that the O1 cluster is energetically more stabilized in comparison to that of the O4 cluster [18]. Deprotonation energy (DE) was calculated using the formula ZeolOH→ZeolO$^-$ + H$^+$. From the value of DE it is observed that O1 will be better proton donor than O4.

The transfer of molecules through the crystal facet is still not very well understood mainly due to the fact that experimentation of the surface-mediated process is difficult and the surface termination of zeolites is not well known. To address these

TABLE 7.2

Total Energy, Deprotonation Energy, Net Charge on Oxygen, and O–H Stretching Frequency for Both the Oxygen O1 and O4 in Cluster Model 2.1 of H–Y Zeolite

Type of Oxygen in H–Y Zeolite	Total Energy (kcal/mol)	Deprotonation Energy (kcal/mol)	Net Charge on Oxygen	O–H Stretching Frequency (cm^{-1})
O_1	−663,059.19	−339.56	−0.58	3712
O_4	−663,055.5	−335.87	−0.56	3619

issues, Whitmore et al. [19] used an atomistic simulation methodology to probe the physical chemistry of the external surface. Previously, Catlow et al. [20] established the fact that for faujasite-type zeolites (111) surface was the most stable surface, which has also been verified from the experimental crystal morphology. They constructed the model by cleaving (1 1 1) plane and found out seven possible distinct cuts to satisfy the dipole moment convergence rule. Of the seven theoretical surfaces identified in that study, only two have been proved to exist experimentally, the type I surface as identified by Terasaki [21] and displayed in Figure 7.5a and Figure 7.5b. The type I termination results in a surface with an opening of a twelve-membered ring to the entrance of surfacial supercages, which is shown in Figure 7.5b and labeled as C. Additionally, two other potential adsorption sites can be identified from the modeled figure: a six-membered ring (labeled as A) and a half sodalite cage (containing one six-membered ring), which is demarcated by hydroxyl groups (indicated as B), were employed in this study. Using the surface simulation code MARVIN [22], the surface was modeled by using a finite total of eight cell layers, containing 1,986 atoms. Within this strategy, the lower four layers were held fixed, and the remaining upper four layers were relaxed explicitly. To determine the number of cell layers to use in each region we reproduced the bulk electrostatic potential within the fixed layers and derived a converged surface energy. The Coulomb sum was performed in two dimensions using a variant of the Ewald approximation as described by Heyes et al. [23]. To represent the zeolite framework, the parameterization of Sanders et al. [24] was employed with additional parameters formulated by Schroder et al. [25] to represent framework hydroxyl groups. To describe inter- and intramolecular benzene molecule and guest–host interactions, the force field derived by Henson et al. [26] was selected because this has been fitted directly to experimental adsorption energy data and has been shown to yield excellent reproduction of adsorption isotherms. Following the simulated annealing strategy, three distinct adsorption sites at the external surface were located and are represented pictorially in Figure 7.6a, Figure 7.6b, and Figure 7.6c. In Figure 7.6a, the guest is adsorbed strongly to the six-membered ring and Figure 7.6b represents the adsorption of benzene approximately 3 Å above the half sodalite (SOD) cage and binds less strongly than the six-membered ring cases because the accessible volume is somewhat less in comparison. Finally, in

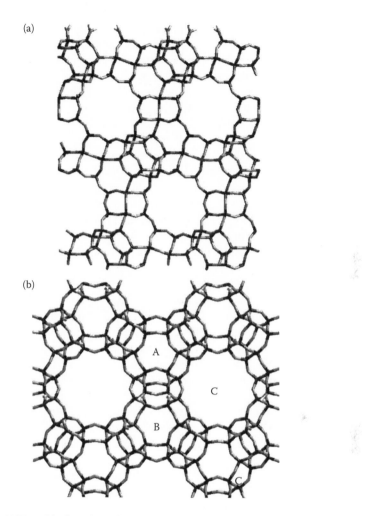

(a)

(b)

FIGURE 7.5 (a) Type-I hydroxylated faujasite(111) surface viewed along the [110] direction; (b) type-I hydroxylated faujasite(111) surface viewed along the [111] direction.

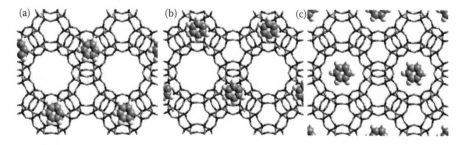

FIGURE 7.6 (a) External surface six MR adsorption of benzene; (b) external surface half SOD cage adsorption of benzene; (c) external surface twelve MR adsorption of benzene.

Figure 7.6c, the benzene molecule can be seen in the center of the twelve-membered ring, although the center of mass of the guest lies approximately 1 Å above the plane perpendicular to the twelve-membered ring. These calculations reveal two distinct results of fundamental importance in understanding the external surface and transport properties of faujasite. Firstly, three favorable sites for physisorption have been identified on the external surface for the type I termination. The magnitude of the adsorption energies is similar to that found in the bulk. The second result provides information about the mode of transport from vacuum to crystal or vice versa, which is only possible through the twelve-membered ring for an ideal (111) facetted crystal. The energetics of this process studied by a series of constrained minimizations to estimate the energy barrier reveal that the most favorable pathway is comparable with that found previously in the bulk calculations of Henson et al. [26].

The surface chemistry of mesoporous materials is very interesting. In contrast to the crystalline structure of the microporous zeolites, mesoporous materials are amorphous. These materials have a tunable pore structure and pore size within the range of 2–10 nm and large specific surface area from 600 m^2 g^{-1} to 1300 m^2 g^{-1}. The surface property can be easily modified by the interaction with guest molecules. Unlike zeolites, the structural study of mesoporous materials through computer simulation is limited. Most of the models are based on the adsorption characteristics of the material. Sonwane et al. [12] described a model of MCM-41 based on molecular simulations that gives an idea of the surface area and pore size as described in Chapter 6. Several models mimic the structural features of MCM-41. Most modeld consider that the surface of MCM-41 is not always represents uniform array of hexagonal channel but they are rough and heterogeneous. Fenelenov and coworkers [27] provided a quantitative estimate of surface roughness by defining a roughness coefficient, β. The value $\beta = 1$ corresponds to the smooth surface, whereas values higher than 1 indicate increased roughness. For MCM-41, they used gas-adsorption analysis and synchrotron X-ray diffraction and found that $\beta = 1.1$–1.2. Next is the model of Denoyel and coworkers [28], which consists of an array of oxygen atoms in a cylindrical annulus with each atom interacting with an adsorptive atom via a Lennard-Jones (LJ) 6–12 potential. For the adsorption of Ar and Kr modeled via the GCMC, they found that the energy site distribution determines low-pressure adsorption, whereas the structural distribution affects multilayer adsorption. In contrast to the previously described model, a cylindrical pore model to represent MCM-41 was described by Cao and coworkers [29]. The results of GCMC and density functional theory (DFT) over their model proposed a good comparison between simulation and experiments for the adsorption isotherm of nitrogen at 77 K, as well as that for carbon tetrachloride and methane at 273 K. For both GCMC and DFT models, the MCM-41 pores were assumed to be smooth cylindrical tubes consisting of one atomic layer.

Therefore, from the above discussion it is obvious that the preparation of a realistic model is the most important factor to make a good correlation between experiment and the simulation. Several factors related to the chosen property need to be considered carefully, which then can mimic the authentic scenario by simulation.

7.3 SURFACE CONFINEMENT

In addition to the relevance of zeolites as catalysts, other important applications are purification, separation, and use as molecular sieves. Size dependence of diffusivity has attracted considerable attention since the discovery of nanoporous materials in a wide variety of systems. Size-dependent self-diffusivity of guests confined to the zeolitic pores is of interest because of its fundamental importance as well as its utility to applications in petroleum and petrochemical industries. The isomers of alkanes in zeolites provide an example where the size of the diffusant varies over a wide range. In the processing of crude oil with the help of zeolites, the diffusivity of the molecules of hydrocarbons is determined by their size and shape. Separation of such isomers of alkanes and alkenes is one of the challenges faced by the petroleum industry. In fact, the well-known term *molecular sieve* refers to the separation achieved through size, although in this case, the sieving property has its origin in the repulsive interactions and not the dispersion interactions. In the processing of crude oil with the help of zeolites, the diffusivity of the molecules of hydrocarbons is determined by their size and shape. Solute diffusion in solution has attracted extensive attention over the last hundred years. Einstein investigated the dependence of diffusivity of a solute on its radius. The relation obtained is the well-known Stokes-Einstein relationship:

$$D = \frac{K_B T}{6\pi \eta r_u}$$

where D is the self-diffusivity, $è$ is the viscosity of the solution, T is the temperature, k_B is the Boltzmann constant, and ru is the solute radius. Diffusion of these neutral solutes has been well-studied both experimentally and theoretically. Influence of mass, density, temperature, viscosity, and many other factors on the Stokes-Einstein relationship has been extensively investigated [23].

Self-diffusivity, D, of diffusants in extensively differing mediums such as liquids (e.g., solution), porous solids (e.g., guests in zeolites), or ions in polar solvents exhibit strong size dependence. We discuss the nature of the size dependence observed in these systems. Altogether, different theoretical approaches have been proposed to understand the nature of size dependence of D not only across these widely differing systems but even in just one medium or class of systems such as ions in polar solvents. In the past decade, molecular dynamics investigations have shown that the size dependence of self-diffusion in guest-porous solids could have origins in the mutual cancellation of forces that occurs when the size of the diffusant is comparable to the size of the void. This effect leads to the maximum in D known as the *levitation effect* (LE). Such a cancellation is a consequence of symmetry, which exists in all porous solids irrespective of the geometrical and topological details of the pore network provided by the solid. Recent studies show that the LE and size-dependent diffusivity maximum exist for uncharged solutes in solvents. One of the consequences of this is the breakdown in the Stokes-Einstein relationship over a certain range of solute–solvent size ratio. Experimental measurements of ionic conductivity over the past hundred years have found the existence of a size-dependent diffusivity maximum leading to violation of the Walden's rule for ions in polar solvents. Molecular dynamics simulations and experimental data suggest that even this maximum has its origin in LE. Simulation studies of impurity atom diffusion in close-packed solids as well as ions in superionic and other solids suggest the existence of a size-dependent diffusivity maximum in these materials as well. The LE is a universal phenomenon leading to a maximum in diffusivity of a diffusant in a variety of condensed matter phases. The only condition for its existence appears to be the presence of van der Waals or electrostatic interactions.

In this regard, an exhaustive work has been carried out by Yashonath et al. [30, 31] that allowed a low loading of one guest per α cage to zeolites Y and A as their structure is of importance to understand the transport properties. Both of these zeolites consist of large voids, called cages, interconnected via narrower windows or bottlenecks. In Na–Y or faujasite, the α cages have a diameter of about 11.8 Å and are interconnected via narrower twelve-ring windows of approximate free diameter 8 Å. Again, each cage is connected tetrahedrally to four other α cages. In zeolite A, each cage of approximate diameter 11.8 Å is linked octahedrally to six other cages via eight-ring windows of diameter 4.5 Å and are shown in Figure 7.7. Average force

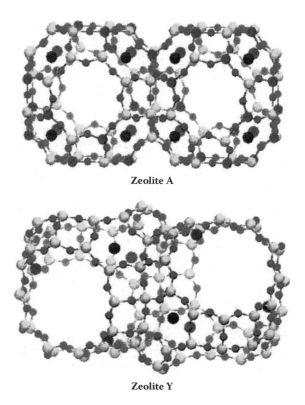

Zeolite A

Zeolite Y

FIGURE 7.7   (For color version see accompanying CD) Two α cages of zeolite A and Y are shown [32] along with the narrower interconnecting window. Each cage in zeolite A is connected to six other cages placed octahedrally, whereas each cage in zeolite Y is connected to four other cages in a tetrahedral fashion. [31] With permission.

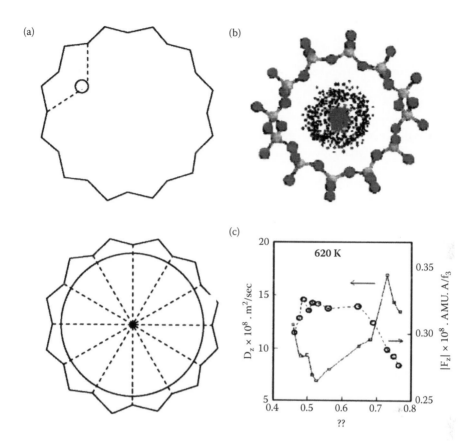

FIGURE 7.8 (For color version see accompanying CD) (a, top) Schematic diagram illustrating the large force on the guest due to the confining medium, zeolite, for a relatively small-sized guest from the linear regime. (bottom) In contrast, the large guest from the anomalous regime, which is comparable in size to the neck diameter, experiences equal force from diagonally opposite directions leading to negligible force on the guest. (b) Twelve-ring window along with positions obtained from molecular dynamics trajectory indicating the position of the guest during the passage through this bottleneck. Blue points indicate the place in the plane of the window that is sampled by the linear regime particle during its passage through the window. Green points are for the anomalous regime particle, which sample the central region of the cage. [31] With permission.

obtained from molecular dynamics simulations when plotted against size of the diffusant exhibits a minimum for that size where self-diffusivity is maximum locally and shown in Figure 7.8. It provides the direct evidence regarding the force, which is indeed lower for the guest in the anomalous regime. The trajectories of guests through the twelve-ring window in zeolite-Y obtained from molecular dynamics are also depicted in Figure 7.8. The oxygen and silicon atoms of the twelve-ring window are shown along with points in the window plane at which guests from linear and from the anomalous regime pass through the window. Size-dependent diffusivity maximum exists in porous solids irrespective of the geometrical and topological characteristics of the pore network. Just as interaction between the atoms is responsible for the deviation from ideal gas behavior, the deviation from the Stokes-Einstein behavior of self-diffusivity occurs in interacting systems. The deviation occurs only if there is either van der Waals interaction or electrostatic interaction between the diffusant and the medium in which it is diffusing. Because the van der Waals interaction is essentially present, there is really no system in nature that does not exhibit a diffusivity maximum, provided that other conditions such as $|U_{dm}|/k_B T \gg 1$ are satisfied.

Disorder appears to only slightly reduce the height of the diffusivity maximum. The diffusivity maximum is highly pronounced in the presence of electrostatic interactions. This is attributed to the long-range nature of these interactions. The main reason why long-range interaction leads to the existence of the diffusivity maximum even at high temperatures or a more pronounced maximum as compared with a system with purely van der Waals interaction is that the total interaction energy between the diffusant and the medium is higher because of the interaction extending to large distances.

Diffusion and adsorption are two crucial phenomena occurring in zeolite. Several experimental [33] and theoretical studies [34] have been done to make a proper understanding of diffusion of adsorbed molecules for practical application. From the macroscopic point of view, diffusion can be transport diffusion (nonequilibrium process) and self-diffusion (equilibrium process). Zeolites maintain a complicated pore topology and are restricted to the small molecule. Thus, confinement (sufficiently larger molecules diffused into zeolites cannot pass each other) of the adsorbate molecule occurs frequently, which plays an important role in the diffusion and is called *anomalous diffusion* [35]. The diffused molecule interacts more strongly with the environment that matches with their size and shape. The curved surface of intercrystalline free volume of zeolite provides a variety of environments and the molecules inside the pores are in strong interaction with their environment because of the confinement effect [36]. Because the size of the guest molecule is similar to that of the pores, the confinement effect increases. The first confinement effect is proposed as a "nest effect" based on the results of the cracking of *n*-pentane over protonated zeolite [37]. It states that the open framework structure possessing channels and cages can originate some remarkable properties, which affect the shape-selective properties of zeolites. Due to this effect, the adsorbed molecules and zeolite framework reciprocally may optimize their respective structures to maximize the van der Waals interaction. The confinement effect can occur inside the intracrystalline volume of the zeolite or on the external surface with "crater" and "hills" [38] and because of the confinement effect, zeolites can also be considered as solid solvents. Many examples such as accommodation of benzene (van der Waals radius =0.6 nm) is easily happened inside the pores of ZSM-5 (pore diameter = 0.55 nm) confirmed the phenomenon of solid solvent. Blum et al. proposed another model of confinement based on the Gaussian curvature of zeolite structures [39]. It accounts for the adsorption property of zeolites and correlates heat of adsorption of a given molecule to a given zeolite. Monte Carlo simulation on the confinement of the Ar_3 cluster to an α-cage of zeolite-A predicted that the confined cluster has different geometry than free one, which is attributed to the heterogeneous environment provided by the zeolite [40]. Cwiklik et al. [41] described the confinement effect on the diffusion and adsorption of n-butane in silicalite-1 using dynamic Monte Carlo simulation. Their study suggests a strong influence of confinement of the channel structure on the dynamics and the steady-state property of adsorption of n-butane in silicalite-1. According to their model the simulated adsorption isotherms are in quite a good agreement with experiments. The catalytic activity of zeolite not only originates from the active sites on the inner surface but is also related to the environment of the reactant, intermediate, and product in the pore system of the zeolite; i.e., confinement. The reactant molecule must experience the influence of the electric field exist in the pores and sometimes strongly affect the product distribution of a reaction. A Monte Carlo integration method employing a classic molecular mechanics force field is used to probe the impact of confinement on the selectivity of n-hexane/3-methylpentane cracking reaction using twelve different zeolites. This model has proven to be predictive for small to medium pore zeolites. For instance, the cracking selectivity of MTT- and FER-type zeolites arises from the adsorption selectivity of reactants, whereas restricted transition-state selectivity is associated with the MFI- and BEA-type zeolites [42]. Moreover, adsorption and dynamics of acetonitrile molecules in mordenite zeolites have been studied by a combination of molecular dynamics and electronegativity equalization methods, which predict the adsorption of the molecules in both the main channels and the side pockets of the structure. Adsorption in the side pockets is favored but is hampered by a sterical hindrance and an energy barrier between main channels and side pockets. Increase of the temperature as well as of the adsorbate loading lead to an increase of the occupation of the side pockets [43]. Marquez et al. suggested that the confinement phenomenon of the zeolite is related to the molecular level; the presence of solid strongly perturbed the molecular orbital [44]. An *ab initio* calculation of adsorption of toluene inside zeolite beta ZSM-12 and ZSM-5 has been modeled using all silica clusters. The increase in HOMO energy is observed as the cavity size and Si/Al decreases. One therefore propose that in terms of reactivity, proton transfer will be favored in a confined system [45].

Novel zeolite ITQ-2, the first member of the family of delaminated zeolites, exhibits unique behavior because of its large accessible pockets. In contrast to the other conventional zeolites like ZSM-5, it tunes the cation that best fits within its geometry and is an example of potential energy well, which is able to generate and stabilize the organic cation [46].

In this context, mesoporous materials, another example of nanoporous materials, must be discussed. It is well known that mesoporous materials contain a larger pore system than microporous materials and require precise characterization. As an inherent property of the molecular ensembles in mesopores, the interplay of the fluid–pore wall and the fluid–fluid interactions may give rise to various specific properties of the confined fluids. Among them is adsorption hysteresis, which is critical to fully understand. The adsorption inside the mesoporous material mainly follows the capillary condensation and is described by the Kelvin equation:

$$\ln\left(\frac{p}{p_0}\right) = -\frac{2\gamma V_m}{rRT}$$

where γ represents surface tension, V_m is molar volume of the condensed phase, r is pore radius, R is the gas constant, and T is the temperature. However, the Kelvin equation underestimates the pore size in the range up to 7.5 nm even when it is corrected for statistical film thickness [47]. Hence, a modification of the Kelvin equation (shown below) has been proposed to obtain the exact scenario of the porous structure of the mesoporous material:

$$r\left(\frac{p}{p_0}\right) = -\frac{2\gamma V_m}{RT\ln\left[\left(\frac{p_0}{p}\right)\right]} + t\left(\frac{p}{p_0}\right) + 0.3nm$$

where the term t represents the statistical film thickness [48]. Experimentally, many techniques such as ESR, NMR, X-ray, and neutron diffraction have been used to study the effect of confinement on mesoporous materials and came to observation that the density of the confined fluid is higher than that of the bulk, which varied regularly with the pore size [49]. Mesoporous materials are widely used as catalyst support. How does this confinement effect influence the catalytic activity of the material? Unfortunately, theoretical studies on catalytic effects are unavailable in this field, but some experimental observations exhibit a strong impact of confinement on the catalytic activity. For example, Tanchoux and coworkers [50] recently observed a correlation between the pore size dependence of the activity of mesoporous aluminosilicates in the catalytic isomerization of 1-hexene and the variation of absorption heat of hexane on the corresponding nonacidic mesoporous silicates. An enhanced activity of MPV reduction of cyclic ketone has also been suggested over aluminum isopropoxide grafted MCM-41, due to the surface confinement [51]. Because the surface confinement of mesoporous materials provides surprising results, it is believed that computer simulation will offer a fruitful scenario in this promising field.

The recent work of Sonwane and Li [52] provides the first study of ordered mesoporous materials SBA-15 coated with microporous zeolites ZSM-5 using molecular simulations. Several model structures with characteristics such as periodic arrangement of mesopores, randomly arranged micropores, surface hydroxyls, and bulk deformations of SBA-15 were used. Cartesian coordinates of ZSM-5 unit lattice were obtained from the literature and the 100 face of H-ZSM-5 unit cell was then placed on the surface of SBA-15 and the entire structure was equilibrated to obtain a final configuration. The resulting structure was characterized using simulated small angle and wide angle X-ray diffraction, Connolly surface area (to compare BET area), accessible pore volume for nitrogen molecules (to compare with t-plot volume of micro and mesopores), and methane adsorption at 303 K. The orientation of ZSM-5 on the SBA-15 had no effect on the surface area, pore volume, or adsorption capacity. In order to find out whether the addition of microporous ZSM-5 should increase the total methane adsorption capacity due to addition of micropores, they studied adsorption on bare and coated SBA-15. However, total adsorption capacity was found to decrease, whereas the number of methane molecules adsorbed per unit cell of the SBA-15 structure increased. An existing experimental method [53] to synthesize hybrid ZSM-5/SBA-15 structure was studied using accessible micropore volume (by *t*-plot). It was found that the procedure made all the micropores inaccessible. A modification of the method or use of other host materials is suggested to use the benefits of narrow micropore distribution in ZSM-5. In a recent work, Ahunbay [54] performed the adsorption and diffusion of trichloroethylene (TCE) and tetrachloroethylene (PCE) in ZSM-5-type zeolite to calculate adsorption isotherms and heats of vaporization using GCMC techniques. The results demonstrated that the *Pnma-P*212121 symmetry transition of the zeolite framework has no significant effect on the TCE adsorption capacity of the silicalite, but it causes an increase of the PCE adsorption capacity. Simulations using a silicalite framework with *Pnma* symmetry showed that the adsorption capacity of the silicalite was limited to five molecules per unit cell. However, when a framework with *P*212121 symmetry was used in the simulations, the capacity reached eight molecules per unit cell, which is the actual adsorption capacity. To calculate intracrystalline diffusion coefficients of these compounds, molecular dynamics simulations were performed at different temperatures and loadings using the sorption module. The results show that zeolite symmetry has a significant impact on diffusion coefficients of the sorbate molecules. The preliminary adsorption simulations indicated that the condensed-phase optimized molecular potentials for atomistic simulation studies (COMPASS) force field [55] overpredicted the equilibrium loadings of the chlorinated alkenes by several orders of magnitude in the low-pressure range. As a result, one knew to use the consisted valence force field (CVFF) [56] rather than COMPASS force field to calculate sorbate–sorbate and sorbate–zeolite interactions. Periodic boundary conditions and a cutoff radius of 10 Å were adopted for the calculation of the van der Waals interactions. The electrostatic interactions were calculated by the Ewald method. Grand canonical Monte Carlo (GCMC) simulations were run at constant temperature T, chemical potential μ, and volume V. The value of μ was imposed through the fugacity of the sorbate molecules, which was assumed to be equal to their pressure at the simulated low-pressure conditions. The Metropolis algorithm [57] was used in the calculation. Details of the GCMC method can be found elsewhere [58]. Averages were collected over 3,000,000 iterations to obtain good statistics. MD simulations for the diffusion of TCE and PCE in the zeolites were performed using the Discover module of the simulation package of Accelrys Inc. The CVFF force field [59] was used for both types of simulations as in the case of the adsorption simulations. Simulations were carried out in the NVT ensemble using the velocity Verlet algorithm [60] and the temperature was kept constant using the Nosé method [61]. Both the zeolite framework and sorbate molecules were considered flexible in the simulations. Simulations were performed using both *ortho* and *para* structures of the silicalite. Other details are similar to those of the adsorption simulations. A time step of 1 fs was adopted for all simulations, with the total simulated time being around 2,000 ps. The energy iso-surfaces corresponding to possible locations of TCE and PCE molecules inside silicalite at different

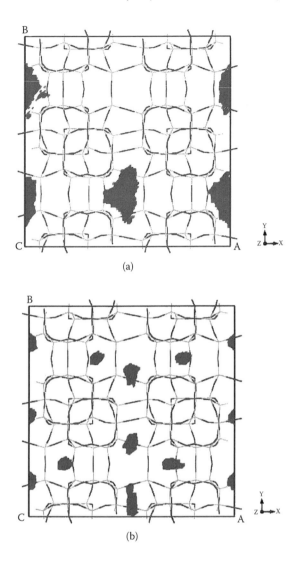

(a)

(b)

FIGURE 7.9 (For color version see accompanying CD) The energy isosurfaces of TCE in silicalite frameworks of *ortho* and *para* symmetries at 298 K. [54] With permission.

loadings at 298 K are presented in Figure 7.9 and Figure 7.10, respectively. Figure 7.9 shows that for loadings of one and ten TCE molecules per unit cell, both the channels (straight and sinusoidal) and the channel intersections are occupied similarly for both the *ortho* and *para* structures of the silicalite. On the other hand, as can be seen in Figure 7.10, PCE molecules show strong site preference below the loading of four molecules per unit cell, where the zeolite framework is of the *ortho* symmetry, and sorbate molecules reside only at the channel intersections. In particular, simulation results on the PCE adsorption showed that the symmetry transition of the ZSM-5

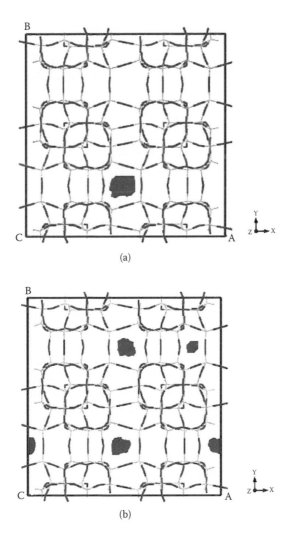

(a)

(b)

FIGURE 7.10   (For color version see accompanying CD) The energy isosurfaces of PCE in silicalite frameworks of *ortho* and *para* symmetry at 298 K. [31] With permission.

zeolite due to loading has a strong effect on the calculated isotherms; thus, this transition has to be taken into account to predict the isotherms correctly.

   Metal organic framework materials possess a large surface area among the nanoporous materials. GCMC simulations were used with a Lennard-Jones (LJ) fluid modeled on methane confined in nanospace with jungle-gym-like (JG) cubic structure, which is typically found in porous coordination polymers. The jungle-gym-like structure of MOF consists of pillar, which composed the cubic structure and modeled as structure-less smooth, solid rods made of LJ carbon. The simulations clarified that the condensation pressure and adsorption amount in the JG structure were influenced by pore size LJ and rod potential, whereas the transition type was determined by rod thickness. The characteristics of the JG structure lie in the sensitivity to the slight changes in pore size, rod thickness, and rod potential owing to the combination of the packing effect of molecules and the superposition effect of rod potentials [62]. This method is still hypothetical and depends on the quality of the parameter set, which is yet to derive atomistically.

7.4   SURFACE ACTIVITY

Zeolites are widely recognized as outstanding heterogeneous catalysts in several industrial processes. The prerequisite condition to heterogeneous catalysis is to have adsorption at the first place followed by the molecular dissociation usually preceded by the diffusion [63]. The fundamental concept behind the adsorption phenomenon is the adsorption isotherm. It is the equilibrium relation between adsorbed quantity of the adsorbate molecule and the pressure or concentration in the bulk fluid phase at constant temperature. Adsorption in porous materials is dependent on the pore size. In microporous materials, due to the comparable sizes of pores the adsorbed molecule can interact with the adsorbate species. Generally, the adsorption in the micropore is the pore-filling type; thus, pore volume is the main controlling factor. For mesoporous materials, mono- and multilayer adsorption can occur and is considered as capillary condensation. Therefore, the basic parameters characterizing the morphology of the mesoporous structure are surface area, pore size, pore volume. The hysteresis loop is predicted due to the deviation between adsorption and desorption curve. An evaluation of the specific surface area and the pore size is the fundamental problem in adsorption and the subject of many experimental and theoretical studies to be discussed below. Species with a kinetic size that makes them too large to pass through a zeolite pore are effectively "sieved." This sieve effect can be utilized to produce sharp separations of molecules by size and shape. The particular affinity a species has for an internal zeolite cavity depends on electronic considerations. The strong electrostatic field within a zeolite cavity results in very strong interaction with polar molecules such as water. Nonpolar molecules are also strongly adsorbed due to the polarizing power of these electric fields. Thus, excellent separations can be achieved by zeolites even when no steric hindrance occurs. Adsorption based on molecular sieving, electrostatic fields, and polarizability is always reversible in theory and usually reversible in practice. This allows the zeolite to be reused many times, cycling between adsorption and desorption. This accounts for the considerable economic value of zeolites in adsorptive applications. Knowledge of the adsorption behavior in the pores of zeolites is very important for understanding the performance of zeolites in various fields. The adsorption behavior is generally quantified by the adsorption isotherm at a given temperature and pressure. Experimental determination of the adsorption isotherm is sometimes time consuming (depending on the nature of the adsorbate), requires expensive instrument setup, and makes the system more complicated when a mixture of adsorbate is used. Furthermore, the adsorption and diffusion of sorbate molecules in zeolites are complicated and the intracrystalline diffusivity is affected by many parameters like size and shape of the sorbate relative to the zeolite pore, connectivity, and the dimension of the zeolite pore structure. Therefore, computer simulations can provide a valuable tool to the adsorption in porous materials like zeolites. Many techniques such as molecular mechanics [64], molecular dynamics [65], and Monte Carlo [66] simulation are widely applied. Among those methods, GCMC and molecular dynamics are used to predict the adsorption behavior and diffusivity of various sorbate molecules in zeolites. The surfaces of any accepted structure can be generated and fed into the computer so that one can obtain much richer information than the experimental methods. Thus, it is possible to carry out computational measurements of adsorption isotherms, adsorption isobars to calculate all types of adsorption heats and the heat capacity, to evaluate the local density profiles, radial distribution functions, structure factors, etc. The methods under consideration give a truly unique possibility to observe how the system evolves in time and changes its inner structure and what happens with every single particle. The main reason for using computer simulation is that it eliminates the inaccuracies that result from the approximate statistical thermodynamics. GCMC methods are used to study the adsorption isotherms, interaction energies, entropies, and density distributions of nanopores of varying sizes and shapes. In terms of the interaction energy between adsorbate–adsorbate, adsorbate–pore, and density profile, the isotherms are different according to the shape and size of the pores. The relationship between adsorbed amounts and pore size and shape are found to be a strong function to correlate the external potential and accessible pore volume [67].

7.4.1 ADSORPTION OF ALKANES

Adsorption of single, binary, and mixtures of alkanes is an industrially important issue for the separation of gas mixtures. The main advantages of adsorption are its high selectivity compared to other separation techniques and relatively high capacity for the adsorbent, even at lower pressure. Experimental measurement of the multi-component system is complicated and time consuming. So advanced techniques like computer simulation are used to predict the adsorption as well as the mixture of the components. There are several models to describe the adsorption of single component, but a Langmuir model can also be applied for the multicomponent system in a way that the number of carbon atoms corresponds to the number of adsorption sites:

$$N = \sum_{i=1}^{n} N_{mi} \frac{b_i P}{(1 + b_i P)},$$

where i is the number of carbon atoms in the alkane.

For alkane–alkane interaction, the united atom model was used. It is a good approximation to simulate a molecular system in which intermolecular motion is much more important than the intramolecular motion. In the adsorption of alkane, CH_4, CH_3, and CH_2 groups are considered as single interaction centers and described with the Lennard-Jones potential as follows:

$$U_{non\text{-}bonded} = U_{LJ} = \sum 4\varepsilon_{ij} \left[\left(\frac{\sigma_{ij}}{r_{ij}} \right)^{12} - \left(\frac{\sigma_{ij}}{r_{ij}} \right)^{6} \right]$$

where r_{ij} is the distance between pseudo atoms i and j, ε_{ij} is energy parameter, and σ_{ij} is the size parameter. The Lennard-Jones parameters are given in Table 7.3.

TABLE 7.3
**Parameters for the Lennard-Jones Potential Describing
the Interactions Between Pseudo-Atoms of a Branched
Alkane as Developed by Wang et al.**[a]

	$\dfrac{(\varepsilon k_B)}{K}$	σ (Å)
CH_2-CH_2	59.38	3.905
CH_3-CH_3	88.06	3.905
CHb_3-CHb_3	80.51	3.910
$CH-CH$	40.25	3.850
CH_3-CH_2	72.31	3.905
CH_3-CHb_3	84.20	3.9075
CH_3-CH	59.53	3.8775
CH_2-CHb_3	69.14	3.9075
CH_2-CH	48.89	3.8775

Notes: A CH_3 group connected to a CH group is denoted by $-CHb_3$. This
group is given a different set of interaction parameters. The interactions are truncated at Rc = 9.626 Å.

[a] Wang et al. [121].

The configurational-bias Monte Carlo technique is applied to simulate the adsorption of long-chain alkanes in zeolites. This simulation technique is several orders of
magnitude more efficient than conventional methods that can be used to simulate
the adsorption of long-chain alkanes. The calculated heats of adsorption are found
to be in excellent agreement with experimental data. The results show a surprising
chain length dependence of the heats of adsorption. This dependence has a simple
molecular explanation in terms of preferential sitting of the long-chain alkanes [67].
Similarly, an anisotropic united atoms potential AUA-4 has been developed to predict
the equilibrium properties of various hydrocarbons. The use of combination rules
enabled prediction of the adsorption properties of other alkanes (methane, ethane,
propane, butane, isobutane, pentane, 2-methylbutane, hexane, 2-methylpentane, hep-
tane, 2-methylhexane, octane, and nonane) without any further readjustment of the
force field parameters and correlates well with the experimental data at 277 and 374 K
[69]. Adsorption of the binary mixture like C-4 and C-7 alkane isomers in MOR- and
MFI-type zeolites is simulated using GCMC and configurational biased MC techniques and it was observed that MFI behaved quite differently [70]. Fox et al. [71]
carried out Monte Carlo simulations using new potential parameters for cyclohexane
to model the properties of C-6 alkanes (linear, branched and cyclic) in silicalite-1. The
vapor–liquid coexistence curves for the three types of alkanes, the heats of adsorption,
and Henry coefficients are in good agreement with the experimental data [71]. Several
other examples in the related field will also be provided throughout this chapter.

7.4.2 ADSORPTION AND DIFFUSION OF AROMATICS

The increasing industrialization and development in transportation require great concern about the environmental problem. Concerning the environmental issue, the adsorption of aromatics in molecular sieves like zeolites is an obvious choice and an extensively studied topic [72]. Moreover, the adsorption of aromatics is the foundation of many industrially important reactions like BTX (benzene, toluene, xylene) separation, toluene disproportionation, etc. The adsorption of benzene has been studied over different types of zeolites starting from silicalite to ITQ-2 mainly by GCMC simulation methods, MD, force field, and lattice models. According to the lattice model of the adsorption of benzene in silicalite, there are various adsorption sites for benzene and a transformation of silicalite structure from *ortho-* to *para-* due to the adsorption of the benzene has been predicted [73]. A comparison was made between all silica Y and Na–Y zeolites using GCMC calculations and it has been suggested that the amount benzene adsorbed is larger in Na–Y compared to the all-silica zeolite because of the interaction between the cation and the zeolite [74]. Because computer simulation of benzene in HY zeolites is quite complicated in comparison to siliceous or even cationic faujasite, force field–based determination of the adsorption site of benzene was conducted by Jousse et al. [75]. A variety of molecular modeling techniques like molecular docking, equilibrium, nonequilibrium, molecular dynamics, and Monte Carlo umbrella sampling were applied. These methods revealed multiple adsorption sites and a creeping motion of the adsorbate molecule both for intercage and intracage diffusion. Moreover, nonequilibrium molecular dynamics simulation showed that the majority of the molecules were relaxed after the final stage of adsorption. The interaction between zeolite and benzene is modeled by simple 6-12-1 potential between each guest–zeolite atom pair:

$$U = \sum_{iJ} \left(-\frac{A_{iJ}}{r_{iJ}^6} + \frac{B_{iJ}}{r_{iJ}^{12}} + \frac{q_i q_J}{r_{iJ}} \right)$$

where i and J refer to the guest and zeolite atoms, respectively [76]. In the case of large-pore zeolites like ITQ-2, the adsorption and diffusion of benzene can be modeled by the combination of GCMC and molecular dynamics studies. A strong interaction of benzene molecule at S2 S3 site (represents the position near the 10 MR to 6 MR in supercages, while S3 lies near the centre of the 12 MR) as well as mobility of benzene mainly predicted in twelve-membered ring supercages. GCMC methods have also been applied to understand the mechanism behind the adsorption of other aromatics like *p*-xylene, *m*-xylene, or a mixture of aromatics [77], and these methods can also be extended to multicomponent adsorption. So, from the above discussion it seems that the adsorption of aromatics in zeolites is a complex system irrespective of the zeolite structure and number of simulation techniques necessary to get a complete picture of the adsorption.

7.4.3 ADSORPTION OF OTHERS, MAINLY CO$_2$ AND H$_2$

The increasing concentration of CO_2 in the atmosphere due to the combustion of the fossil fuel has led to the concern of global warming. Most common methods of CO_2 capturing are gas adsorption and are commonly evaluated by determining the adsorption isotherm via gravimetric or volumetric methods, Henry's constant, and enthalpies of adsorption [78]. Zeolites are most suitable due to their properties such as pore size, crystalline structure, and chemical composition influencing the adsorption performance. In addition, zeolites have a strong affinity toward carbon dioxide, it is stable in a vacuum, and it can be used as a most suitable material for PSA (pressure swing adsorption)-type applications. FTIR spectra show that the adsorption of carbon dioxide on the NaX, KX, and LiX is the chemisorption. This is confirmed by the position of the band, which lies in the region of 1,250 to 1,750 cm^{-1} and corresponds to the bond between carbon atom and the oxygen atom of the zeolite skeleton. On the contrary, physisorption of CO_2 on NaA zeolites has been reported at various temperatures between 178 and 423 K [79]. From the theoretical point of view, GCMC combined with molecular dynamics is used to study the adsorption of CO_2 inside the pores of the ion-exchanged X-type zeolite, like Na-4A, and Na-ZSM-5 system using interatomic potential by parameter fitting with the experimental data [80].

Another important procdess to study is the adsorption is H$_2$. Widespread use of the H$_2$ as energy carrier demands a practical need to design a material with enhanced capacity of H$_2$ storage and release at ambient temperature and moderate pressure. Regarding the H$_2$ storage material, research generally revolves around mesoporous carbons [81] and metal-organic frameworks [82]. However, the disadvantages of those materials are that the solid–gas interaction energy is too small and does not allow much gas H$_2$ to retain at ambient temperature. Zeolites are not considered good candidates for H$_2$ storage or delivery because of their high-density aluminosilicate structures. But it is well known that ion-exchanged zeolites are the ideal materials for H$_2$ bonding and have the potential to be a model to portray the insight and could be helpful for designing other prospective materials. The adsorption of molecular hydrogen on model zeolites has been simulated employing GCMC procedures. The hydrogen adsorption can be affected mainly by the available volume and surface area per gram of zeolite at the same temperature and pressure. Increase of temperature results in the decrease of sorption intensity and capacity. The adsorption capacity correlates well with the pressure with high linearity at room temperature. Organic zeolites with larger available volume show larger adsorption capacity [83]. Barbosa et al. [84] studied the adsorption and dissociation of H$_2$ on Zn-exchanged zeolites theoretically by density functional theory. The Zn(II) cation is most exposed to probe molecules when situated on 4T ring of zeolites and H$_2$ is prefered to adsorb physically on that site and the dissociative process requires an activation energy of 50 kJ/mol [84].

The adsorption of nitrogen was also studied over various zeolites such as NaA- and CaA-type zeolites, where the selectivity of nitrogen results from the electrostatic interaction between the nitrogen molecule, which possesses a quadrupole moment, and the extraframework cations. The pore size of zeolites in combination with the location, size, and charge of an extraframework cation determine a zeolite's effectiveness as an adsorbent [85]. During the elution chromatographic study, a steep increase in the heat of adsorption after a particular level of calcium exchange in zeolite NaA was observed experimentally [86]. However, a linear increase in the heat of adsorption with increasing calcium content is expected on the basis of the electrostatic interaction of the N_2 molecule with higher charge density Ca^{+2} ions. To understand this anomalous behavior of N_2 adsorption in NaCa-containing zeolite A, a detailed adsorption study employing the volumetric adsorption method and GCMC simulations was performed by Nicholson & Fuchs et al. [87]. GCMC simulations are suitable for establishing a correlation between the microscopic behavior of zeolites and an adsorbate system with macroscopic properties such as the adsorption isotherm and the heat of adsorption that are measured experimentally [88]. The absolute isotherms were then computed using a GCMC algorithm via the sorption module in the Cerius^2 software, which allows displacements, creations, and destructions of adsorbate species. All of these simulations were performed with fixed pressures at 303 K using one unit cell of each model with the typical number of Monte Carlo steps ranging from 4 to 5 million. The evolution of the total energy over the Monte Carlo steps was plotted in order to monitor the equilibration conditions. The zeolite structure was assumed to be rigid during the sorption process, and the extraframework cations were maintained fixed in their initial optimized positions. The Ewald summation method [89] was used to calculate the electrostatic interactions and the short-range interactions, with a cutoff distance of 12 Å. The studied method also allows insertion of molecules throughout the zeolite framework regardless of the physical diffusion pathways. Sorbate molecules such as N_2, O_2, and Ar, because of their larger size, cannot be adsorbed into the sodalite cages; therefore, dummy atoms of zero mass and charge were placed at the center of the sodalite cages to prevent the introduction of these molecules inside these cages. The heat of adsorption for N_2, O_2, and Ar at various adsorbate loadings in NaCaA were calculated using the fixed-loading Monte Carlo simulation method. Figure 7.11 shows both experimental and simulated adsorption isotherms of nitrogen, oxygen, and argon in zeolite A at 303 K, exchanged with Ca^{2+} cations at different exchange levels. The sample nomenclature is depicted the weight percentage of Ca exchanged in the sample. It is clear from the results that the simulation of N_2, O_2, and Ar isotherms predicts experimental values very well. The slightly higher values for the simulated adsorption isotherm could be due to the loss of crystalline structure during cation exchange, inaccurate parameter selection for the simulation studies, or the cation locations used for the simulations that may not exactly match the exact position inside the adsorbent samples used for the adsorption experiment. However, up to 35% exchange of Ca ions in zeolite A was obtained from experiments. However, simulated adsorption isotherms of oxygen show higher values than experimental values due to the selection of the same LJ parameters for O_2 in both low- and high-Ca-content zeolite A. Furthermore, the presence of impurity of other phase in the experimental sample and the loss of crystallinity during the cation exchange process may also contribute to the slight difference between the simulated and experimental adsorption isotherms. Simulations of the adsorption of N_2, O_2, and Ar in zeolite A having different calcium ion exchange levels were carried out with the framework structure given in the crystallographic data [90] with Ca^{2+} ions located in the extraframework positions using the cation locator module of Cerius^2 software (Cerius^2, v. 4.2; Accelrys, Inc., San Diego). As the percentage of Ca^{2+} was increased,

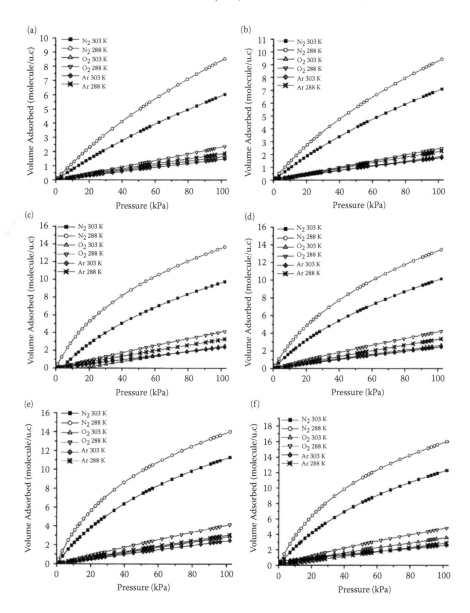

FIGURE 7.11 Experimental and simulated adsorption isotherm of N_2, O_2, and Ar in (a) NaA, (b) NaCaA25, (c) NaCaA35, (d) NaCaA60, (e) NaCaA75, (f) NaCaA90, and (g) NaCaA97 at 303 K. [86] With permission.

their occupation of site I was on par with the crystallographic data. All the calculations were performed with Cvff_aug force field. Both experimental and simulation studies showed that there was a sudden increase in the heat of adsorption of N_2 after a particular exchange level of Ca^{2+}, as can be seen from Figure 7.12. All of the main cavities in CaA were accessible to these adsorbate molecules because there are no

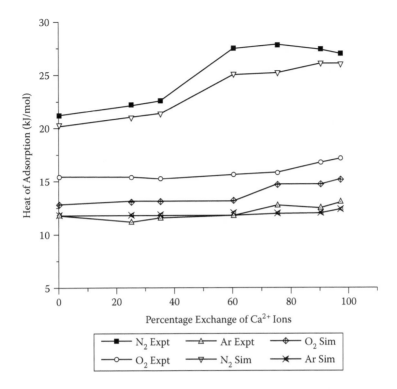

FIGURE 7.12 Heat of adsorption of N_2, O_2, and Ar in zeolite A with different percentages of Ca^{2+} exchange levels. [86] With permission.

blocking cations at the entrance of 8 MR. As the percentage of calcium exchange increases, the Na^+ ions at the entrance of the 8 MR decrease, leading to the rapid accessibility of N_2 into the main cavities of higher exchanged Ca^{2+} zeolite A. At lower exchange levels, Ca^{2+} ions are located very close to the centroid of the 6 MR, but at higher exchange levels, they are shifted more into the main cavity and hence are easily accessible for the adsorption as shown in Figure 7.13. The snapshots of the simulation of N_2 adsorption at 101 kPa pressure and 303 K temperature in zeolite A

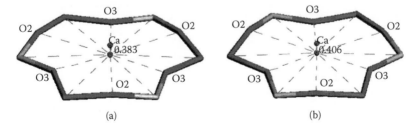

FIGURE 7.13 (For color version see accompanying CD) Molecular graphics showing the distance between an extraframework Ca^{2+} ion and the centroid of six MR in (a) NaCaA35 and (b) NaCaA60 (green = centroid, yellow = Si, pink = Al, red = O, and cyan = Ca^{2+}). [86] With permission.

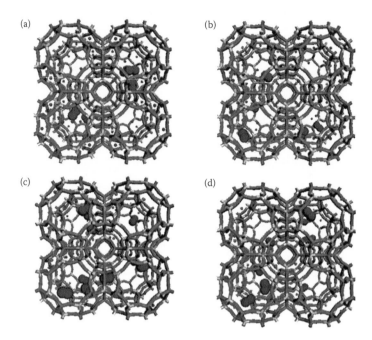

FIGURE 7.14 (For color version see accompanying CD) Molecular graphics snapshot of nitrogen adsorption at 101.3 kPa and 303 K in (a) NaA, (b) NaCaA35, (c) NaCaA60, and (d) NaCaA97 (yellow = Si, pink = Al, red = O, cyan = Ca^{2+}, violet = Na^+, and blue = N_2 molecule). [86] With permission.

with different Ca^{2+} exchange levels are shown in Figure 7.14. In the case of NaA, N_2 molecules are sitting close to the Na^+ ions located at sites I–III.

Therefore, one can find N_2 molecules well inside the main cavity of NaA, but in the case of Ca-exchanged zeolite A, the N_2 molecules are inside the supercage of the zeolite close to the calcium cations located at site I at the center of the six-membered ring. This is one of the best examples to show the capability of simulation in combination with the experiment to rationalize the process and usage of a particular nanoporous material.

7.5 CHEMISORPTION/PHYSISORPTION

The study of adsorption of different molecules in the porous materials has significant importance in various fields of applications. Basically there are two types of adsorption: (1) physisorption, where a molecule is adsorbed without undergoing a significant change in the electronic structure, it is a weak bonding based on van der Waals forces, which can be easily broken, and this process is nonselective; and (2) chemisorption, where the electronic structure of the molecules is significantly changed, a strong bonding type, which leads to longer equilibrium time and selective. It has been observed that the energies for chemisorption (20–40 kcal/mol) are higher than physisorption (0.1–1.0 kcal/mol). Among the variety of intrazeolite phenomena, physisorption and chemisorption are the first elementary steps in most of the acid-catalyzed

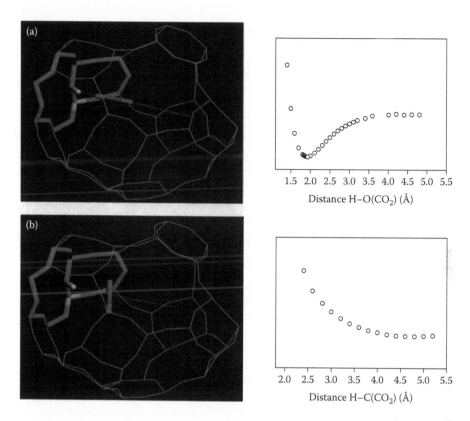

FIGURE 7.15 (For color version see accompanying CD) Schematic representation of the CO_2/H-zeolite geometry for the calculation of the potential energy curve corresponding to the interactions: (a) H–O(CO_2) and (b) H–C(CO_2). On the right is a schematic representation of their respective potential energy curves obtained by our hybrid calculation. [104] With permission.

hydrocarbon conversion reactions and have been investigated by several experimental [91] and theoretical [92] methods. Chemical heterogeneity of zeolites (arising from the substitution of Si by Al) and surface curvature give rise to chemisorption and physisorption sites, respectively. The adsorption of argon and nitrogen in silicalite-1 and xenon in Na–Y faujasite are examples of physisorption in zeolites and have been thoroughly studied experimentally [93] and theoretically [94]. Hydrocarbon conversion is an important reaction studied over zeolites. Theoretical findings predicted that the physisorption of hydrocarbon in zeolites proceeds through the interaction between hydrocarbon atoms and the highly polarizable O-site of the zeolite wall arising from the dispersive van der Waals force. Benco et al. [94] studied a series of hydrocarbons to investigate the adsorption; for example, the conversion of hydrocarbons over gemilinite zeolites using static and molecular dynamics DFT. The result shows that the conversion of olefin proceeds through the chemisorption inside zeolites because the chemisorbed species are more stable than the physisorbed one. In the case of paraffin and olefin physisorption, they found two different types of bonding mechanisms, while their adsorption energy is same. For paraffin, the interaction is dominated by the dispersive forces, whereas for olefin the presence of a double bond leads to a weak interaction between the Brønsted acid site and π-electron density of the molecule. The situation chages in the chemisorption of alkene. It occurs via protonation and the formation of an isoalkoxy species bonded through a secondary carbon atom. The energy of the chemisorption depends on the zeolite O-site and the length of the olefin chain. As the chain length increases, the chemisorption energy decreases. Much stronger deformation of the zeolite structure can be expected from the chemisorption on the O1 and O3 sites. Rozanska et al. studied the flexibility of the zeolite framework on the chemisorption of hydrocarbons. The activation energy and the heat of reaction for the transformation of physisorbed propylene into chemisorbed alkoxy species in chabazite are strongly affected due to the gradual relaxation of the environment [95]. The extended DFT parameterized force field has been combined with B3LYP and MP2 quantum mechanical methods to simulate the physisorption and chemisorption of a series of alkene over all-silica FAU and aluminum-containing H-FAU zeolite. It suggests more stable chemisorption complex than physisorption π-complex and a linear increase of both physisorption and chemisorption energy of 8.7 kJ/mol per carbon atom is derived. In addition, the nature of the zeolite plays a role in stabilizing the physisorbed alkene, which follows the order H-ZSM-5 > H-MOR > H-BEA > H-FAU [96]. DFT with gradient corrections to the exchange correlation functional was applied to study the adsorption of water in mordenite. It is observed that in neutral water molecules strongly physisorbed through two different types of hydrogen bond, the stronger one is between acid sites and water oxygen and the weaker one between is the hydrogen of water and oxygen of framework. Moreover, the *ab initio* molecular dynamics calculation including the chemisorbed species indicates the formation of unstable hydroxonium ion for the low coverage of water molecule per unit cell [97]. Similarly, a comparison between the simulated isotherm of water confined in the pores of Na–Y zeolite is in good agreement with experimental. A simple and transferable force field can reproduce the different aspects of the physisorbed water [97].

Nonlocal DFT has been used to study the capillary condensation of vapor inside the cylindrical channel of mesoporous materials. The results are comparable with the thermodynamic approach based on the Kelvin-Cohan and Derjaguin-Broekhoff-de Boer equations, but the results differ significantly as the pore size becomes smaller than ≈4 nm. It is observed that the adsorption isotherm predicted from the NLDFT calculation matches well with the adsorption of nitrogen (77 K) and argon (87 K) when the pore size is larger than 5 nm. Thus, for smaller pore sizes the desorption data could be used to determine the pore size distribution analysis as the adsorption is really hampered due to the lack of theoretical description of the metastable adsorption states in pores smaller than 5 nm [98]. GCMC molecular simulation on nanoporous silica shed some light on the confinement, pore morphology, and surface texture of mesoporous materials. Confinement phenomena are restricted to the pore size and are not observed for pores larger than 10 nm. Similarly, the amount of adsorption is greater in the rough pore surface than the smoother one [99].

Industrial process demands material designing, and simulation can be the best tool to design new material cost effectively. As we know, carbon dioxide, typically present in natural gas streams, drastically reduces its energy content of the gas [100] and is extremely corrosive in the presence of water, which is detrimental for the life of natural gas pipelines. Hence, removing the carbon dioxide from natural gas appears to be an important energy-saving parameter. The current technologies applied on an industrial level are mainly based on the absorption of CO_2 in amine or hot potassium carbonate aqueous solutions [101], which does not represent a convenient solution because of high cost and environmental hazards. The SAPO materials are synthesized by introducing silicon atoms into electrically neutral $AlPO_4$ frameworks [102] and are characterized by the presence of Brønsted acid sites that prefer specific adsorption and catalytic properties to these materials [103]. A recent work in search for a matrix that can best adsorb CO_2 figured out that the silicon content governs the concentration of Brønsted acid sites affect the thermodynamics properties and as well the architecture of the SAPO material, which consequently controls the adsorption process of CO_2 [104] on the SAPO matrix. The procedure is first to elucidate the structure and its interaction with adsorbate; the formulation is based on fitting and testing potential parameter followed by GCMC simulation to compare the adsorption isotherm. Once the adsorbate position was known, then the structures were optimized for two types of adsorbate. The final structures were then further studied to monitor to resolve the stability and the architectural issue with various loadings of Si. in the matrix These behaviors were again compared with the obtained pure aluminophosphate, $AlPO_4$-18, presenting an analogous structure of the desired matrix, to estimate the influence of the Brønsted acid sites on the adsorption process. The crystal structures of both materials were modeled as follows: the $AlPO_4$-18 structure was built from Rietvelt refined crystallographic data followed by a geometry optimization using an energy minimization technique implemented in GULP. For SAPO STA-7 material, the structure of the framework was extracted from previous X-ray diffraction data as obtained from the literature [105]. A DFT calculation was used with the PW91 functional and the double numerical basis set with polarization function, implemented in the $DMol^3$ program ($DMol^3$, v. 4.0, Accelrys, Inc., San Diego). During the optimization, the CO_2 molecule, the hydroxyl group, and both of the adjacent rings (six- and four-membered) were allowed to relax, and the rest of the cluster was kept fixed. This optimized arrangement provides a suitable starting geometry to generate a potential energy curve, by calculating a series of single-point energies. In this way, we translated the CO_2 molecule on the straight line defined by the position of the proton and those of the nearest oxygen atom of the CO_2 molecule (Figure 7.15a), within limits of 1.0–5.0 Å, using an increment of 0.1 Å. for each geometry created, a single-point energy calculation was performed. A similar strategy was adapted to derive the energy profile for the interaction $H–C(CO_2)$, starting from a configuration where the carbon dioxide molecule is perpendicularly orientated to the hydroxyl group (Figure 7.4b). Furthermore, by contrasting the experimental adsorption enthalpy profile with the simulated ones obtained for different silicon atom arrangements in the SAPO material, the authors [104] were able to discriminate the contribution of each type of silicon arrangement possibly present in the sample with respect to the global adsorption behavior of the material. The location of CO_2 as observed by the GCMC calculation is typically in the vicinity of a four-membered or an eight-membered ring (Figure 7.16a, Figure 7.16b). It was found that the CO_2 molecules are probably situated in the center of the cavities; that is, more or less homogeneously distributed within the cages, with characteristic distances separating the oxygen of the CO_2 molecule and the oxygen of

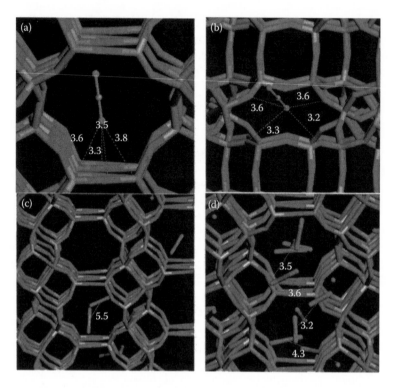

FIGURE 7.16 (For color version see accompanying CD) Representation of the preferential adsorption sites for CO_2 in $AlPO_4$-18: (a) four-ring and (b) eight-ring window. Typical arrangement of the CO_2 molecules in $AlPO_4$-18 at (c) low loading (0.5 bar) and (d) high loading (30 bar). Corresponding distances are reported in angstroms. [104] With permission.

the framework ranging from 3.2 to 3.8 Å. Calculations also showed that the average distance between the oxygen of the framework and that of the carbon dioxide remains almost unchanged. By contrast, the distance between the carbon dioxide molecules, $d(C–C)$, becomes significantly shorter. The results were compared and are shown as snapshots in Figure 7.16c and Figure 7.16d, which show the arrangements of CO_2 molecules at low and high loadings, respectively. Different types of material have different silica loadings and the architectures vary as well. Once the material is designed in terms of the architecture, loading, and silica stability, the next step is to check the influence of the various parameters, which will affect the CO_2 adsorption. Based on the physisorption and chemisorption study, it is possible to design new materials depending on the situation.

GCMC simulation can be performed to study the adsorption of gases in metal-organic frameworks (MOFs). Because the GCMC simulation will help to plot the adsorption isotherm, one can therefore predict the pore architecture and the amount of gas loading within the material domain. It is possible to calculate the binding energy between adsorbent and the adsorbate to predict the nature of adsorption (physisorption or chemisorption). Calculation of intermolecular distance is also another measure to check the nature of adsorption. Adsorption energy can be utilized to know the active site present in the MOF type materials. Now, surely this will be a physisorption and one can then further extend the model to do some quantitative calculations to determine the binding energy for the proposition of chemisorption. Another possibility is to predict the adsorbate–adsorbent distance and the most active site present in the MOF through GCMC simulation. With the adsorption energy spectrum it is also possible to predict the influence of pressure on adsorption. Recent work by Liu et al. [106] has shown that anions play a significant role in the methane adsorption behavior in heterometallic MOFs and are the preferential adsorption sites at low pressure. However, this kind of priority is not pronounced at high pressure. Methane was adsorbed closer to the linking units containing Ag atoms than to the bridging structures containing Fe atoms. Based on the simulations, this work indicates that an appropriate selection of anions at a given pressure may control the methane adsorption sites in heterometallic MOFs, which is useful for understanding the function of these novel porous materials in practical applications.

MOFs are becoming very popular nowadays for adsorption and storage usage to find a better application in comparison with zeolites. In the nanoporous domain, MOFs have to compete with existing materials, including carbon nanotubes. The best way to compare the adsorption reaction for a separation usage is to look at the diffusion rate. Simulation is preferred to describe the diffusion within the porous structure of MOFs. As mentioned before, it is the most advanced technique to gain insight into the molecular feature occurs in the backstage and difficult to predict from the experimental results only. To run the diffusivity calculation, it is essential to have a clear idea of the force field; otherwise, it would be complicated to see what traps the molecule and what releases inside a nanopore any of the three materials mentioned. It is therefore demanding to look into the mechanism of diffusion. The mechanism of diffusion calculations were performed in order to assess similarities and/or differences with the behavior observed in carbon nanotubes, with a comparable pore structure and where transport is known to occur very rapidly. For instance, the calculation of gas diffusion in two-dimensional covalent organic frameworks shows that it is one order of magnitude more rapid compared to MOFs or zeolites but still not as fast as in carbon nanotubes. The adsorption and diffusion characteristics of these materials are related to the peculiar structure of the solid–fluid potential energy surface [107]. The diffusion behavior has been rationalized through the diffusivity calculation and diffusion plot to compare the energy and the diffusion activity of the material towards the adsorbate, respectively.

Metal-organic frameworks, with their unique structures, are thought to be a promising hydrogen storage material. The thermal stability of this material is low at temperatures above 673 K. Among these metal-organic frameworks, MOF-5 is prescribed as the most stable MOF [107]. It is observed that MOF-5 is stable under 673 K in a nitrogen environment and its stability decreases to a lower temperature range in the presence of air to about 573 K. It has a surface area of 2,900 m^2 g^{-1} with pore volume of 0.36 cm^3 g^{-1}. Experimentally it does not show any hysteresis during physisorption and hence it is thought to be a good adsorbent for hydrogen storage applications. But the problem is that not much is known about the structure or the structure property correlation has been standardized. To rationalize the structure property relationship the adsorption mechanism of MOFs has been studied by

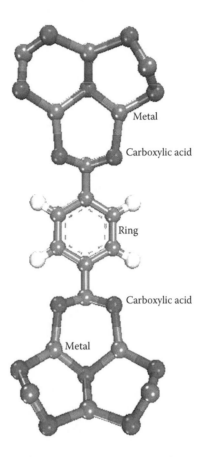

FIGURE 7.17   (For color version see accompanying CD) MOF model to represent three different regions for hydrogen adsorption.

quantum chemical calculations. In MOFs there are a couple of major areas to govern the adsorption process like any other material. There are three distinct regions of interest: (1) the metal region located around the metal center, (2) the site around the carboxylic acid, and (3) the sites around the ring moiety. An example model is shown in Figure 7.17. The regions are labeled to show that position really affects the adsorption situations. The adsorption energies for all the structures with different possible orientations were calculated using the DMol^3 program of Accelrys, details of which are given elsewhere [95]. The adsorption energy scenario is prescribed for a whole range of MOF materials along with IRMOFs (isoreticular MOFs) [109]. One can therefore determine the best binding site for hydrogen adsorption and be able to predict which is the detrimental factor in enhancing the adsorption. Once the regular structure adsorption analysis is performed, one can see that there are two key issues with adsorption: one is to change the metal cation, but it is not as simple as a metal substitution in zeolites, because it will involve the change in the ligand network and hence the structure. It will be much simpler to try the other option, which is to change the ligand or organic linker; once the ligand is changed, the ligand surface increases

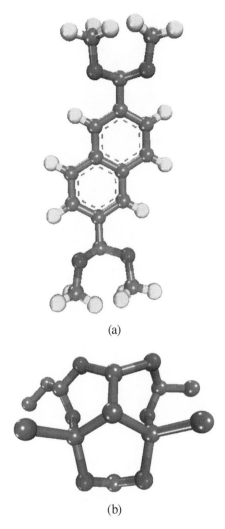

(a)

(b)

FIGURE 7.18   (For color version see accompanying CD) (a) An organic linker model example with a functional group where the attachment may occur at the terminals; (b) an organic linker model example with a functional group where the attachment may occur at the middle, which looks similar to a cluster.

and hence the surface area. So it is plausible to study different MOFs with varying ligands and varying surface areas to find an *a priori* rule for which MOFs can be a better adsorber for hydrogen. If the binding site increases, the adsorption energy changes and pictures are shown in Figure 7.18 to compare different linker sites. Two different linkers are shown in Figure 7.18a and Figure 7.18b, respectively, that can be adopted in the MOF domain. Figure 7.18a represents an example of organic linker model with a functional group where the attachment may occur at the terminals, and Figure 7.18b depicts an organic linker model with a functional group where the attachment may occur at the middle and looks similar to a cluster. With that

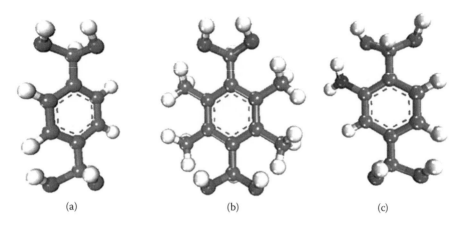

FIGURE 7.19   (For color version see accompanying CD) The representative organic linker structures as present in (a) IRMOF-1, (b) IRMOF-3, and (c) IRMOF-18, respectively.

background it will be a good idea to probe the functionalization of MOFs [110]. One can compare the known structure with variation in the functional group and then with that confidence gained can propose a new structure. Three structures of different fictionalization were tested: IRMOF-1, IRMOF-3, and IRMOF-18. After synthesis, followed by structural analysis or characterization by experiments, one can look into the local geometry by DFT calculations to know the local bond and angles to see the structural correlations. The three structures are labeled and are shown in Figure 7.19 as (a) IRMOF-1, (b) IRMOF-3, and (c) IRMOF-18, respectively. It can be seen that there is some commonness in the structure and some differences in terms of the functionalization. This matters a lot in MOF structures. Once the geometry optimization is performed, one then can probe the surface area of these molecules and the solvent accessible surface area as if a reaction occurs or even adsorption the surface area surely will matter. One may expect from the local geometry that IRMOF-18 probably has the larger surface area but it does not; IRMOF-3 has the largest surface area. This can be further validated by the electrostatic potential map to predict the electronics structure in terms of activity of the molecule. It is observed from the electrostatic potential map in comparison to solvent accessible surface area that the IRMOF-3 surface area is larger and that can be experimentally verified. To compare the adsorption behavior of hydrogen, a GCMC simulation was performed using the Sorption module of Accelrys, the details of which are described elsewhere [57]. The adsorption isotherm results in comparison to the experimental results confirm the fact that IRMOF-3 has the largest loading. One can expand this work further to interpenetrating MOFs for a more complicated system where the systems have a combination or mixtures of MOFs and the pore architecture is hierarchical [111]. MOF as a matrix is very challenging and it has a bright future with many more applications in optical properties and electronics where zeolitic materials failed, but MOFs are more complicated matrixes to study and modeling will surely be a helpful tool to rationalize structure property correlations.

7.6 SURFACE CHARACTERIZATION — A COMPARISON WITH EXPERIMENTAL TOOLS

Porous materials are widely used in several scientific and technological fields. An actual characterization of those materials is essential for optimizing their applications. For instance, measurement of the surface area, porosity, and pore size distribution is crucial to designing the catalyst depending on the nature of the reaction. In addition, calculation of total surface area is important because it determines the accessibility to the active site and consequently is related to the catalytic activity. The shape and size of the pore have direct relation with the transport phenomenon and dictate the product distribution of solid catalyzed reaction. There are several experimental techniques to correlate pore architecture and structural characterization such as X-ray and neutron scattering, gas adsorption, mercury porosimetry, NMR, electron microscopy, calorimetry (thermoporometry), etc., but each method has applicability that is limited to a definite range of pore size. Gas adsorption is one of the experimental methods used to characterize porous materials that is convenient and can be used in the pore size range of 0.35 to 100 nm, from micropores to macropores. However, experimental observation of adsorption isotherms is time consuming and it is extremely difficult to gain molecular-level understanding of the behavior of the molecules inside the zeolite. Molecular simulation offers an attractive idea about macroscopic adsorption thermodynamics and the visualization of microscopic sitting of the molecules. Thus, a combination of experimental and theoretical methods will provide valuable insight into the adsorption desorption phenomena and the characterization of the surface. A combination of experimental and theoretical (GCMC) work on adsorption of a liquid-phase mixture of linear alkane in silicalite was carried out by Chempath et al. [112] Excellent agreement between the experimental results and simulation was observed and it was predicted that if the force field parameters were optimized for single-component adsorption, then it would work well with the multicomponent system. The shorter alkane pushed to the zigzag channel, whereas the longer alkane occupied the straight channel [113]. A comparison has been made on the adsorption of methane, ethane, and argon in Na-mordenite at ambient and cryogenic conditions. At ambient conditions, experiment and theory match well but differ considerably in the case of argon adsorption at cryogenic conditions. Hence, inclusion of cation and framework aluminum with a realistic charge distribution is required to match with the experimental data [113]. Accurate reproduction of experimentally obtained Henry coefficient, heats of adsorption, and adsorption isotherms is possible in the adsorption of cycloalkanes and a mixture of cycloalkanes and n-hexane in MFI-type siliceous zeolites via Monte Carlo simulation [114]. Due to the crystalline structure and their wide range of applications, the adsorption of different molecules in zeolites, the experimental results are not only compared with theoretical calculations but a combination of experimental methods and theory has also been used to derive a complete understanding of the adsorption process. Calorimetric data and MC docking calculations provide a complete scenario of the interaction between aromatics and siliceous zeolites. From the calorimetric measurement, the heat of adsorption is obtained as 55 kJ/mol at density lower than 22 molecules/unit cell, whereas MC calculations predicted three different binding sites near to four-ring (58.4 kJ/mol), six-ring (50.0 kJ/mol), and twelve-ring windows (43.7 kJ/mol). In addition, the motion of the benzene molecule can be described depending on the temperature; at low temperature confined to a single supercage but at high temperature through twelve-ring windows adjacent to the supercage [115]. Monte Carlo along with calorimetric measurements also applied to the adsorption of chlorocarbons, chloroform, and trichloroethylene in a series of fauzasite-type zeolite, Na–X, Na–Y, and siliceous form, where experiment suggests a correlation between the heat of adsorption and polarity of zeolite host (siliceous < NaY < NaX). Simulation construct an important bridge between the structural and thermodynamic features, which are impenetrable by conventional diffraction methods [116]. However, neutron diffraction and computer simulation together applied to the adsorption of $CFCl_3$ in Na–Y zeolite. The result of neutron diffraction shows the importance of cation migration due to the adsorbate and simulation makes an exact picture of the position of the cation [117]. The location of 4-4' bi-pyridine in the straight channel of the silicalite and the extraframework cation can easily be confirmed by different spectroscopic techniques such as diffuse reflectance UV and Raman scattering, which is not far behind to the surface characterization of zeolite material and simulation [118].

In contrast to the crystalline zeolite materials, mesoporous materials are amorphous and problematic considering their characterization. However, the large surface area and pore size make them an attractive model to study the surface phenomena like adsorption and diffusion. Capillary condensation hysteresis in nanopores has been studied by Monte Carlo simulations and nonlocal density functional theory. Comparing the theoretical results with the experimental data on low-temperature sorption of nitrogen and argon in cylindrical channels of mesoporous siliceous molecular sieves of MCM-41 type, four qualitatively different sorption regimes have been predicted depending on the temperature and pore size. A quantitative agreement is found between the modeling results and the experimental hysteresis loops formed by the adsorption–desorption isotherms [119–120].

Thus, to achieve the ability to fabricate the complex structure of nanoporous materials tailored for specialized applications, it is necessary to employ a combination of experiment and theoretical calculations that complement each other.

REFERENCES

1. Cerius² modeling software, available at: http://www.msi.com/cerius2, Accelrys Inc., San Diego, CA (2001).
2. E. C. Moloy, L. P. Davila, J. F. Shackelford, A. Navrotsky, *Microporous and Mesoporous Materials*, 54 (2002) 1.
3. A. B. Mukhopadhyay, C. Oligschleger, M. Dolg, *Phys. Rev. B*, 67 (2003) 014106.
4. H. Jobic, *Phys. Chem. Chem. Phys.*, 1 (1999) 525; S. Yasonath, P. Demontis, M. L. Klein, *J. Phys. Chem.*, 95 (1991) 5881.
5. J. A. Martens, W. Souverijns, W. Verrelst, R. Parton, G. F. Fermont, P. A. Jacobs, *Angew. Chem. Int. Ed.*, 34 (1995) 2528.
6. A. Corma, V. Fornes, M. S. Galletero, H. Garcia, C. J. Gomez-Garcia, *Phys. Chem. Chem. Phys.*, 3 (2001) 1218.
7. L. Whitmore, B. Slater, C. R. A. Catlow, *Phys. Chem. Chem. Phys.*, 2 (2000) 5354.

8. B. Slater, J. D. Titiloye, F. M. Higgins, S. C. Parker, *Curr. Opin. Solid State Mater. Sci.*, 5 (2001) 417.
9. W. M. Meier, D. H. Bearlocher, *Atlas of Zeolite Structure Type*, Elsevier, London (1995).
10. S. Brunauer, P. H. Emmett, E. Teller, *J. Am. Chem. Soc.*, 60 (1938) 309.
11. S. J. Gregg, K. S. W. Sing, *Adsorption, Surface Area and Porosity*, Academic Press, London (1982).
12. C. G. Sonwane, C. W. Jones, P. J. Ludovice, *J. Phys. Chem. B*, 109 (2005) 23395.
13. B. P. Feutson, J. B. Higgins, *J. Phys. Chem.*, 16 (1994) 4459.
14. S. Bhattacharya, B. Cosane, F. R. Hung, K. E. Gubbins, *Langmuir*, 25, (2009) 10.
15. (a) A. F. Wells, *Three-Dimensional Nets and Polyhedra*, Wiley, New York (1977); (b) J. V. Smith, *Chem. Rev.*, 88 (1988) 149.
16. M. Trombetta, G. Busca, M. Lenarda, M. Storaro, M. Pavan, *Appl. Catal. Gen.*, 182 (1999) 225.
17. M. Trombetta, G. Busca, *J. Catal.*, 187 (1999) 521.
18. A. Chatterjee, T. Iwasaki, T. Ebina, H. Tsuruya, T. Kanougi, Y. Oumi, M. Kubo, A. Miyamoto, *Appl. Surf. Sci.*, 130 (1998) 555.
19. L. Whitmore, B. Slater, C. R. A. Catlow, *Phys. Chem. Chem. Phys.*, 2 (2000) 5354.
20. C. R. A. Catlow, J. D. Gale, D. H. Gay, D. W. Lewis, Computer Modeling of Sorption in Zeolite in *Access in Nanoporous Materials*, T. J. Pinnavaia, M. F. Thorpe (eds.), Plenum Press, New York (1995).
21. O. Terasaki, *J. Electron Microsc.*, 43 (1994) 337.
22. D. H. Gay, A. L. Rohl, *J. Chem. Soc. Faraday. Trans.*, 91 (1995) 925.
23. D. M. Heyes, M. Barber, J. H. R. Clarke, *J. Chem. Soc. Faraday Trans.*, 73 (1977) 1485.
24. M. J. Sanders, M. Leslie, C. R. A. Catlow, *J. Chem. Soc. Chem. Comm.*, (1984) 1271.
25. K. P. Schroder, J. Sauer, M. Leslie, C. R. A. Catlow, J. M. Thomas, *Chem. Phys. Lett.*, 188 (1992) 320.
26. N. J. Henson, A. K. Cheetham, M. Stockenhuber, J. A. Lercher, *J. Chem. Soc. Faraday Trans.*, 94 (1998) 3759.
27. V. B. Fenelonov, A. Yu. Derevyankin, S. D. Kirik, L. A. Solovyov, A. N. Shmakov, J. Bonardet, A. Gedeon, V. N. Rommanikov, *Microporous and Mesoporous Materials*, 44 (2001) 33.
28. B. Kuchta, P. Llewellyn, R. Denoyel, L. Firlej, *Colloid. Surf.*, 241 (2004)137.
29. D. Cao, Z. Shen, J. Chen, X. Zhang, *Microporous and Mesoporous Materials*, 67 (2004) 159.
30. M. Sharma, S. Yashonath, *J. Phys. Chem. B*, 110 (2006) 17207.
31. S. Yashonath, P. K. Ghorai, *J. Phys. Chem. B*, 112 (2008) 665.
32. W. Humphrey, A. Dalke, K. Schulten, *J. Mol. Graphics*, 14 (1996) 33.
33. (a) D. M. Ruthven, M. F. P. Post, H. van Bekkum, E. M. Flaningen, P. A. Jacobs, J. C. Jansen, *Introduction to Zeolite Science and Practice*, Elsevier, Amsterdam (2001); (b) H. Jobic, *J. Mol. Catal. Chem.*, 158 (2000) 135; (c) L. Song, L. V. C. Rees, *Microporous and Mesoporous Materials*, 41 (2000) 193.
34. (a) S. M. Auerbach, *Int. Rev. Phys. Chem.*, 19 (2000) 155; (b) R. A. van Santen, X. Rozanska, A. Chakraborty, *Molecular Modeling and Theory in Chemical Engineering*, Academic Press, San Diego (2001); (c) R. Krishna, D. Paschek, *Chem. Eng. J.*, 87 (2002) 1; (d) H. Takaba, T. Suzuki, S. Nakao, *Fluid Phase Equil.*, 219 (2004) 11.
35. (a) J. Karger, *Adsorption*, 9 (2003) 29; (b) J. Karger, D. M. Ruthven, *Diffusion in Zeolites and Other Microporous Solids*, Wiley & Sons, New York (1992).
36. E. G. Deroune, *CATTECH*, 6 (2006) 11.
37. E. Kikuchi, H. Nakano, K. Shimomura, Y. Morita, *Sekiyu Gakkaishi*, 28 (1985) 210.
38. E. G. Deroune, *J. Mol. Catal. Chem.*, 134 (1998) 29.
39. Z. Blum, S. T. Hyde, B. N. Ninham, *J. Phys. Chem.*, 97 (1993) 60.

40. R. Chitra, S. Yashonath, *Indian Acad. Sci. (Chem. Sci.)*, 109 (1997) 189.

41. B. J. Cwiklik, L. Cwiklik, M. Frankowicz, *Appl. Surf. Sci.*, 252 (2005) 699.

42. M. C. Macedonia, E. J. Maginn, *AIChE J.*, 46 (2000) 2504.

43. K. S. Smirnov, F. T. Starzyk, *J. Phys. Chem. B*, 103 (1999) 8595.

44. (a) F. Marquez, H. Garcia, E. Palomares, L. Fernandez, A. Corma, *J. Am. Chem. Soc.*, 122 (2000) 6520; (b) F. Marquez, C. Zicovich-Wilson, A. Corma, E. Palomares, H. Garcia, *J. Phys. Chem. B*, 105 (2001) 9973.

45. A. Corma, H. Garcia, G. Sastre, P. M. Viruela, *J. Phys. Chem. B*, 101 (1997) 4575.

46. M. S. Galletero, A. Corma, B. Ferrer, V. Forns, H. Garcia, *J. Phys. Chem. B*, 107 (2003) 1135.

47. C. Lastoskie, K. E. Gubbins, N. Quirke, *J. Phys. Chem.*, 97 (1993) 4780, and references therein.

48. M. Kruk, M. Jaroniec, *Langmuir*, 13 (1997) 6261.

49. R. Guegan, D. Morineau, C. Alba-Simionesco, *Chem. Phys.*, 317 (2005) 236.

50. S. Pariente, P. Trens, F. Fajula, F. Di Renzo, N. Tanchoux, *Appl. Catal. Gen.*, 307 (2006) 51.

51. R. Anwander, C. Palm, G. Gerstberger, O. George, G. Engelhardt, *Chem. Comm.*, (1998) 1811.

52. C. G. Sonwane, Q. Li, *J. Chem. Soc. Chem. Comm.*, (2007) 3261.

53. T. On Do, A. Nossov, M. Springuel-Huet, C. Schneider, J. L. Bretherton, C. A. Fyfe, S. Kaliaguine, *J. Am. Chem. Soc.*, 126 (2004) 14324.

54. M. Göktug Ahunbay, *J. Chem. Phys.*, 127 (2007) 044707.

55. H. Sun, *J. Phys. Chem. B*, 102 (1998) 7338.

56. P. Dauber-Osguthorpe, V. A. Roberts, D. J. Osguthorpe, J. Wolff, M. Genest, A. T. Hagler, *Protein. Struct. Funct. Genet.*, 4 (1988) 31.

57. N. Metropolis, A. W. Rosenbluth, M. N. Rosenbluth, A. H. Teller, E. Teller, *J. Chem. Phys.*, 21 (1953) 1087.

58. D. Frenkel, B. Smit, *Understanding Molecular Simulations*, Academic, London (1996).

59. P. Dauber-Osguthorpe, V. A. Roberts, D. J. Osguthorpe, J. Wolff, M. Genest, A. T. Hagler, *Protein. Struct. Funct. Genet.*, 4, (1988) 31–47.

60. L. Verlet, *Phys. Rev.*, 159 (1967) 98.

61. S. Nosé, *Mol. Phys.*, 52 (1984) 255.

62. S. Watanabe, H. Sugiyama, M. Miyahara, *Adsorption*, 14 (2008) 165.

63. K. I. Zamaraev, *Chem. Sustain. Dev.*, 1 (1993) 133.

64. (a) J. A. Horsley, J. D. Fellmann, E. G. Derouane, C. M. Freeman, *J. Catal.*, 147 (1994) 231; (b) H. Klein, C. Kirschhock, H. Fuess, *J. Phys. Chem.*, 98 (1994) 12345.

65. (a) P. Demontis, G. B. Suffritti, A. Alberti, S. Quartieri, E. S. Fois, A. Gamba, *Gazz. Chim. Ital.*, 116 (1986) 459; (b) S. Yashonath, *Chem. Phys. Lett.*, 177 (1991) 54; (c) S. Yashonath, P. Demontis, M. L. Klein, *J. Phys. Chem.*, 95 (1991) 5881; (d) C. R. A. Catlow, C. M. Freeman, B. Vessal, S. M. Tomlinson, M. Leslie, *J. Chem. Soc. Faraday Trans.*, 87 (1991) 1947; (e) P. Demontis, G. B. Suffritti, E. S. Fois, S. Quartieri, *J. Phys. Chem.*, 96 (1992) 1482; (f) R. L. June, A. T. Bell, D. N. Theodorou, *J. Phys. Chem.*, 96 (1992) 1051.

66. (a) Q. S. Randall, T. B. Alexis, N. T. Doros, *J. Phys. Chem.*, 97 (1993) 13742; (b) P. A. Van Tassel, H. T. Davis, A. V. McCormick, *J. Chem. Phys.*, 98 (1993) 4173.

67. D. Keffer, H. T. Davis, A. V. McCormick, *Adsorption*, 2 (1996) 9.

68. (a) B. Smit, J. I. Siepmann, *Science*, 264 (1994) 1118; (b) Z. D. G. Manos, T. J. H. Vlugt, B. Smit, *AIChE J.*, 44 (2004) 1756.

69. (a) P. Ungerer, C. Beauvais, J. Delhommelle, A. Boutin, B. Rousseau, A. H. Fuchs, *J. Chem. Phys.*, 112 (2000) 5499; (b) P. Pascual, P. Ungerer, B. Tavitian, P. Pernot, A. Boutin, *Phys. Chem. Chem. Phys.*, 5 (2003) 3684.

70. L. H. Lu, Q. Wang, Y. C. Liu, *J. Phys. Chem. B*, 109 (2005) 8845.

71. J. P. Fox, V. Rooy, S. P. Bates, *Microporous and Mesoporous Materials*, 69 (2004) 9.

72. (a) N. J. Henson, A. K. Cheetham, A. Redondo, S. M. Levine, J. M. Newsam, *Zeolites Relat. Microporous Mater.*, 84 (1994) 2059; (b) Q. Snurr, A. Bell, D. Theodoru, *J. Phys. Chem.*, 98 (1994) 5111; (c) B. Grauert, K. Fiedler, H. Stach, J. Janchen, *Zeolites: Facts, Figures, Future*, (1989) Elsevier, New York; (d) C. Raksakoon, J. Limtrakul, *J. Mol. Struct.*, 631 (2003) 147.

73. Q. Snurr, A. Bell, D. Theodoru, *J. Phys. Chem.*, 97 (1993) 13742.

74. Y. Zeng, S. Ju, W. Xing, C. Chen, *Ind. Eng. Chem Res.*, 46 (2007) 242.

75. F. Jousse, S. M. Auerbach, D. P. Vercauteren, *J. Phys. Chem.. B*, 104 (2004) 2360.

76. (a) M. M. Laboy, I. Santiago, G. E. López, *Ind. Eng. Chem. Res.*, 38 (1999) 4938; (b) V. Lachet, S. Buttefey, A. Boutin, A. H. Fuchs; *Phys. Chem. Chem. Phys.*, 3 (2001) 80.

77. (a) J. S. Lee, J. H. Kim, J. T. Kim, J. K. Suh, J. M. Lee, C. H. Lee, *J. Chem. Eng. Data*, 47 (2002) 1237; (b) V. R. Choudhary, S. Mayadevi, A. Pal Singh, *J. Chem. Soc. Faraday Trans.*, 91 (1995) 2935; (c) V. R. Choudhary, S. Mayadevi, *Zeolites*, 17 (1996) 501; (d) R. M. Barrer, R. M. Gibbons, *Trans. Faraday Soc.*, 61 (1965) 948; (e) K. S. Walton, M. B. Abney, M. Douglas LeVan, *Microporous and Mesoporous Materials*, 91 (2006) 78; (f) J. A. Dunne, R. Mariwala, M. Rao, S. Sircar, J. Gorte, A. L. Myers, *Langmuir*, 12 (1996) 5896; (g) S. S. Khvoshchev, A. V. Zverev, *J. Colloid Interface Sci.*, 144 (1991) 571.

78. Y. Delaval, E. Cohen de Lara, *J. Chem. Soc. Faraday Trans.*, 77 (1981) 869.

79. (a) K. Mizukami, H. Takaba, Y. Konayashi, Y. Oumi, R. V. Belosludov, S. Takami, M. Kubo, A. Miyamoto, *J. Membr. Sci.*, 188 (2001) 21; (b) S. M. Gheno, S. Damyanova, B. A. Riguetto, C. M. P. Marques, C. A. P. Leite, J. M. C. Bueno, *J. Mol. Catal. Chem.*, 198 (2003) 263; (c) E. D. Akten, R. Siriwardane, D. S. Sholl, *Energ. Fuel.*, 17 (2003) 977; (d) A. Hirotani, K. Mizukami, R. Miura, H. Takaba, T. Miya, A. Fahmi, A. Stirling, M. Kubo, A. Miyamoto, *Appl. Surf. Sci.*, 120 (1997) 81.

80. (a) Z. Yang, Y. Xia, R. Mokaya, *J. Am. Chem. Soc.*, 129 (2007) 167; (b) M. Choi, R. Ryoo, *J. Mater. Chem.*, 17 (2007) 4204; (c) M. Armandi, B. Bonelli, C. Otero Areán, E. Garrone, *Microporous and Mesoporous Materials*, 112 (2008) 411.

81. (a) J. L. C. Roswell, O. M. Yaghi, *Angew. Chem. Int. Ed.*, 44 (2005) 4670; (b) D. J. Collins, H. C. Zhou, *J. Mater. Chem.*, 17 (2007) 3154.

82. M. Song, K. Tai, *Catal. Today*, 120 (2007) 374.

83. L. A. M. M. Barbosa, G. M. Zhidomirov, R. A. van Santen, *Catal. Lett.*, 77 (2001) 55.

84. (a) W. Breck, *Zeolites Molecular Sieves: Structure, Chemistry and Use*, Wiley-Interscience, New York, (1974); D. M. Ruthven, S. Farooq, K. S. Knaebel, *Pressure Swing Adsorption*, Wiley-VCH, New York (1994); (b) C. C. Chao, J. D. Sherman, J. T. Mullhaupt, C. M. Bolinger, U.S. patent 5,174,979 (1992); (c) C. G. Coe, J. F. Kirner, R. Pierantozzi, T. R. White, U.S. patent 5,-152,813 (1992); (d) S. A. Peter, J. Sebastian, R. V. Jasra, *Ind. Eng. Chem. Res.*, 44 (2005) 6856.

85. N. V. Choudary, R. V. Jasra, S. G. T. Bhat, *Ind. Eng. Chem. Res.*, 32 (1993) 548.

86. R. S. Pillai, S. A. Peter, R. V. Jasra, *Langmuir*, 23 (2007) 8899.

87. (a) D. Nicholson, R. Pellenq, *Adv. Colloid Interface Sci.*, 179 (1998) 76; (b) A. H. Fuchs, A. K. Cheetham, *J. Phys. Chem. B*, 105 (2001) 7375.

88. (a) J. J. Pluth, J. V. Smith, *J. Am. Chem. Soc.*, 102 (1980) 4704; (b) J. J. Pluth, J. V. Smith, *J. Am. Chem. Soc.*, 105 (1983) 1192.

89. A. J. Richards, K. Watanabe, N. Austin, M. R. Stapleton, *J. Porous Mater.*, 2 (1995) 43.

90. (a) N. W. Cant, W. K. Hall, *J. Catal.*, 25 (1972) 161; (b) H. Stach, J. Jänchen, H. Thamm, E. Stiebitz, R. A. Vetter, *Adsorption Science and Technology*, 3 (1986) 261; (c) F. Eder, M. Stockenhuber, J. A. Lercher, *J. Phys. Chem. B*, 101 (1997) 5414; (d) F. Eder, J. A. Lercher, *Zeolites*, 18 (1997) 75; (e) J. F. Denayer, W. Souverijns, P. A. Jacobs, J. A. Martens, G. V. Baron, *J. Phys. Chem. B*, 102 (1998) 4588; (f) J. F. Denayer, G. V. Baron, J. A. Martens, P. A. Jacobs, *J. Phys. Chem. B*, 102 (1998) 3077; (g) L. A. Clark, G. T. Ye, A. Gupta, L. L. Hall, R. Q. Snurr, *J. Chem. Phys.*, 111 (1999) 1209.

91. (a) C. Tuma, J. Sauer, *Phys. Chem. Chem. Phys.*, 8 (2006) 3955; (b) B. Jansang, T. Nanok, J. Limtrakul, *J. Phys. Chem. B*, 110 (2006) 12626; (c) L. Benco, T. Demuth, J. Hafner, F. Hutschka, H. Toulhoat, *J. Chem. Phys.*, 114 (2001) 6327; (d) T. Demuth X. Rozanska, L. Benco, J. Hafner, R. A. van Santen, H. Toulhoat, *J. Catal.*, 214 (2003) 68; (e) L. A. Clark, M. Sierka, J. Sauer, *J. Am. Chem. Soc.*, 125 (2003) 2136; (f) L. A. Clark, M. Sierka, J. Sauer, *J. Am. Chem. Soc.*, 126 (2004) 936; (g) S. Kasuriya, S. Namuangruk, P. Treesukol, M. Tirtowidjojo, J. Limtrakul, *J. Catal.*, 219 (2003) 320; (h) S. Namuangruk, P. Pantu, J. Limtrakul, *J. Catal.*, 225 (2004) 523; (i) M. Boronat, P. M. Viruela, A. Corma, *J. Am. Chem. Soc.*, 126 (2004) 3300; (j) S. P. Bates, W. J. M. van Well, R. A. van Santen, B. Smit, *J. Am. Chem. Soc.*, 118 (1996) 6753; (k) L. Dixit, T. S. R. Prasada Rao, *J. Chem. Inform. Comput. Sci.*, 39 (1999) 218; (l) D. Dubbeldam, S. Calero, T. J. H. Vlugt, R. Krishna, T. L. M. Maesen, B. Smit, *J. Phys. Chem. B*, 108 (2004) 12301; (m) P. Pascual, P. Ungerer, B. Tavitian, P. Pernot, A. Boutin, *Phys. Chem. Chem. Phys.*, 5 (2003) 3684; (n) P. Pascual, P. Ungerer, B. Tavitian, A. Boutin, *J. Phys. Chem. B*, 108 (2004) 393; (o) X. Rozanska, R. A. van Santen, T. Demuth, F. Hutschka, J. Hafner, *J. Phys. Chem. B*, 107 (2003) 1309; (p) X. Rozanska, L. A. M. M. Barbosa, R. A. van Santen, *J. Phys. Chem. B*, 109 (2005) 2203; (q) R. Rungsirisakun, B. Jansang, P. Pantu, J. Limtrakul, *J. Mol. Struct.*, 733 (2005) 239; (r) J. O. Titiloye, S. C. Parker, F. S. Stone, C. R. A. Catlow, *J. Phys. Chem.*, 95 (1991) 4038; (s) C. Tuma, J. Sauer, *Angew. Chem. Int. Ed.*, 44 (2005) 4769.

92. (a) U. Muller, H. Reichart, E. Robens, K. K. Unger, Y. Grillet, F. Rouquerol, J. Rouquerol, D.-F. Pan, A. Mersmann, *Fresenius' Z. Anal. Chem.*, 333(1989) 433; (b) H. Reichart, U. Muller, K. K. Unger, Y. Grillet, F. Rouquerol, J. Rouquerol, J. P. Coulomb, *Stud. Surf. Sci. Catal.*, 62 (1991) 535; (c) N. J. M. Tosi-Pellenq, Adsorption in Zeolite. Inter molecular interaction and computer Simulation Thesis, Université de Provence, Marseille (1994); (d) T. T. P. Cheung, C. M. Fu, S. Wharry, *J. Phys Chem.*, 92 (1988) 5170.

93. (a) R. J.-M. Pellenq, D. Nicholson, *Stud. Surf. Sci. Catal.*, 87 (1994) 31; (b) R. J.-M. Pellenq, A. Pellegatti, D. Nicholson, C. Minot, *J. Phys. Chem.*, 99 (1995) 10175; (c) P. Llewellyn, J. P. Coulomb, Y. Grillet, J. Patarin, H. J. Lauter, H. Reichart, J. Rouquerol, *Langmuir*, 9 (1993) 1846; (d) G. B. Woods, J. S. Rowlinson, *J. Chem. Soc. Faraday Trans.*, 85 (1989) 765.

94. B. A. DeMoor, M. F. Reyniers, M. Sierka, J. Sauer, G. B. Marin, *J. Phys. Chem. C*, 112 (2008) 11796.

95. L. Bencol, J. Hafner, F. Hutschka, H. Toulhoat, *J. Phys. Chem. B*, 107 (2003) 9756.

96. (a) X. Rozanska, T. Demuth, F. Hutschka, J. Hafner, R. A. van Santen, *J. Phys. Chem. B*, 106 (2002) 3248; (b) X. Rozanska, R. A. van Santen, T. Demuth, F. Hutschka, J. Hafner, *J. Phys. Chem. B*, 107 (2003) 1309.

97. T. Demuth, L. Bencol, J. Hafner, H. Toulhoat, *Int. J. Quant. Chem.*, 84 (2001) 110.

98. A. D. Lella, A. Desbiens, A. Boutin, I. Demachy, P. Ungerer, J. P. Bellar, A. H. Fuchs, *Phys. Chem. Chem. Phys.*, 8 (2006) 5396.

99. A. V. Neimark, P. I. Ravikovitch, *Microporous and Mesoporous Materials*, 44–45 (2001) 697.

100. B. Cosane, R. J. M. Pellenq, *J. Chem. Phys.*, 120 (2004) 2913.

101. S. Li, G. Alvarado, R. D. Noble, J. L. Falconer, *J. Membr. Sci.*, 251 (2005) 59–66.

102. H. Lin, B. D. Freeman, *J. Mol. Struct.*, 739 (2005) 57.

103. B. M. Lok, C. A. Messina, R. L. Patton, R. T. Gajek, T. R. Cannan, E. M. Flanigen, *J. Am. Chem. Soc.*, 106 (1984) 6092.

104. U. Olsbye, M. Bjorgen, S. Svelle, K. P. Lillerud, S. Kolboe, *Catal. Today*, 106 (2005) 108.

105. I. Deroche, L. Gaberova, G. Maurin, M. Castro, P. A. Wright, P. L. Llewellyn, *J. Phys. Chem. C*, 112 (2008) 5048.

106. D. Liu, Q. Yang, C. Zhong, *Mol. Simulat.*, 35 (2009) 213.

107. G. Garberoglio, R. Vallauri, *Microporous and Mesoporous Materials*, 116 (2008) 540.

108. Y. Kang, Y. Yao, Y. Qin, Z. Zhang, Y. Chen, Z. Li, Y. Wen, Z. Cheng, R. Hu, *Chem. Comm.*, (2004) 1046.
109. Q. Yang, C. Zhong, *J. Phys. Chem. B*, *109* (2005) 11862.
110. D. Kim, T. B. Lee, S. B. Choi, J. H. Yoon, J. Kim, K. Choi, *Chem. Phys. Lett.*, 420 (2006) 256.
111. D. H. Jung, D. Kim, T. B. Lee, S. B. Choi, J. H. Yoon, J. Kim, K. Choi, S. Choi, *J. Phys. Chem. B*, *110* (2006) 22987.
112. S. Chempath, J. F. M. Denayer, K. M. A. De Meyer, G. V. Baron, R. Q. Snurr, *Langmuir*, 20 (2004) 150.
113. P. A. Wright, M. J. Maple, A. M. Z. Slawin, *J. Chem. Soc. Dalton Trans.*, 8 (2000) 1243.
114. M. D. Macedonia, D. D. Moore, E. J. Maginn, M. M. Olken, *Langmuir*, 16 (2000) 3823.
115. M. Schenk, B. Smit, T. L. M. Maesend, T. J. H. Vlugte, *Phys. Chem. Chem. Phys.*, 7 (2005) 2622.
116. N. J. Henson, A. K. Cheetham, M. Stockenhube, J. A. Lercher, *J. Chem. Soc. Faraday Trans.*, 94 (1998) 3759.
117. C. F. Mellot, A. K. Cheetham, S. Harms, S. Savitz, R. J. Gorte, A. L. Myers, *J. Am. Chem. Soc.*, 120 (1998) 5788.
118. C. Mellot-Draznieks, J. Rodriguez-Carvajal, D. E. Cox, A. K. Cheetham, *Phys. Chem. Chem. Phys.*, 5 (2003) 1882.
119. A. Moissette, C. Bremard, *Microporous and Mesoporous Materials*, 47 (2001) 345.
120. A. V. Neimark, P. I. Ravikovitch, A. Vishnyakov, *Phys. Rev. E*, 62 (2000) R1493.
121. Y. Wang, K. Hill, J. G. Harris, *J. Phys. Chem.*, 100 (1994) 3276.

8 Application of Nanoporous Materials

Nanoporous materials combine the advantages of porous materials. This tiny material provides huge surface area, controllable pore sizes, morphology, and distribution capable of any surface-related applications. As discussed in previous chapters, due to their considerably small size porous structure, material properties have been increased compared to their bulk counterpart. Therefore, it is obvious that the revolutionary properties of nanoporous materials make them a strong contender for a wide range of applications. The main interest of this chapter is to provide some promising applications of nanoporous materials, such as (a) photonic crystals, (b) bio-implants, (c) sensors, and (d) for separation. It is well known that nanoporous materials have two basic areas of applications, of which one is as bulk material and another as membrane. Photonic crystals, bio-implants, and sensors are related to its bulk nature, whereas separation is in the category of membrane-related applications. Before going into detail on the above-mentioned purposes of nanoporous materials, the essential criteria for easy access to the applicability of nanoporous material in every aspect of life must be known. Nanoporous materials are porous, but control of pore size and uniformity as well as a fundamental understanding of the structure–property relationship and availability of the suitable raw material are still challenges in material chemistry. From the technological point of view, there is still a lack of meaningful equipment for quick characterization and inadequate software. Another important point that has hindered its wide applicability is the lack of understanding of the industrial requirements. Despite the above-mentioned difficulties, nanoporous materials are reigning in every field of everyday life.

8.1 PHOTONIC CRYSTALS

Photonic crystals are a new class of dielectric or metallo-dielectric material that affects the propagation of electromagnetic waves. The underlying physical phenomenon is based on diffraction. Therefore, the lattice constant of the photonic crystal structure has to be in the same length scale as half the wavelength of the electromagnetic wave. The photonic crystals are actually periodic nanostructures with repeating internal regions of high and low dielectric constant. They have most promising capability to move from regular electronics to photonic technology in the information processing. Generally, colloidal crystal arrays, artificial opal, and inverse or hollow spherical micelles are used for photonic crystals [1]. Significant research has been dedicated to the improvement of the photonic crystal in combination with the organic material or hybrid material to enhance its interesting applicability as light emitting diodes (LEDs), solar cells, organic lasers, electro-optic switches, etc. [2]. Crystalline macroporous materials are a true candidate for the preparation of photonic crystals; that is, three-dimensional dielectric composites with lattice spacing of the order of wavelengths of light (about 500 nm) [3]. These crystals can be used to create photonic bandgaps (frequency ranges that will not propagate light because of multiple Bragg reflections) [4] that induce useful optical properties, such as inhibition of spontaneous emission or photon localization [5]. To achieve bandgaps for the visible and infrared spectrum, several challenges exist: both constituent materials of the crystal should be topologically interconnected [6] and the ratio of their refractive indices n should at least be 2. Wijnhoven et al. synthesized a new class of photonic crystals of air spheres in titania with radii between 120 and 1,000 nm using artificial opals by precipitation from liquid phase. Different spectroscopic techniques demonstrated that the crystals are strongly photonic [7]. The sensitivity of a photonic crystal optical biosensor is greatly enhanced through the incorporation of low porous dielectric material into the device structure. Computer models are used to predict the reflectance spectra and sensitivity performance of a one-dimensional photonic crystal. Tay et al. developed a simple yet effective melt processing method for the preparation of infiltrant nano-composites as photonic crystals. The infiltrated device showed tuning of the photonic bandgap, which is controllable with nanoparticle loading level [8]. Moreover, block copolymers (BCPs), which self-organize into ordered one-, two-, and three-dimensional periodic equilibrium structures, can exhibit photonic bandgaps (PBGs). Sun et al. describes the treatment of cylinder microdomain nanoporous films with a new kind of two-dimensional BCP-based photonic crystal. The minor component of the nanoporous films has been removed chemically with only pores left in order to enhance their dielectric constant contrast, which provides a new solution to achieve necessary PBG properties with BCPs. The finite-difference time-domain (FDTD) method is used to investigate band features of this kind of photonic crystal theoretically. In addition, theoretical predictions of photonic properties of nanoporous copolymer films as photonic bandgap materials using FDTD has also been discussed [9]. The molecular simulation method has also been developed to understand the preparation of various photonic materials. In that case, molecular dynamics (MD) simulation techniques are applied to study the formation of liquid crystalline nanodroplets, starting from an isotropic and uniform binary solution of spherical Lennard-Jones (solvent) and elongated ellipsoidal Gay-Berne (solute) rigid particles in low (<10%) concentration. The study also can be extended to the characterization of the resulting nanodroplets assessing the effect of temperature, composition, and specific solute–solvent interaction on the morphology, structure, and anisotropy. They find that the specific solute–solvent interaction, composition, and temperature can be adjusted to tune the nanodroplet growth and size [10]. Mesoporous materials like MCM-41 and SBI-15 have shown promise in their usage as photonic materials in the form of thin films apart from their regular usage as porous silica. Porous silicon-type crystals have immense potential as label-free chemical and biological sensing trans-ducers owing to the ease of fabrication, high-quality optics, and a sensitive optical response to changes in refractive index. A major advantage of nanoporous materials, especially mesoporous materials for photonic crystals, is to protect the active site from ambient air and an aqueous environment because of the surface roughness. Photonic crystal design can be performed by molecular modeling by measuring the band width to determine the feasibility of the material or looking at the interaction of the matrix for oxidation or hydroxylation to know the capability of these matrices. This is an emerging field, and nothing concrete has developed yet in terms of as an active material and its simulation.

8.2   BIO-IMPLANTS

Bio-implant materials are generally defined as those materials that can be used for clinical purposes to replace a part of a living system or with the intimate contact of the living tissue. For the traditional material scientist, the main challenge with bio-implant materials is biocompatibility, i.e., the acceptance of the material by the surrounding living tissues to elicit the specific cellular response. Hence, the synthesis of proper materials, especially for surgical implants, is still a great challenge to the material scientist. There are very few materials that can be used as bio-implant materials because the surface of the implants that contact the body might be made of a biomedical materials like titanium or silicon. In fact, the enormous growth in the field of nanoporous materials is extending its hand to bio-implant materials. A very new study shows that the nanoporous carbon membrane may be used to resolve the issue of dialysis and that can be implanted in the human body [11]. Those biomaterials are designed to be used in the biological interface. Thus, it is necessary to have a molecular-level understanding of those materials. The relevance of computer simulations at the molecular level in material science is recognized because they offers significant insight on the bulk and surface properties of the material. A recent review on computer simulation of the adsorption model of biomaterials by Ganazzoli and Raffaini describes why computer simulations play an important role in predicting the nature of biomaterials [12]. From their approach it is evident that when biomaterials are inserted in a biological environment, for instance, within the body proteins do quickly adsorb on the exposed surface of implants. Such a process is of fundamental importance, because it directs the subsequent cell adhesion. Thus, a rationalization of the mode of adsorption of protein on biomaterials is of utmost significance. So, coarse-grained models that can provide important general results have long been recognized in polymer science. The hierarchical structure of a very complex copolymer such as a protein, together with the nature of the biomaterial surface, suggests that atomistic models are better suited to investigate these phenomena. An extensive study has been carried out through molecular dynamics simulations on the adsorption of a γ-chain fragment of fibrinogen over self-assembled monolayers (SAMs) with five different terminal functionalities in explicit water with salt added. The simulations showed different adsorption behaviors on different SAMs, but the fibrinogen fragment did not show major conformational rearrangements over a 5-ns simulation time but only rotational and translational motions. At present, the latter study appears to be stretching the limits of what can be realistically done in this field with computer simulations [13]. Other features, such as the simulation of the material wettability, hydration of the adsorbed fragments, kinetics of spreading, and sequential adsorption of two protein fragments on top of each other, highlighting the results of general interest have also been considered. Segmented polyurethane (PU) is seen as a critical biomaterial for clinical applications due to the excellent combination of mechanical and elastic properties with biocompatibility. However, surface modification of the material is necessary for its application to bio-implant. It has been found that the pulsed laser deposition (PLD) has been useful to deposit titanium nitride (TiN) on PU due to possible deposition without PU substrate heating, leading to thermal degradation. The formation of the hard and brittle ceramic TiN coating can influence the rigid properties of the bulk material and the physicochemical properties, which are related to the thickness of the deposited layer. A computer simulation study has been carried out to simulate the contribution of the TiN coatings to the rigid properties of implant focusing on the short contacts with surgery tools [14]. A hydroxyapatite crystal that gives rigid (high Young's modulus) and shock-resistant tissues in bone [15], behaves like an elastomer with low rigidity and high deformation to rupture in tendons [16], or shows optical properties such as transparency in cornea [17] is an important biomaterial. Molecular dynamics (MD) simulations were employed to study hydroxyapatite/biopolymer interface interactions in composites for biomedical applications. The study analyzed the binding energies between hydroxyapatite (HA) and three polymers: polyethylene (PE), polyamide (PA), and polylactic acid (PLA). The interactions of polymers on HA crystallographic planes (0 0 1), (1 0 0), and (1 1 0) were simulated and the effects of the silane coupling agent (A174) on interfacial binding energies were also examined. The results exhibit the presence of silane coupling agent A174, increasing the binding energy between PE and HA but not for the PA/HA and PLA/HA systems. The MD results can be used to guide the design of polymer/HA composites and to select proper coupling agents [18]. In hard tissue replacements, bioglass is an example of interest for bioactive materials that stimulate tissue mineralization. Bioglass is a composite of selenium, calcium, and sodium oxides favoring apatite hydroxyl-carbonate crystallization but also contributing to the cell cycle implied in tissue formation. An *ab initio* B3LYP calculation was used to asses the structure of the material after the addition of modifiers like sodium in terms of the vibrational spectrum. The initial glass structure unit cell envisaging seventy-eight atoms was generated through a melt quench process by means of classical molecular dynamics simulations. The molecular mechanics optimized unit cell was then fully reoptimized (both unit cell parameters and internal coordinates) at the B3LYP level in a periodic approach using Gaussian basis sets by means of the CRYSTAL06 code. Although long-range structural properties cannot be modeled by using this *ab initio* approach because of the intrinsic amorphous nature of the glass, the quantum mechanical simulation proved to be extremely effective in predicting and analyzing the vibrational features of this biomaterial [19]. Despite the efforts made this past decade to elaborate bio-inspired materials, characterize their structural and physicochemical properties, and understand their structure–function relationships and most of all their formation steps; many unexplored mechanisms remain to be investigated. As human longevity increases, this domain becomes economically significant and a major challenge of the biology/material interface.

8.3   SENSORS

Nanoporous materials have large surface areas and are highly sensitive to change in physicochemical conditions like temperature, light, humidity, etc. Therefore, nanoporous materials are widely used as sensors and actuators. Gas sensor is the reply on the detection of the change in the gas concentration. The underlying principle is the interaction of the gas species with the surface of the sensing material, which will change the electronic conductivity in the near surface region. Hence, a large surface-to-volume ratio is necessary for gas sensors. Apart from the large surface area, porosity is also an important criterion to improve the quality of the sensor because the diffusion of respective gas molecules is strongly correlated with the pore sizes and pore structures. Deliberate control of uniform porosity is necessary for gas detection. Generally, semiconducting nanoporous oxides such as SnO_2, ZrO_2, WO_3, and In_2O_3 are of interest due to their effective use as gas sensors for detecting hazardous gases in factories, automobile emissions, and air. Dickey et al. studied the effect of pore size and pore uniformity for sensing NH_3 and humidity at room temperature and observed that the response of NH_3 and humidity to the material is a strong function of pore size [20]. Zeolites are also used as gas sensing materials because of their excellent porous structure and chemical and thermal stability. The framework of zeolites can easily be varied by changing the composition; consequently, the properties are varied. Table 8.1 shows a few examples of zeolites and their use as sensors for a particular gas or mixture. In addition to zeolites, mesoporous materials can also be used as sensors. A mesoporous silica material has also been developed that contains low-molecular-weight organic compounds. This material claimed to be used as chemical sensor aiming the development of new and practical analytical technique. It has been observed that the material can successfully detect benzene gas at ppm levels in car exhaust containing interference components [21]. Again, novel nanostructured materials, such as aluminum oxide (Al_2O_3), silicon oxide (SiO_2), or zirconium oxide (ZrO_2) embedded into PVA, were investigated as potential matrices to incorporate

TABLE 8.1
Examples of Zeolite Used as Gas Sensing Materials

Material	Sensing Material	Reference
Ru-bipy in FAU	O_2	[80]
Zeolite as filter + $Na_3Zr_2Si_2PO_{12}$ + Li_2CO_3	CO_2, CO	[81]
Polyaniline + KA or NaA or LTA-type zeolite	CO	[82]
Proton-conductive zeolite	NH_3	
	Hydrocarbon	
	Methanol, 2-propanol	[83–85]
Ytria- and zirconia-stabilized zeolite	NO, SO_2	[86]
Ag/zeolite	Hydrocarbon, organic molecules	[87]
SnO_2/zeolite	H_2	[88]
TiO_2/zeolite	H_2, CO, O_2	[89]

organometallic compounds (OMCs) for the development of optical oxygen-sensitive sensors that make use of the principle of luminescence quenching [22].

It is well known that considerable effort has been devoted to gas sensing methodology. However, a major problem in choosing materials for selective gas sensors is that the sensing mechanism is not yet fully understood. Therefore, computer simulation is an essential technique for designing the sensor materials. The main advantage of computer simulation is to provide design optimization by varying geometry, layer dimension, and materials of the device without actual fabrication. This systematic approach can savethe time and cost of device fabrication and experiments. *Ab initio* density functional theory and pseudopotential approaches were carried out for the study of NO_2 adsorption on a transition metal-exchanged zeolite (M = Zn, Cu, Ni, Co, Fe). A tritetrahedral model (T3) was used to represent the structure of the zeolite. The density functional calculations predicted that the bonding energy would follow the order Zn > Ni > Cu > Fe > Co. The optimization results showed that there was a charge transfer from the complex M-T3 to HOMO orbital of NO_2 [23]. Computer simulations provided accurate predictions of sensor behavior of a novel photonic crystal biosensor incorporating a surface-patterned, low-index material that exhibits up to a fourfold sensitivity increase over similar sensors that use a patterned higher index polymer. The increased sensitivity of the porous glass device will enable more accurate characterization of smaller molecules at lower concentrations [24].

The topic of biosensors must be mentioned as an application for nanoporous materials. Due to the large surface area, biological molecules can be immobilized to the surface and serve as a biological detection system or used as a transducer when microscale piezoelectronics are used. The incorporation of nanoporous silica in between the biological molecule and the cantilever increases the surface-to-volume ratio and consequently the sensitivity is enhanced due to the shift in resonance frequency. Instead of silica, Ryu et al. [25] studied the resonant characteristics and the sensitivity of a cantilever-shaped $Al/PZT/RuO_2$ resonant biosensor for the application of micro-endoscope by computer simulation techniques. The resonant frequency and sensitivity of the sensor were more dependent on the sensor length than other geometric variations such as width and thickness. The sensing area should be confined near the end of the cantilever sensor for fixed loads to maximize the sensitivity and to enable quantitative analysis of the analytic mass. Good agreement between the experimental and the simulated resonant frequencies was observed [25]. Thus, biosensors are an important tool in the health care industry, where nanoporous silica can serve as a versatile platform because several features make it especially attractive for chemical and biological sensors, including a very high surface area–to-volume ratio, simple and inexpensive fabrication techniques, and suitability for integration with silicon electronics for optical and electrical detection.

The emerging field of functionalized material resulting from the combination of the structural aspect of inorganic lattice with the intrinsic chemical reactivity of organic components is demanding attraction for its various possible applications. Colorimetric sensors are one of the functional materials whose design is important for innumerous applications. Generally, mesoporous materials are used as inorganic lattices and tetra-alkoxysilane or organoalkoxysilanes are typically used to synthesize these materials [26]. The synthesis of mesoporous materials with large surface

area like SBA-15, with well-defined pore size and pore shapes [27], have great potential in advanced materials designing. However, applications such as adsorption, sensing, ion exchange, and catalysis require the materials to have specific attributes such
as binding sites, stereochemical configuration, charge density, and acidity, which are
discussed in a review by Sayari [28]. In analytical science, there is always the need to
use different techniques for selective separation of metal ions. Colorimetric sensors
(naked eye sensors) is one of such method. The sensor do not require the use of any
sophisticated instrumentation and are particularly attractive due to the simplicity and
quick analysis in measuring the presence of analytes by colorimetry. Here, we wish to
explore the interaction between these chromophores and the silylating agents responsible for the binding of specific metals to design a sensor for a particular usage. The
binding of silylating agent with mesopores is well studied, but the interaction of the
silylating agent and the chromophore remains rare in the literature. If that interaction
is proposed right in terms of the active site of the interacting species, this can foresee
the binding nature of the chromophore with the silylating agent and the metal ions
to complete designing of a specific sensor. Solvent plays role in the sensing behavior.
We believe that in cases like this where intra- and intermolecular interactions probability can prescribe the selective binding between the centers, reactivity descriptors can be the best method to rationalize these issues. Reactivity descriptors are
a well-established methodology, which was discussed in Chapter 4. Previously, a
reactivity index scale for heteroatomic interaction with zeolite framework was
proposed [29] and it was used for the first time in gas sensing over mesopore matrices [30], which really can be reproduced by experimentation. The game plan here
is therefore a multistep simulation methodology, first to perform the localized reactivity of the silylating agent N-trimethoxysilylpropyl N,N,N-trimethyl ammonium
chloride (TMAC), which is used for the organic monolayer formation over mesoporous material SBA-15, and two representative chromophores 4-(2-pyridylazo)-re-
sorcinol (PAR) and 2-[1-(2-hydroxy-5-sulphonyl)-3-phenyl-5 formazano] benzoate
(ZINCON), respectively, both in the gas phase and in the presence of solvent were
chosen. From the localized reactivity descriptors, a probable interaction is to be postulated. These proposed interactions from the localized reactivity index were then
validated by the interaction energy calculations for the TMAC-chromophore complex in blank as well as in the presence of metal ions and they were validated with the
experimental results. An *a priori* rule was formulated to design the sensors.

The Fukui functions were calculated using the simplest equation of

$$s_x^+ = [q_x(N+1) - q_x(N)S]$$

$$s_x^- = [q_x(N) - q_x(N-1)S]$$

where s is the local softness, N is the total number of electrons, and q is the amount
of charge with S defined as the global softness. S can be approximated as

$$S = \frac{1}{(IE - EA)}$$

where *IE* and *EA* are the first ionization energy and electron affinity of the molecule, respectively.

$$s(r) = f(r)S$$

Thus, local softness contains the same information as the Fukui function $f(r)$ plus additional information about the total molecular softness, which is related to the global reactivity with respect to a reaction partner, as stated in the hard–soft acid–base (HSAB) principle. In this way, the frontier-orbital theory of reactivity of Fukui [31] can be easily incorporated into the theory. The function $f+(r)$ is associated with the lowest unoccupied molecular orbital LUMO and measures reactivity toward a donor reagent, and the function $f-(r)$ is associated with the highest occupied molecular orbital HOMO and measures reactivity toward an acceptor reagent.

Condensed Fukui functions for chromophore molecules and the silylating agent at gas phase and in solvents have been calculated in the framework of density functional theory (DFT) within the helm of the HSAB principle. The calculations were performed using the DMol^3 program [32] of Accelrys Inc. The geometry of all the interacting molecules in all cases was optimized using the double numerical with polarization basis set DNP [33], which is equivalent to the 6-311G** basis set of Gaussian and with BLYP as the exchange correlation functional [34]. This has also been used for cadmium considering the d electrons. Basis set superposition error (BSSE) was also calculated for the current basis set in nonlocal density approximation (NLDA) using the Boys-Bernardi method [35]. Single-point calculations of the cation and anion of the chromophore molecule and the silylating agent, both in gas phase and in solvent medium, at the optimized geometry of the neutral molecule were also carried out to evaluate Fukui functions and global and local softness. The condensed Fukui function and atomic softness were evaluated using the above equations. The gross atomic charges were evaluated using the technique of electrostatic potential (ESP)-driven charges. In order to consider the effect of solvent, the conductor-like screening model (COSMO) salvation method within the DFT formalism as in the program DMol^3 of Accelrys Inc. was introduced [36]. In this method, the solute molecules form a cavity within the dielectric continuum of permittivity _ that represents the solvent. The charge distribution of the solute polarizes the dielectric medium. The response of the dielectric medium is described by the generation of screening (or polarization) charges on the cavity surface. The cavity surface is obtained as a superimposition of spheres centered at the atoms, discarding all parts lying on the interior part of the surface. The spheres are represented by a discrete set of points, the so-called basic points; eliminating the parts of the spheres that lie within the interior part of the molecule thus amounts to eliminating the basic grid points that lie in the interior of the molecule. The radii of the spheres are determined as the sum of the van der Waals radii of the atoms of the molecule and of the probe radius. The surviving basic grid points are then scaled to lie on the surface generated by the spheres of van der Waals radii alone. The basic points are then collected into segments, which are also represented as discrete points on the surface. The screening charges are located at the segment points. The interaction energy calculations were performed with all electrons relativistic, which includes all electrons explicitly and introduces some relativistic effects into the core. This is to handle the 5d metal system of cadmium.

The important aspect is to find out first the affinity of these molecules with solvent and here the solvent matrix is water as proposed by experiments. The best method is to observe the global affinity of the molecule in the presence and absence of the solvent as detailed in [30]. To rationalize the molecular scenario one needs to monitor the inter and intra molecular interaction. The dipole moment, global softness values for the independent chromophore molecules, and the silylating agent are shown in Table 8.1. It shows that the global softness of the silylating agent TMAC is higher than the values for the chromophore molecules both in the solvated and unsolvated form. There is a consistent increase in the global softness after solvation and the trend is that same as that for the unsolvated situation. The order of activity remains ZINCON < PAR, whereas TMAC has the highest global softness value. A larger dipole moment means that the solvent molecules can interact favorably with charged solute molecules by screening their charges. Consequently, a high dipole moment usually implies a high dielectric constant. A high dielectric constant, such as that found in water, is important because the forces between charges are attenuated. The values of dielectric constant are again related with the hydrophobicity and the hydrophilicity of a system. The results therefore propose that chromophore PAR is hydrophobic and the other chromophore ZINCON together with the silylating agent TMAC is hydrophilic, which needs further verification in terms of the localized interaction. The grafting and elucidation of that grafted structure is beyond the scope of this study. TMAC was chosen mainly for the presence of N^+, which can be used for the ion pair interaction with the chromophores. The $-NH_3$ groups of the aminopropyl segments present in TMAC were converted into the $-NH_3^+$ groups on the SBA-15 surface showing that the SiO/H^+ NH_3-type structure was predominantly stabilized on the surface of silica gel. This is more about the binding of TMAC on SBA-15, which is again beyond the scope of this study. Experimentally, it is established that for a 1:1 loading of SBA-15 (mesoporous material) and TMAC, the pore size decreases from 7.1 to 5.4 nm and the surface area decreases from 897 to 620 m^2g^{-1}. Hence, the silylating agents TMAC must be present over or inside SBA-15. The optimized geometry of all the three components is shown in Figure 8.1, Figure 8.2, and Figure 8.3 with the order of (a) unsolvated and (b) solvated, respectively. The atoms are all labeled for easy understanding. In terms of global softness the order of activity for the chromophores both when solvated and unsolvated condition is PAR > ZINCON. PAR shows much higher global softness than ZINCON, even compared to that of TMAC. In terms of dipole moment, ZINCON exhibits greater hydrophilicity compared to TMAC when unsolvated, but TMAC is more hydrophilic than ZINCON when solvated and at the same time it shows greater hydrophobicity for PAR. Comparing the atomic center of the silylating agent with highest relative electrophilicity/nucleophilicity to match with the counteractive atom from chromophore with highest nucleophilicity/electrophilicity for pseudo-bond formation, the interaction of TMAC with PAR and ZINCON, bonded with the SBA-15 surface was considered. From the results it was observed that when the molecules are unsolvated, C139 of TMAC interacts with C16 of PAR and C34/O37/O42 of ZINCON. In reality, the reaction is carried out in solvated medium. So in the solvated situation, Si1 of TMAC interacts with the same C16 of PAR and

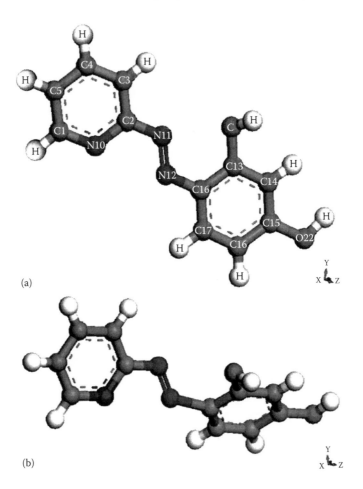

(a)

(b)

FIGURE 8.1 (For color version see accompanying CD) (a) Optimized geometry of 4-(2-pyridylazo)-resorcinol (PAR) at unsolvated condition with all the constituent atoms labeled. (b) Optimized geometry of 4-(2-pyridylazo)-resorcinol (PAR) at solvated condition.

may interact with C23/O45 of ZINCON. Among the possible interaction sites, the second option, O45, looks more reasonable apart from that of PAR. But the results of dipole moment suggest that PAR is hydrophobic, so the solvent does not influence the binding, whereas O45, which is the free oxygen present in ZINCON, will guide itself toward Si1 of TMAC. These results were obtained in terms of the highest electrophilic site of TMAC interacting with the highest nucleophilic site of the chromophore molecule, considering that the interaction takes place between two neutral centers with the delocalized charge inside. From the reactivity descriptors results it is observed that when the molecules were unsolvated, Cl39 of TMAC interacted with C16 of PAR and C34/O37/O42 of ZINCON. So these models were constructed and optimized the optimized geometry of these two structures in the presence and absence of solvent as shown in Figure 8.4 and Figure 8.5, respectively. At the starting geometry, the

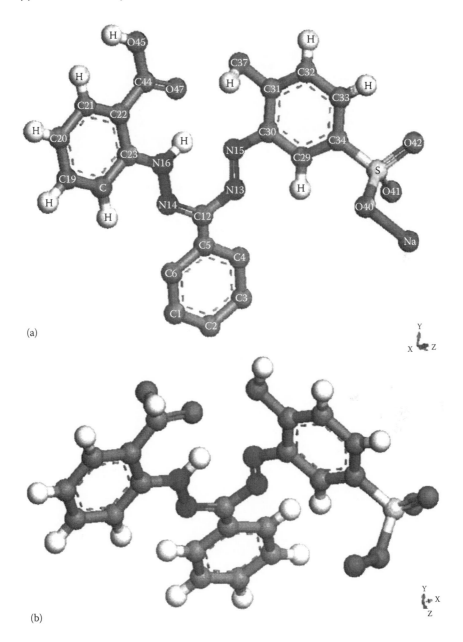

FIGURE 8.2 (For color version see accompanying CD) (a) Optimized geometry of 2-[1-(2-hydroxy-5-sulphonyl)-3-phenyl-5 formazano) benzoate (ZINCON) at unsolvated condition with all the constituent atoms labeled. (b) Optimized geometry of 2-[1-(2-hydroxy-5-sulphonyl)-3-phenyl-5 formazano) benzoate (ZINCON) at solvated condition.

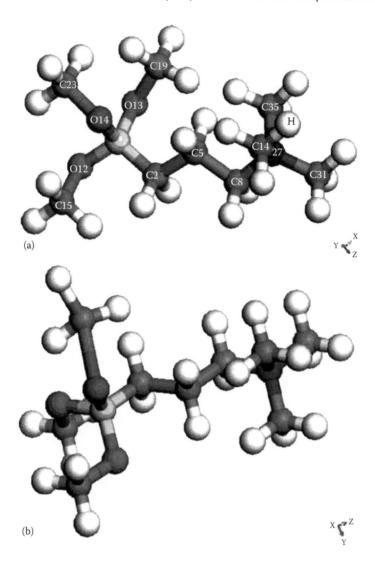

FIGURE 8.3 (For color version see accompanying CD) (a) Optimized geometry of N-trimethoxysilylpropyl *N,N,N*-trimethyl ammonium chloride (TMAC) at unsolvated condition with all the constituent atoms labeled. (b) Optimized geometry of N-trimethoxysilylpropyl *N,N,N*-trimethyl ammonium chloride (TMAC) at solvated condition.

interacting species were kept at a distance of 2.5 Å. The calculations were repeated for various initial distances starting from 2.0 to 4.0 Å, and from the potential energy surface generated thereforth, 2.5 Å distance was chosen for the interaction energy calculation. In addition, the variation of interacting centers with comparable activity has also been taken care of. The results show that unsolvated TMAC + PAR complex is more stable than that of TMAC + ZINCON complex by 47.23 kcal/mol, whereas ZINCON complex is more stable than the solvated PAR complex with TMAC by

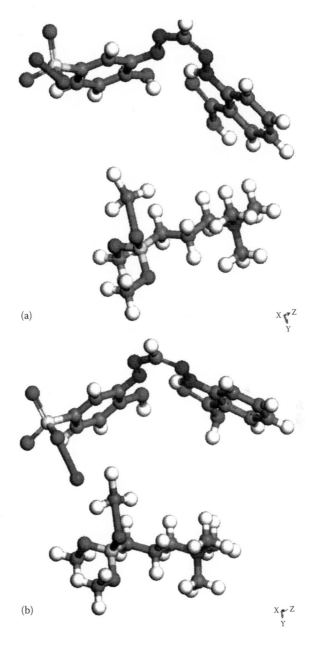

(a)

(b)

FIGURE 8.4 (For color version see accompanying CD) Interaction between N-trimethoxysilylpropyl*N,N,N*-trimethylammonium chloride (TMAC) and 2-[1-(2-hydroxy-5-sulphonyl)-3-phenyl-5 formazano) benzoate (ZINCON). (a) Unsolvated; (b) solvated.

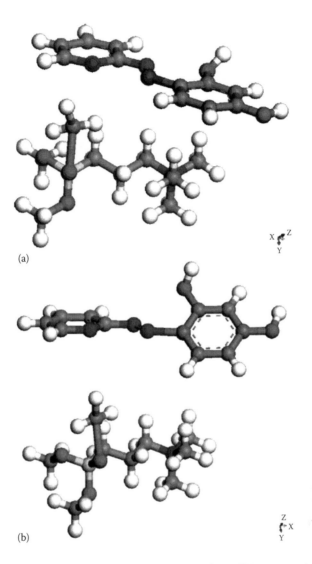

(a)

(b)

FIGURE 8.5 (For color version see accompanying CD) Interaction between N-trimethoxysilylpropyl*N*,*N*,*N*-trimethylammoniumchloride(TMAC)and4-(2-pyridylazo)-resorcinol (PAR). (a) Unsolvated; (b) solvated.

23.17 kcal/mol in the solvated condition. The O45 center of ZINCON binds more favorably with the Si1 center of TMAC while solvated, whereas in the unsolvated situation Cl39 of TMAC binds strongly with PAR and ZINCON binds more strongly with TMAC when solvated. Now what could be the scenario if a metal ion binds with the chromophores? To study this interaction, Cd was chosen and placed over PAR and ZINCON at a distance of 2.5 Å. The situation did not change even when the molecules were solvated. After salvation, Cd–ZINCON complex is less stable than the complex of Cd and PAR (17.09 kcal/mol). The geometry of the optimized structures for the

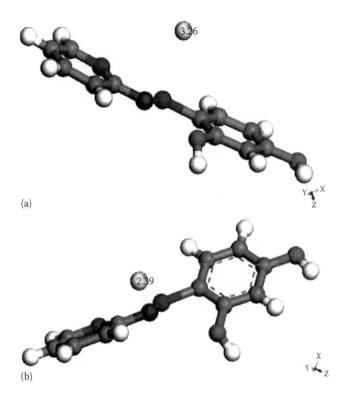

(a)

(b)

FIGURE 8.6 (For color version see accompanying CD) (a) Optimized geometry of the complex formed between Cd and 4-(2-pyridylazo)-resorcinol (PAR) at unsolvated condition. (b) Optimized geometry of 4-(2-pyridylazo)-resorcinol (PAR) and Cd at solvated condition. The numbers are the distances between Cd and the nearest neighboring atom.

Cd–PAR complex and the Cd–ZINCON complex for both the unsolvated and solvated situations are shown in Figure 8.6 and Figure 8.7, respectively. It was observed that there was a dramatic change in the location of the metal centers. The metal center in the solvated situation came closer to the PAR surface within a bonding distance of, say, 2.39, from 3.26 in the unsolvated case. In case of ZINCON, the nature of bonding explains the lower stability of Cd; in the unsolvated case, the cadmium is closer to the double-bonded oxygen present on the ZINCON surface (2.72) but goes further away from the nitrogen centers due to the steric repulsions resulting from the proximity of other centers. The situation worsens in the presence of solvent as the cadmium atom moves behind the surface with a change in Cd–N distance of about 3.56. This forces us to look at the situation when TMAC, chromophores, and cadmium all are present in the reaction. The result is that the unsolvated case TMAC + cadmium + PAR is more stable than the respective ZINCON complex by 18.76 kcal/mol. The situation for ZINCON deteriorates when it is solvated; the complex is less stable than the PAR complex and matches well with experimental observations. According to the experimental outcome, PAR is the best performer for complex formation with cadmium in comparison with ZINCON and is attributed to the fact that the cadmium–PAR

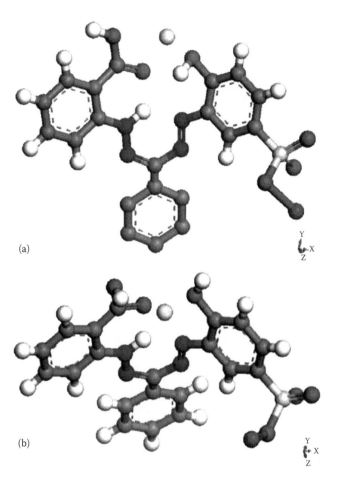

FIGURE 8.7 (For color version see accompanying CD) (a) Optimized geometry of 2-[1-(2-hydroxy-5-sulphonyl)-3-phenyl-5 formazano) benzoate (ZINCON) at unsolvated condition with Cd. (b) Optimized geometry of 2-[1-(2-hydroxy-5-sulphonyl)-3-phenyl-5 for-mazano) benzoate (ZINCON) and Cd at solvated condition. The numbers are the distances between Cd and the nearest neighboring atom.

distribution coefficient is much higher than that of ZINCON–cadmium. Thus, it cannot bind cadmium strongly enough but its binding with TMAC when solvated is stronger than the metal. The circumstances may change for interaction with other metal ions. Hence, chormophores can be selectively used for metal identification or separation and a reactivity index can be successfully used to design colorimetric sensor to address a specific analytical problem.

Fluorophores are any molecules capable of exhibiting emitting fluorescence in the excited state and are widely used as sensor materials. Depending on the kind of substitution on the aromatic ring (the electronic excitation), the proton affinity undergoes highly localized changes and results fluorescence. The influence of the ground-state geometries on the photophysical properties is quite complicated. So we have calculated the reactivity index from the nonprotonated moiety to propose their activity in the protonated form, resulted due to the intramolecular hydrogen bonding. The intramolecular hydrogen bonding is the consequence of the interaction between the nucleophilic site with an electrophilic site as a combination of electron donor and acceptor site to favor a pseudo-bond formation process inside the molecule. On the other hand, if the situation arises when there is a possibility to interact with other molecules, then these molecules may interact with their active centers to form intermolecular bonding. Hence, an interaction between the center with highest nucleophilicity and the center with highest relative electrophilicity takes place. Depending on the activity of the respective centers, electron donor acceptor property will vary. We have used this methodology successfully to explain the intermolecular and intramolecular H-bonding for fluorophore sensors. Anthracenes bearing aliphatic or aromatic amino substituents, which behave as molecular sensors, have shown their potential to act as photoinduced electron-transfer (PET) systems. In this PET, the fluorophore moieties are responsible for electron release during protonation and deprotonation. The HSAB principle deals with both intra- and intermolecular electron migration.

It is possible to calculate the localized properties in terms of Fukui functions in the realm of DFT and thus establish a numerical matchmaking procedure that will generate an *a priori* rule for choosing the fluorophore in terms of its activity. A qualitative scale is proposed in terms of the feasibility of intramolecular hydrogen bonding. To investigate the effect of the environment of the nitrogen atom on protonation going from mono- to diprotonated systems, the results show that the location of the nitrogen atom in an aromatic ring does not influence the PET, but for aliphatic chains the effect is prominent and the reactivity indices can scale the activity of fluorophore molecules for the PET process [37].

Computer simulation thus can be a useful technique to identify the capability of a material to be a sensor and then tune the material to make it selective for a specific application. Sensing needs localized interaction, and simulation can help rationalize that fact.

8.4   SEPARATION

Nanoporous materials, particularly zeolites, are widely used in separation and purification processes involving a mixture of gases. Zeolites are crysatalline porous materials and thus the separation process is mainly based on the difference in the rate of transportation, their rate difference of the adsorption into the pore, or their difference in shape and molecular size. The three-dimensional framework structure provides an ideal environment for selective adsorption of different molecular species. The accessible pore size of the zeolite in combination with the location, size, and charge of any cations that may be present determine the zeolite's effectiveness. However, it is necessary to have a clear understanding of the transportation process. In addition, those materials that are used for the separation process of exceedingly small pores, so the interaction between mixture molecules and the pore walls cannot be ignored. Hence, molecular modeling of these materials and transport processes that takes place in the pores have attracted considerable attention to rationalize the understanding. It is well known that zeolites are widely used for shape-selective separation. Shape-selective separations can be catagorized into three regimes: (1) molecular sieving, (2) size-based separation, and (3) shape-based separation, and an interesting example is separation of xylene isomer. In fact, the way the methyl group is connected to the benzene ring is mainly the consideration of shape and has been studied in detail by several authors [38]. Mohanty and McCormick reviewed the prospects of shape-selective separation based on theoretical and experimental results [39]. In a very recent example, GCMC and configurational bias Monte Carlo simulations were performed for the adsorption of butane isomers binary mixtures in nine types of full-silica zeolites such as MFI, MEL, BEA, BOG, TER, MOR, ISV, TON, and CFI at 300 K. It appears that the channel size and structure control the selectivity of alkane isomers and there is a critical pore size, which is a ten-membered-ring about 5.6 Å. The channel sizes, larger than critical pore size, prefer i-butane over n-butane, whereas TON, with smaller channel size than the critical pore size, prefers n-butane over i-butane. The channel size determines the location of alkane molecules in zeolites. The adsorption of the quaternary mixture suggests that chain-length dependence of n-alkane adsorption presents at low pressure, but the adsorption is controlled by the pore size and structure with the enhancement of the pressure [40]. However, in any case it is necessary to understand in depth the single-component adsorption through the study of fundamentals of adsorption. It has been mentioned before that the intention of this book is to highlight those approaches that correlate between the theory and the experimental results to provide molecular-level information on a practical basis. For example, the adsorption of pure carbon dioxide and methane and their equimolar mixtures was explored in a model zeolite Na–Y by combining GCMC simulation and an original experimental approach based on gravimetry/manometry and manometry/microcalorimetry devices. Both experimental and calculated adsorption isotherms and enthalpy profiles obtained for the single gas and a 50/50 binary mixture were provided for pressures up to 30 bars. In agreement with experiments, the result shows a strong correlation with simulation attributing a high selectivity of CO_2 over CH_4 in Na–Y for the whole range of pressures. The study also shows that the preferential site and nature of the interaction of two adsorbates in the single-component adsorption are very different because carbon dioxide molecules mainly interact with the extraframework Na^+ ions via predominantly electrostatic interaction, whereas van der Waals type of interactions are evident for nonpolar methane molecules. This difference in interaction leads to the selective separation of carbon dioxide from methane [41]. Molecular modeling has also been employed to study the interactions between nitrous oxide and nitric oxide with mordenite, before and after charge equilibration. The results indicate that the charge equilibration leads to an adsorption that is independent of pressure. Comparison with gas-chromatographic analysis indicates that charge equilibration is not allowed in acid mordenite. Therefore, in the case of polar molecules, their adsorption depends on the charge in the zeolite micropores and the separation factor obtained by calculation matches well with experiments [42]. The separation of carbon dioxide and nitrogen on all-silica DDR- and MFI-type zeolites was also studied using GCMC simulations and compared with experimental results. The simulated values of sorption capacity, isosteric heat of adsorption, and Henry's constant are in good agreement with experimental data. Furthermore, simulation was used to assess the adsorption properties and selectivity of carbon dioxide/nitrogen mixture gas as a function of gas-phase composition and pressure [43]. Gallo et al. [44] studied the separation of binary mixture hydrogen/methane and hydrogen/carbon dioxide, over silicalite and titanosilicalite type of molecular sieve ETS-10. This is the first molecular simulation study that presents mixture adsorption isotherms of these components in silicalite and ETS-10 and determines selectivities based on simulation results. From GCMC simulation results the study indicated that the separation of hydrogen from methane or from carbon dioxide in silicalite would be successful, because hydrogen in a 50% bulk mixture does not adsorb unless the pressure is very high, ca. 500 bar. In contrast, in ETS-10, hydrogen in a 50% bulk mixture adsorbs at a pressure near 10 bar. Simulations of adsorption in ETS-10 show a higher selectivity for the separation of carbon dioxide from hydrogen than the separation of methane from hydrogen, independent of pressure. On the other hand, silicalite exhibits a higher selectivity for the separation of carbon dioxide from hydrogen than the methane/hydrogen separation at high pressures only. Moreover, isosteric heat of adsorption information indicates that silicalite is energetically homogeneous with the adsorbates but ETS-10 is energetic heterogeneous [44]. The separation discussion will not be complete without mentioning the application of reactivity index theory, which incorporates the phenomenon of intermolecular and intramolecular interaction behaviors through the prediction of the affinity of each site of a molecule to another molecule or in between. We have used this method exhaustively to predict the selectivity of a specific molecule to be adsorbed within a framework of nanoporous material and prescribed an *a priori* rule to design new materials of interest by carefully monitoring the interaction scenario. We have investigated [45] the separation of individual component gases from a mixture of carbon dioxide, nitrogen, methane, ethane, and sulfur hexafluoride along with the selective permeation of carbon dioxide from a mixture of carbon dioxide and nitrogen through Na-zeolite-Y membranes using reactivity descriptors and interaction energy calculations by DFT. To locate the active site of zeolite-Y, three different cluster models were considered to propose the role of a zeolite framework in the selectivity of gaseous molecular separation. All the interacting molecules were optimized with respect to the framework cluster to compare the stability of the adsorption complex. We analyzed the effect of affinities of gas molecules for the membrane wall on the permeation to predict the optimal affinity strength; e.g., for higher selectivity of carbon dioxide. We investigated the local softness of the interacting species with the zeolite framework cluster models. The order of activity as obtained from reactivity index values of individual molecules was compared with interaction energy calculations using DFT to validate the proposition. The result successfully predicts the experimental observation of selective permeation of gaseous molecules through Na–Y-zeolite pore in the order ethane < methane < sulfur hexafluoride < carbon dioxide < nitrogen. This is the first study to rationalize the understanding between selective permeation of gaseous molecules through zeolite membranes in terms of reactivity index and is validated by interaction energy calculations using DFT. Though this is not a quantitative study, we claim that the findings, which are qualitative in terms of their numerical output, are a novel approach in solving the permeation process of a smaller molecules through zeolite membranes. The novelty lies in the simplicity of the approach, which is so far the only plausible way to explain the mechanism of the permeation process. The results validate our earlier proposition [46] that molecules with one active site — i.e., a nucleophilic site — can preferentially interact with most electrophilic sites of the interacting material (e.g., zeolite); the reactivity descriptors can conclusively predict the selectivity of the interacting molecules with respect to a particular interacting matrix. Experimentally, the selectivity of carbon dioxide over nitrogen is clearly understood. The current results also propose the mechanism of permeation, which logically defers from the experimental prediction of nonadsorptivity of nitrogen. The results show that nitrogen forms the strongest adsorption complex, which further helps in the migration of other gases through zeolite pores. Zeolite clusters were rationally modeled to consider the environment of the active cation, which realistically considers the experimental situation. The optimistic result paves a novel way of explaining reactive centers in a particular reaction, which can eliminate tedious experimentation.

The removal of sulfur in the petroleum industry is a great problem because it is responsible for the pollution of the environment. The Environmental Protection Agency (EPA) has announced new regulations that mandate refineries to reduce the sulfur level down to 20 ppm weight in gasoline and 5 ppm weight in diesel by 2016. The removal of organic sulfur compounds from transportation fuels has become an increasinly important issue in recent years because of the possibility that these fuels can be reformed onboard or onsite to produce hydrogen-rich gas as a fuel for fuel cells for different types of applications. Zeolites can also be successfully used for the separation of sulfur compounds. The GCMC simulation on the investigation of the adsorption of thiophene and benzene on MFI- and MOR-type zeolites shows that the distribution of thiophene over various channels, for MFI adsorption capacity of the intersections of straight and zigzag channels appears preferable than the other channels. The calculated amount of absorbed thiophene on MFI is in good agreement with the experimental data at high ratios of Si/Al for a single-component system. However, for a thiophene/benzene binary mixture, the system obeys the competition classification, because both species prefer the intersections and thiophene adsorbs more strongly, pushing benzene into the zigzag channels [47].

Molecular modeling is a powerful tool in material science as well as in environmental applications. Hydrogen sulfide is a pollutant present in the biogas and it is necessary to remove this pollutant to exploit the better performance. Different zeolites such as FAU, LTA, and MFI have been used for the removal of hydrogen sulfide. As it is a polar compound, so high-aluminum-containing hydrophilic zeolites would be more effective for separation. GCMC and quantum mechanics procedures adopted have been validated by comparison with experimental data available for hydrogen sulfide removal from the atmospheric environment. The result shows that FAU Na–Y type appears to be the best choice over Na–X due to the different Si/Al ratio, but hydrogen sulfide is scarcely adsorbed on LTA- and MFI-type zeolites when a mixture is considered [48].

Metal-organic framework (MOF) materials are a class of nanoporous materials that have many potential advantages over traditional nanoporous materials for adsorption and other chemical separation technologies. Because of the large number of different MOFs that exist, efforts to predict the performance of MOFs using molecular modeling can potentially play an important role in selecting materials for specific applications. The current state-of-the-art in the molecular modeling and quantum mechanical modeling of MOFs has been reviewed. Quantum mechanical calculations were used to examine the structural and electronic properties of MOFs and the calculation of MOF–guest interactions and to study pure and mixed fluid adsorption in MOFs. Similar calculations have recently provided initial information about the diffusive transport of adsorbed fluids in MOFs. Cu–BTC (Cu(II) benzene 1, 3, 5 tricarboxylate) is a metal-organic compound with a microporous structure. GCMC simulation was conducted to systematically evaluate the adsorption and separation of three binary mixtures, ethane/carbon monoxide, ethylene/carbon dioxide, and ethylene/ethane, over Cu–BTC at room temperature. The simulated results showed that Cu–BTC could be potentially used for the purification of carbon monoxide, capture of carbon dioxide, and separation of olefin/paraffin caused by the side pockets of the material. The selective adsorption behaviors between gas components and its effects were more evident at low adsorption loadings [49].

In the last ten years, interest in developing thin zeolite films or zeolite membranes has grown enormously [50]. Zeolite membranes are generally used as catalytic membrane reactors because they can combine separation [51] and catalytic activity efficiently due to the significantly different diffusivities in the uniform, molecular-sized pores and the presence of its catalytic sites. The selective sorption properties with their catalytic activity and thermal stability make zeolites an ideal candidate for the inorganic catalytic membrane. Most of the zeolite membranes reported in the literature are of the MFI type because of its pore architecture, which consists of straight channels that are interconnected by the zigzag channels. This amazing pore structure leads to highly anisotropic diffusivity and MFI membranes have been widely used in gas separation. A review on zeolite membrane preparation and characterization by Caro and Noack provides a clear scenario on the progress of the material [52]. Here we provide some selective examples of the application of various types of zeolite membranes.

To perform an accurate simulation of steady-state pressure-driven diffusion of the permeation of Lennard-Jones gases across a zeolite model membrane, Pohl et al. [53] developed a dual-control volume grand canonical molecular dynamics technique. Due to the molecular sieving nature of microporous zeolites, the permeation of helium, hydrogen, and methane were favored selectively and the results compared very well with experimental results of permeation through ZSM-5 polycrystalline membranes [53]. Based on the GCMC method, Kobayashi et al. [54] developed a novel simulation technique called *dual ensemble* to treat the nonequilibrium system. This technique has also integrated into two programs, dual ensemble Monte Carlo (DEMC) and dual ensemble molecular dynamics (DEMD) programs. The DEMC method treats the movement of molecules in a stochastic manner and therefore it can calculate the distribution of the molecules on the membrane surface or on the pores. The DEMD program rectifies the movement of the molecules using molecular dynamics techniques and therefore the permeability can be calculated directly from the simulation result. This program has been used to study the separation of carbon dioxide/nitrogen using Na–Y membranes. Carbon dioxide exhibits greater permeability than nitrogen and temperature dependence has also been observed [55].

Under the Clean Air Act, the depletion of the ozone layer is an environmentally hazardous problem. It mainly generates from the emission of chlorocarbons or chlorofluorocarbons (CFCs) in air. Because those molecules are polar, polar adsorbents like zeolites, silica gel, or alumina may be preferred. However, the hydrophilicity of zeolites makes the situation complicated because of the presence of water molecules in the air, which prevents the adsorption of other organic molecules. So, it is necessary to properly design the porous material to achieve the highest separation of toxic material. This issue can be well addressed by the reactivity index theory. The reason for separation lies in the affinity of the reactants toward the nanoporous material, which can be explored and validated through the intra- and intermolecular reactivity concept. Adsorption of CFCs over zeolites has attracted a great deal of attention at present due to its relevance to environmental issues concerning ozone layer depletion [56]. Faujasite has shown its credibility in separation of hydrofluorocarbons (HFCs) during the synthesis of CFC substitutes [57]. Grey et al. have investigated the dehydrofluorination reactions of various HFCs over the basic X and Y zeolites. Their investigation is based on solid-state NMR to rationalize the factors responsible for gas separation [58]. They found that CF_2HCF_2H molecules are known to bind more tightly to the zeolites than CF_3CFH_2 molecules. The lower energy structure of the CF_2HCF_2H molecules with sodium is influenced by both hydrogen bonding and sodium–fluoride interactions. Coordination of two sodium cations maximizes the interactions with the partial negative charge of the F atom, which is located at either end of CF_2HCF_2H molecules. There are very few experimental data concerning the adsorption of fluorocarbons and CFCs in zeolites and other microporous materials. Calorimetric methods have been used [59] to estimate the heats of adsorption of several CFCs in deuterated H–Y zeolite. NMR has been used to probe the host–guest interactions between hydrofluorocarbons and faujasite. In addition, HFC binding on Na–X has been studied using X-ray diffraction and NMR together [60]. Isotherm measurements have been reported for adsorption of dichlorodifluoromethane in Na–, K–, and Cs–Y zeolite [61]. Therefore, it is important to figure out the role of Na in bonding with CFCs depending on the halocarbon concentration, would be an important criteria of the separation mechanism of hydrohalocarbons in a mixture. We have explored this domain by DFT calculations [62]. Again, the role of H–Y in bonding the CFCs and the nature of bonding between chlorine and fluorine centers with the active center of H–Y and Na–Y will help to choose the zeolite matrix for desired separation of molecules with chlorine and/or fluorine centers. The reactivity index was used to solve this intriguing problem of halocarbon separation by choosing a suitable zeolite matrix. The procedure of the calculation is to first locate the reactive site present in the zeolite system as well as in CFC to predict the interacting structure, followed by a periodic optimization to include the neighboring effects of zeolite, which is very important in these nanoporous matrices and computationally demanding. Each of the molecules along with the zeolite framework cluster models representing Na–Y and H–Y zeolites were optimized in their neutral, anionic, and cationic forms to calculate the global and localized properties. All the cluster calculations were performed within the DFT domain using DMol^3 code of Accelrys. $NaSiAlO_7H_6$ cluster was used to represent the zeolite Na–Y and a cluster with the

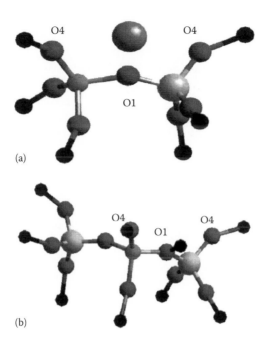

O4   O4

O1

(a)

O4   O1   O4

(b)

FIGURE 8.8 (a) The cluster model representing the Na–Y zeolite with formula $NaSi_2Al_2O_{13}H_{11}$. (b) The cluster model representing the H–Y zeolite with formula $Si_2AlO_{10}H_9$.

formula of $Si_2AlO_{10}–H_9$ represented the H–Y zeolite structure. The adjacent silica and aluminum atoms occurring in the zeolite lattice were replaced by hydrogen to preserve the electroneutrality of the model as shown in Figure 8.8a and Figure 8.8b. Terminal hydrogen was kept at a distance of 1.66 Å which is the O–Si distance. This is to locate the most nucleophilic and electrophilic site of the interacting species from which a favorable pseudo-bond formation is postulated. Now keeping in mind two constraints of cluster models, (1) if fully relaxed overestimates the freedom of relaxation and (2) fixing boundary atoms will result in restricted optimization, it will be more appropriate to consider a periodic model. Payne et al. used first principle methodology for methanol adsorption in zeolites to model the methanol-to-gasoline conversion process [63]. The theory of reactivity descriptor has been discussed in length in Chapter 4 and also needed here to rationalize the CFC adsorption phenomenon with zeolite. The aim is to study the effect of chlorine and fluorine concentration on the adsorption phenomenon to simplify the separation behavior. The relative nucleophilicity and electrophilicity were also calculated as mentioned earlier [64]. The need for the relative scale is to know simultaneously the tendency of the atom center to be more nucleophilic or electrophilic as the philicity is a site-dependent behavior for heteromolecular systems. *Ab initio* pseudopotential calculations were performed using Cambridge Serial Total Energy Package (CASTEP) and associated programs for symmetry analysis were used [65]. In this code, the wave functions of valence

TABLE 8.2

Global Softness, Local Softness, Relative Electrophilicity in Terms of Both Na–Y and H–Y for all the Interacting Halocarbons

Molecule	Global Softness (a.u.)	Local Softness	s_x^+/s_x^-	s_x^-/s_x^+	Interaction Energy (kJ/mol)
$NaSiAlO_7H_6$	2.9188	O1 0.291			
		O4 0.175			
		Na 0.350			
$Si_2AlO_{10}H_9$	2.9069	O1 0.055			
		O4 0.029			
		H 0.898			
CF_4	1.5901	C 0.063	1.361 (3.494)	0.734 (0.286)	−25.78
		F 0.257			
CF_3Cl	1.9138	C 0.202			
		F 0.213	1.643 (4.215)	0.608 (0.237)	−29.12
		Cl 1.351	0.259 (0.664)	3.860 (1.504)	−19.38
CF_2Cl_2	2.1440	C 0.397			
		F 0.391	0.895 (2.296)	1.117 (0.435)	−19.35
		Cl 0.879	0.398 (1.021)	2.511 (0.978)	−26.90
$CFCl_3$	2.3180	C 0.183			
		F 0.149	2.348 (6.026)	0.425 (0.165)	−33.43
		Cl 0.784	0.446 (1.145)	2.240 (0.873)	−30.49
CHF_3	1.7015	C 0.005			
		H 0.409	0.855 (2.195)	1.168 (0.455)	
		F 0.432	0.810 (2.078)	1.234 (0.481)	−12.54
$CHCl_3$	2.3036	C 0.103			
		H 0.152	2.302 (5.907)	0.434 (0.169)	
		Cl 0.751	0.466 (1.195)	2.145 (0.836)	−34.23

electrons are expanded in a basis set of plane wave with kinetic energy smaller than a specified cut of energy E_{cut}. We have taken 1 k point in the Brillouin zone. The kinetic energy cutoff used was 600 eV. The exchange correlation contribution to the total electronic energy was treated in the generalized gradient corrected (GGA) [66] form. Faujasite is a unit cell with Fd3m symmetry containing 576 atoms. To compromise between CPU and accuracy we used a rhombohedral cell with 144 atoms, where aluminum replaced one of the 48 silica atoms in the cell and H was attached to one of the O atoms bonded to Al. We first optimized the unit cell and then the complex with the interacting species was optimized. The smaller model can reproduce the local and extended environment for the adsorption complex. The results of global softness for the clusters, interacting molecules, and their local reactivity descriptors are shown in Table 8.2. From the local properties it is observed that in terms of local softness for the clusters the order of activity is H of H–Y > Na of Na–Y.

The order observed for the interacting molecules in terms of chlorine center — i.e., considering chlorine as the active site — is as follows:

$$CF_3Cl > CF_2Cl_2 > CFCl_3 > CHCl_3$$

The order of activity for the interacting species considering fluorine as the active center is

$$CHF_3 > CF_2Cl_2 > CF_4 > CF_3Cl > CFCl_3$$

This molecular-level interaction can be explored within the domain of the HSAB principle. From the results obtained it is possible to predict a qualitative activity order for the interacting halocarbons. Earlier studies [64] also suggested that the relative scales are more accurate than the individual localized scale when compared between the active nucleophile and electrophile. This ratio of s_x^+ and s_x^- and vice versa is a better descriptor. Hence, we have calculated the relative electrophilicity and nucleophilicity for the atom centers of the interacting species with that of the active center present in the zeolite clusters. We generated the qualitative scales in terms of the electrophilic centers present in the interacting CFCs. The scale is made with respect to the nucleophilic centers present in H–Y and Na–Y type zeolites, respectively. This implied the suitable interaction between the nucleophilic centers of zeolites and the electrophilic interacting centers of CFCs. The values are shown in Table 8.2. First in Na–Y zeolites, the trend observed for the relative electrophilicity considering a Cl center,

$$CF_3Cl > CF_2Cl_2 > CFCl_3 > CHCl_3$$

whereas for an F center the trend is

$$CFCl_3 > CF_2Cl_2 > CF_3Cl > CF_4 > CHF_3$$

The trend in terms of the relative nucleophilicity for a Cl center and that for an F center is opposite.

The trend for Cl is as follows:

$$CF_3Cl < CF_2Cl_2 < CFCl_3 < CHCl_3;$$

and for an F center

$$CFCl_3 < CF_2Cl_2 < CF_3Cl < CF_4 < CHF_3$$

Now, the trend for H–Y type zeolites, the relative electrophilicity with chlorine center is same as that observed for Na–Y zeolite, the trend is also the same for the relative nucleophilicity. But for the F center the order in terms of relative electrophilicity for H–Y zeolite is as follows:

$$CFCl_3 > CF_3Cl > CF_4 > CF_2Cl_2 > CHF_3$$

whereas in terms of relative nucleophilicity the trend is opposite. The trend is

$$CFCl_3 < CF_3Cl < CF_4 < CF_2Cl_2 < CHF_3$$

From the results it is observed that in case of Na–Y, both in terms of relative electrophilicity and nucleophilicity, the trend depicts a very nice correlation with the concentration of fluorine as present in the interacting CFC species, whereas for chlorine center the same trend is observed with an opposite direction of the order. In the other words, the chlorine and fluorine concentration correlations run opposite to each other, showing that there is a dependence of fluorine/chlorine concentration on the reactivity indices. For H–Y the trend for fluorine shows no correlation with the fluorine concentration, whereas that for chlorine behavior was the same in terms of concentration as observed in case of Na–Y zeolites. It seems obvious that CHF_3 and $CHCl_3$ are always present at the end of the series due to the presence of hydrogen in the interacting molecules, which will favor H-bonding as well in competition with interaction with other active center present in zeolites (like Na). After finding the trend of interaction of different halocarbon molecules, it is necessary to perform a periodic calculation that incorporates all the components of the zeolite matrixes around the interacting molecule. The supercage in Figure 8.9 shows the crystallographic sitting in faujasite. The structure of Na–Y was taken from the neutron diffraction data by Fitch et al. [67]. The framework is composed of cubooctahedral sodalite cages linked together in a tetrahedral arrangement by six-membered rings

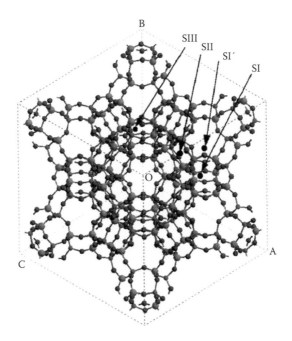

FIGURE 8.9 Zeolite Na–Y structure showing idealized cation positions.

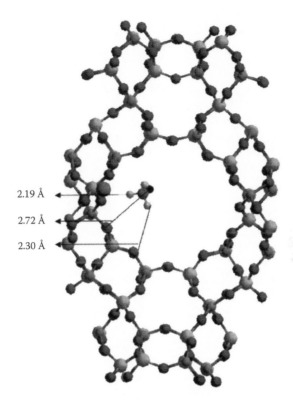

2.19 Å ⟵

2.72 Å ⟵

2.30 Å ⟵

FIGURE 8.10 The optimized geometry of CHF_3 inside Na–Y.

of O1 atoms to form large cavities, called supercages. The supercages were interconnected by windows that were formed by rings consisting of twelve Si/Al and twelve O atoms. The probable locations of cation positions are SI, SI′, and SII as shown in Figure 8.9. All structures with Na–Y zeolite were optimized first and the cation position was shifted from SI to the SII position of the supercage after optimization and is shown in Figure 8.10 as an example that represents the interaction between CHF_3 and Na–Y zeolite. The optimized geometry shows that the F–Na distance is 2.19 Å, whereas the distance between hydrogen to framework oxygen is 2.72 Å. These calculations are repeated for all the molecules within the supercage to reproduce the experimental situation. Only the cation position with which direct interaction is taking place is shown for visual clarity. This has been chosen in terms of its performance observed in relative scales. The interaction energy is shown in Table 8.2. This is calculated by the difference of energy between the complex formed by halocarbon with Na–Y and the bulk before complexation with the halocarbons. The interaction energy has been calculated in terms of both the fluorine and chlorine center of the CFC molecules. The trend obtained for the chlorine center is as follows:

$$CF_3Cl > CF_2Cl_2 > CFCl_3 > CHCl_3$$

And that for the fluorine center is

$$CFCl_3 < CF_2Cl_2 < CF_3Cl < CF_4 < CHF_3$$

The interaction energy as obtained from the periodic calculations shows the same trend as observed using the relative elctrophilicity/nucleophilicity scale. The calculations here indicate that H-bonding plays a very important role in defining the bonding arrangements. It was also observed that Cl and F show a marked difference in the bonding type and nature. This trend is then validated with the experimental observation of Grey et al. [68]. They showed from NMR and XRD that Na–F interactions play an important role for the binding of CF_2HCF_2H molecules on zeolite Na–Y compared to that of the CF_3–CFH_2 molecule. This causes the migration of cations from the SI′ position in the sodalite cages to the SII of the supercage, where they can bind to both ends of the HFC molecule. The combination of ^{23}Na NMR and diffraction data provides evidence that considerable sodium cation migrations occur on CF_2HCF_2H adsorption. The study shows a disappearance of SI′ and sufficient displacement of SII site attributed to local electrostatic interactions involving a combination of H-bonding with the framework and with the Na cations, which controlled the actual binding arrangement of the molecules [69]. However, cation migrations into the supercages were not observed in the earlier studies with structural refinements for the adsorption of chloroform, dichlorobenzene, and 1,4-dibromobutane in Na–Y [70]. Instead, a decrease in the SII occupancy with high loading levels was observed, due to the repulsion between the chloroform molecules and cations. In contrast, for dibromobutane the bromine molecules have a favorable interaction with SI′ and have a coordination with SII to propose a second adsorption site, which has been done with MD simulations. Hence, the reactivity descriptor, a simple methodology, has the capability to explain the complex scenario and is capable of predicting or designing new materials of interest.

Molecular dynamics has been used to study the gas separation efficiencies of FAU, MFI, and CHA zeolite membranes. This allows one to see the diffusion path of the gas molecules for a range of temperature and one can decisively put the molecules in a specific direction to move. This study examined the effect of pore size, pores structure, state conditions, and compositions on the permeation of two types of gas mixtures like oxygen/nitrogen and carbon dioxide/nitrogen. It is evident that for the mixture components with similar sizes and adsorption characteristics, such as oxygen/nitrogen, small-pore zeolites are not suited for separations. In the case of carbon dioxide/nitrogen binary mixture, the separation is mainly governed by adsorption and small-pore zeolites are more efficient. When selective adsorption takes place, for species with low adsorption, the permeation rate is low, even if the diffusion rate is quite high. For small-pore zeolites, such as MFI and chabazite, loading (adsorption) dominates the separation of gas mixtures. On the contrary, for larger-pore zeolites the rate diffusion plays an important role, though adsorption is still a significant parameter. This shows that molecular simulations can serve as useful screening tools to determine the suitability of a membrane for potential separation applications [71]. Again, the permeation of carbon dioxide/nitrogen gas mixture through the zeolite Na–Y membrane suggested that the difference in surface structure could provide an effect on the gas separation. The carbon dioxide/nitrogen separation factor of the (111) surface is much lower than that of the (100) surface. [72]. Molecular dynamics methods have also been extended to the separation of single and mixed gases of iso- and n-butane through MFI-type silicalite membranes and show that it can separate two isomers. The permeation of n-butane at 373 K takes place after the saturation of the zeolite pores, whereas at higher temperature, 773 K, it occurs without significant pore saturation. The calculated permeability of n-butane is close to experimental data [73].

The pervaporation (a method for the separation of mixtures of liquids by partial vaporization through a nonporous or porous membrane) separation of liquid mixtures of water/ethanol and water/methanol using three zeolite (silicalite, NaA, and chabazite) membranes has been examined using the method of molecular dynamics. The simulations correctly exhibited all the qualitative experimental observations for these systems, including the hydrophobic or hydrophilic behavior of zeolite membranes. The simulations showed that for silicalite the separation was strongly influenced by the selective adsorption of ethanol but in NaA and chabazite, pore size was found to play an important role in the separation [74].

The permeation of supercritical fluid through the zeolite membrane is interesting from both a fundamental and a practical point of view. Several experimental studies have already been done on the separation of oil species, homogeneous catalyst from the fluid while maintaining the supercritical condition. Molecular dynamics has been used to study the separation of supercritical water from sodium chloride aqueous solution using ZK-4 zeolite membranes [74]. Results indicate that molecular dynamics is a viable technique for studying such separation processes at the fundamental molecular level. This study is also looked into the effect of the electric field on separation-- one of the key reasons to choose molecular dynamics.

One of the most innovative applications of zeolite membranes is their use as membrane extractor-reactors, which could enhance the activity and selectivity of a reaction. A unique zeolite membrane with a combination of hydrophilic entrance and hydrophobic exit using computer simulation methodology was developed. Based on the Si/Al ratio, a gradient-type membrane reactor was designed in terms of stability and its capability for adsorption and diffusion processes. A gradient-type membrane is defined as a membrane when there is a regular or graded ascent or descent in terms of any measurable parameter like adsorption and diffusion. The stability of a gradient membrane depends on its compatibility with its constituents, which can be measured by the diffusivity and from the amount mismatch between the components. Experimentally, the synthesis of zeolitic gradient-type membranes is critical and challenging due to the surface matching. Moreover, complete interpretation of adsorption and diffusion processes of guest molecules in zeolites only by experiments is quite difficult or impossible. So computer simulation is highly desirable to get the fundamental insight into the nature of the adsorption and diffusion behavior of the zeolite and adsorbate system. A combination of Si/Al 300 (hydrophobic) and Si/Al 30 (hydrophilic) was found to be the best match in terms of percentage mismatch as well as binding energy. The adsorption and diffusion regarding the conversion of methanol to ethylene was examined on the resultant gradient-type membrane. Depending on the Si/Al ratio, a significant difference in the adsorption isotherm and

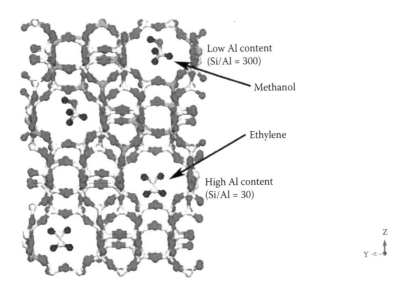

FIGURE 8.11   (For color version see accompanying CD) Diffusion path of methanol through silicalite and the gradient-type membrane.

diffusion behavior of methanol and ethylene molecule was observed. Methanol was adsorbed selectively at high-aluminum-content sites, whereas ethylene passed easily without being adsorbed independent of the Si/Al ratio (Figure 8.11). A plausible mechanism was proposed to justify the product formation and separation on the newly designed membrane material [75].

Since their discovery, mesoporous materials have attracted much attention due to their adjustable pore size, large surface area, and long arrays of ordered cylindrical pores. It is obvious that these types of materials have a wide range of application. Several experiments have been carried out to use mesoporous materials as separating membranes. For example, Park et al. [76] used surface-modified mesoporous silica (silylated mesoporous) membranes for the separation of organics and alcohol and achieved very high permeselectivity. Knowles et al. [77] used HMS (hexagonal mesoporous silica) for carbon dioxide separation. However, compared to the crystalline zeolite material, theoretical studies are limited. Furukawa et al. [78] investigated the effect of surface modification of mesoporous silica by NVT-ensemble molecular dynamics techniques with the melt-quench algorithm for modeling a nonsilylated (an 011 surface model) and a fully silylated mesoporous silica (an FS surface model), and the mu VT-ensemble orientational-bias Monte Carlo method for calculating adsorption isotherms. Good agreement was obtained between simulations and experiments for adsorption of pure acetone and water at 298 K on nonsilylated and silylated forms and a considerably large separation factor of 7300 was obtained for the adsorption from a liquid mixture in the FS surface model [78]. Molecular simulation techniques were also applied to the next-generation membrane materials, which are composed of organic–inorganic nanocomposites. Molecular dynamics simulation was used to gain insight on the relationship between microstructure and separation performance in a prototype solubility-based separation (propane/nitrogen). The permeselectivity data obtained from the simulation corresponds well with experiments. According to the results, the most critical variable is the ratio of the size of the organic group length to pore size, and as the size of the organic group increases, the permeselectivity increases [79].

The isotopic separation factors for T_2/H_2 and D_2/H_2 mixtures adsorbed in various metal-organic frameworks (IRMOF-1, Cu–BTC) have been calculated using path integral Monte Carlo simulations. The results are in fair agreement with the available experimental data and show a very strong dependence on temperature. The selectivity shows a less pronounced dependence on the external pressure, but different behaviors were found for different materials. They were analyzed from the point of view of the solid–fluid potential energy surface [80].

In addition to the above-mentioned applications, nanoporous materials can be used in the electronic industry as electrodes and for different types of biological applications such as in bioseparation. Expanding the range of pore sizes can expand their range of applicability. Considering the clean and green environmental applications, they have tremendous potential to act as storage materials, greenhouse gas reduction catalysts, cost-effective photocatalytic materials, and in nano-imprinting lithography for integrated circuit manufacturers.

There are many challenges stored for future in the application part of the nanoporous material. The most studied part is catalysis, for which it is challenging to obtain the effective catalyst for high selectivity and high yield. Moreover, designing of the new materials with improved pore size, thermal and hydrothermal stability, and structural modification by surface functionalization using the combination of experimental and theoretical techniques is a great challenge to the material scientist. This chapter aimed to prescribe different new applications of zeolites apart from their orthodox usage in heterogeneous catalysis. Within the new applications, zeolites are used in separation, adsorption, chromatographic analysis, and are potential candidates for sensors and other applications in biological fields. The applications are mainly due to the porous structure, which allows shape selectivity as well the acidity both of Brønsted and Lewis type makes of very active species for a reaction. The sensing works well within zeolites for their capability to adsorb and select materials from a mixture. It is unreasonable only to emphasize the application of zeolite materials but it is preferable to extend to nanoporous materials, which found new applications in photonic crystals or bio-implants. In addition, the combination of mesoporous and zeolite materials or combining polymeric structures within the micro mesopore enhances its usage in optical property as a film for diode applications. Zeolites are now used more for composite-type materials where they provide an edge over carbon nanotubes, because it is difficult to synthesize nanotubes with activity similar to zeolites, where the pores and architectures are tunable. The modeling simulation offers to probe the application domain with mainly atomistic simulation like MD to probe the dynamics of the mixture, followed by Monte Carlo-type simulation to probe the adsorption capability and to monitor the permeation process. Finally, DFT remains a potential tool to quantitatively propose the mode of the interaction, adsorption, and selectivity from its versatile functionalization; the merit is visualization and understanding, to give shape to the dream materials.

REFERENCES

1. (a) N. M. Lawandy, R. M. Balachandran, A. S. L. Gomes, E. Sauvain, *Nature*, 368 (1994) 436; (b) B. T. Holland, C. F. Blanford, A. Stein, *Science*, 281 (1998) 538.

2. (a) K. Busch, S. John, *Phys. Rev. Lett.* 83 (1999) 967; (b) S. F. Mingaleev, M. Schillinger, M. Hermann, K. Busch, *Opt. Lett.*, 29 (2004) 2858; (c) M. Schmidt, M. Eich, U. Huebner, R. Boucher, *Appl. Phys. Lett.*, 87 (2000) 121110; (d) K. Inoue, M. Sasada, J. Kawamata, K. Sakoda, J. W. Haus, *Jpn. J. Appl. Phys.*, 38 (1999) L157; (e) R. Lawrence, Y. Ying, P. Jiang, S. H. Foulger, *Adv. Mater.*, 18 (2006) 300; (f) T. Hanic, C. M. de Sterke, M. J. Steel, *Optic. Express*, 14 (2006) 12451; (g) S. K. Lee, G. R. Yee, J. H. Moon, S. M. Yang, D. J. Pine, *Adv. Mater.*, 18 (2006) 2111; (h) S. Furumi, H. Fudoizi, H. T. Miyazaki, Y. Sakka, *Adv. Mater.*, 19 (2007) 2067; (i) Y.-C. Kim, Y. R. Do, *Optic. Express*, 13 (2005) 1598; (j) B. Kannan, K. Castelino, A. Majumdar, *Nano Letters*, 3 (2003) 1729.

3. E. Yablonovitch, *Phys. Rev. Lett.*, 58 (1987) 2059.

4. C.B. Bowden, (a) *Development and Applications of Materials Exhibiting Photonic Band Gaps*, C. M. Bowden, J. P. Dowling, H. O. Everitt (eds.); (b). G. Kuriziki, J. W. Haus, *J. Opt. Soc. Am.* 2 (1993) 10; (c) J. D. Joannopoulos, R. D. Meade, J. N. Winn, *Photonic Crystals*, Princeton University Press, Princeton, NJ (1995).

5. E. Yablonovitch, *Phys. Rev. Lett.*, 58 (1987) 2059.

6. E. N. Economou, M. M. Sigalas, *Phys. Rev. B*, 48 (1993) 13434.

7. J. E. G. J. Wijnhoven, W. L. Vos, *Science*, 281 (1998) 802.

8. S. Tay, J. Thomas, B. Momeni, M. Askari, A. Adibi, P. J. Hotchkiss, S. C. Jones, S. R. Marder, R. A. Norwood, N. Peyghambarian, *Appl. Phys. Lett.*, 91 (2007) 221109.

9. T. Sun, D. Zhu, Z. Yang, Z. Liu, Y. Liu, *Appl. Phys. B*, 82 (2006) 89.

10. R. Berardi, A. Costantini, L. Muccioli, S. Orlandi, C. Zannoni, *J. Chem. Phys.*, 126 (2007) 044905.

11. R. G. Narayan, R. Aggarwal, W. Wei, C. Jin, N. A. Montereo-Riviere, R. Crombez, W. Shen, *Biomed. Mater.*, 3 (2008) 034107.

12. F. Ganazzoli, G. Raffaini, *Phys. Chem. Chem. Phys.*, 7 (2005) 3651.

13. M. Agashe, V. Raut, S. J. Stuart, R. A. Latour, *Langmuir*, 21 (2005) 1103.

14. R. Major, P. Lacki, *Arch. Metall. Mater.*, 50 (2005) 379.

15. S. Weiner, H. D. Wagner, *Ann. Rev. Mater. Sci.,* 28 (1998)271.

16. J. H. Evans, J. C. Barbenel, *Equine Vet. J.,* 7(1975) 1.

17. K. M. Meek, N. J. Fullwood, *Micron,* 32 (2001) 261.

18. H. Zhang, X. Lu, Y. Leng, L. Fang, S. Qu, B. Feng, J. Weng, J. Wang, *Acta Biomaterialia*, 5 (2008) 1169.

19. M. Corno, A. Pedone, R. Dovesi, P. Ugliengo, *Chem. Mater.*, 20 (2008) 5610.

20. E. C. Dickey, O. K. Verghese, K. G. Ong, D. W. Gong, M. Paulose, C. A. Grimes, *Sensors*, 2 (2002) 91.

21. Y. Ueno, *Bunseki Kagaku*, 57 (2008) 871.

22. J. F. Fernandez-Sanchez, R. Cannas, S. Spichiger, R. Steiger, U. E. Spichiger-Keller, *Anal. Chim. Acta*, 566 (2006) 271.

23. A. Sierraalta, R. Anez, M. Brussin, *J. Catal.*, 205 (2002)107.

24. I. D. Block, L. Chan, B. T. Cunningham, *Sensor and Actuator*, 120 (2006) 187.

25. W. H. Ryu, Y. C. Chung, D. K. Choi, C. S. Yoon, C. K. Kim, Y. H. Kim, *Sensor Actuator*, 97 (2004) 98.

26. (a) K. Moller, T. Bein, *Stud. Surf. Sci. Catal.*, 117 (1998) 53; (b) C. E. Fowler, B. Lebeau, S. Mann, *Chem. Comm.*, 17 (1998) 1825; (c) M. C. Burleigh, M. A. Markowitz, M. S. Spector, B. P. Gaber, *J. Phys. Chem. B*, 105 (2001) 9935; (d) T. Asefa, C. Yoshina-Ishii, M. J. MacLachlan, G. A. Ozin, *J. Mater. Chem.*, 8 (2000) 1751.

27. D. Zhao, J. Feng, Q. Huo, N. Melosh, G. H. Fredrickson, B. F. Chmelka, G. D. Stucky, *Science*, 279 (1998) 548.

28. A. Sayari, *Chem. Mater.*, 8 (1996) 1840.
29. A. Chatterjee, T. Iwasaki, T. Ebina, *J. Phys. Chem. A*, 103 (1999) 2489.
30. A. Chatterjee, T. Balaji, H. Matsunaga, F. Mizukami, *J. Mol. Graph. Model.*, 25 (2006) 208.
31. K. Fukui, T. Yonezawa, H. Shingu, *J. Chem. Phys.*, 20 (1952) 722.
32. (a) B. Delley, *J. Chem. Phys.*, 92 (1990) 508; (b) B. Delley, *J. Chem. Phys.*, 94 (1991) 7245; (c) B. Delley, *J. Chem. Phys.*, 113 (2000) 7756.
33. B. Delley, DMol^3 a standard tool for density functional calculation review and advances in *Theoretical and Computational Chemistry*, vol. 2, J. M. Seminario, P. Politzer (eds.), Elsevier Science, Amsterdam (1995).
34. (a) A. D. Becke, *J. Chem. Phys.*, 88 (1988) 2547; (b) C. T. Lee, W. T. Yang, R. G. Parr, *Phys. Rev. B*, 37 (1988) 785.
35. S. F. Boys, F. Bernardi, *Mol. Phys.*, 19 (1970) 553.
36. A. Klamt, G. Schürmann, *J. Chem. Soc. Perkin Trans.*, 2 (1993) 799.
37. A. Chatterjee, T. M. Suzuki, Y. Takahashi, D. A. P. Tanaka, *Chem. Eur. J.*, 9 (2003) 3920.
38. (a) M. Seko, T. Miyake, T. Inada, *Ind. Eng. Chem. Res. Dev.*, 18 (1979) 263; (b) M. Milewski, J. M. Berak, *Separ. Sci. Tech.*, 17 (1982) 369; (c) S. Namba, J. H. Kim, T. Komatsu, T. Yashima, *Microporous Mater.*, 8 (1997) 39; (d) S. Namba, Y. Kanai, H. Shoji, T. Yashima, *Zeolites*, 4 (1984) 77.
39. S. Mohanty, A. V. McCormick, *Chem. Eng. J.*, 74 (1999) 1.
40. L. Lu, X. Lu, Y. Chen, L. Huang, Q. Shao, Q. Wang, *Fluid Phase Equilib.*, 259 (2007) 135.
41. A. Ghoufi, L. Gaberova, J. Rouquerol, D. Vincent, P. L. Llewellyn, G. Maurin, *Microporous and Mesoporous Mater.*, 119 (2009) 117.
42. A. Stefanis, G. Romani, E. Semprini, F. Stefani, A. Tomlinson, G. Perez, *J. Porous Mater.*, 9 (2002) 97.
43. S. Himeno, M. Takenaka, S. Shimura, *Mol. Sim.*, 34 (2008) 1329.
44. M. Gallo, T. M. Nenoff, M. C. Mitchell, *Fluid Phase Equil.*, 247 (2006) 135.
45. A. Chatterjee, F. Mizukami, *J. Phys. Chem. A*, 103 (1999) 9857.
46. A. Chatterjee, T. Iwasaki, T. Ebina, *J. Phys. Chem. A*, 103 (1999) 2489.
47. Y. Zeng, S. Ju, W. Xing, C. Chen, *Separ. Purif. Tech.*, 55 (2007) 82.
48. P. Cosoli, M. Ferrone, S. Pricl, M. Fermeglia, *Chem. Eng. J.*, 145 (2008) 86.
49. S. Wang, Q. Yang, C. Zhong, *Separ. Purif. Tech.*, 60 (2008) 30.
50. (a) J. Caro, M. Noack, P. Kolsch, R. Schafer, *Microporous and Mesoporous Materials*, 38 (2000) 3; (b) A. Tavolaro, E. Drioli, *Adv. Mater.*, 11 (1999) 975.
51. (a) H. H. Funke, M. G. Kovalchick, J. L. Falconer, R. D. Noble, *Ind. Eng. Chem. Res.*, 35 (1996) 1575; (b) H. Kita, T. Inoue, H. Asamura, K. Tanaka, K. Okamoto, *Chem. Comm.*, (1997) 45; (c) K. Kusakabe, T. Kuroda, S. Morooka, *J. Membr. Sci.*, 148 (1998) 13.
52. J. Caro, M. Noack, *Microporous and Mesoporous Materials*, 115 (2008) 215.
53. P. I. Pohl, G. S. Heffelfinger, D. M. Smith, *Mol. Phys.*, 89 (1996) 1725.
54. Y. Kobayashi, S. Takami, Y. Kubo, A. Miyamoto, *Fluid Phase Equil.*, 194 (2002) 319.
55. W. Jia, S. Murad, *J. Chem. Phys.*, 122 (2005) 234708.
56. L. E. Menzer, *Science*, 249 (1990) 31.
57. D. R Corbin, B. A. Mahler, World Patent W.O. 94/02440 (1994).
58. M. F. Ciraolo, J. C. Hansen, B. H. Toby, C. P. Grey, *J. Phys. Chem. B*, 105 (2001) 12330.
59. K. H. Lim, F. Jousse, S. M. Auerbach, C. P. Grey, *J. Phys. Chem. B*, 105 (2001) 9918, and references therein.
60. H. Stach, K. Sigrist, K.-H. Radeke, V. Ridol, *Vacc. Chem. Tech.*, 47 (1995) 55.
61. S. Kobayashi, K. Mizuno, S. Kushiyama, R. Aizawa, Y. Koinuma, H. Ohuchi, *Ind. Eng. Chem. Res.*, 30 (1991) 2340.
62. A. Chatterjee, T. Ebina, T. Iwasaki, F. Mizukami. *J. Mol. Struct.*, 630 (2003) 233.
63. R. Shah, J. D. Gale, M. C. Payne, *J. Phys. Chem.*, 100 (1996) 11688.

64. A. Chatterjee, T. Iwasaki, T. Ebina, *J. Phys. Chem. A*, 106 (2002) 641.

65. (a) M. P. Teter, M. C. Payne, D. C. Allen, *Phys. Rev. B*, 40 (1989) 12255; (b) M. C. Payne, M. P. Teter, D. C. Allen, T. A. Arias, J. D. Johannopoulos, *Rev. Mod. Phys.*, 64 (1992) 1045.

66. (a) J. P. Perdew, Y. Wang, *Phys. Rev. B*, 46 (1992) 6671; (b) J. A. White, D. M. Bird, *Phys. Rev. B*, 50 (1994) 4954.

67. A. N. Fitch, H. Jobic, A. Renouprez, *J. Phys. Chem.*, 90 (1986) 1311.

68. C. P. Grey, F. I. Poshini, A. F. Gualtieri, P. Norby, J. Hansen, D.R. Corbin, *J. Am. Chem. Soc.*, 119 (1997) 1981.

69. K. H. Lim, F. Jouse, S. M. Auerbach, C. P. Grey, *J. Phys. Chem. B*, 105 (2001) 9918.

70. (a) Z. A. Kaszkur, R. H. Jones, J. H. Couves, D. Waller, R. A. Catlow, J. M. Thomas, *J. Phys. Chem. Solid.*, 52 (1991) 1219; (b) Z. A. Kaszkur, R. H. Jones, D. Waller, R. A. Catlow, J. M. Thomas, *J. Phys. Chem.*, 97 (1993) 426.

71. K. Mizukami, Y. Kobayashi, H. Morito, S. Takami, M. Kubo, R. Belosludov, A. Miyamoto, *Jpn. J. Appl. Phys.*, 39 (2000) 4385.

72. H. Takaba, R. Koshita, K. Mizukami, Y. Oumi, N. Ito, M. Kubo, A. Fahmi, A. Miyamoto, *J. Membr. Sci.*, 134 (1997) 127.

73. W. Jia, S. Murad, *Mol. Phys.*, 104 (2006) 3033.

74. S. Murad, J. Lin, *Ind. Eng. Chem. Res.*, 41 (2002) 1076.

75. A. Chatterjee, M. Chatterjee, *Mol. Simulat.*, 34 (2008) 1091.

76. D. H. Park, N. Nishiyama, Y. Egashira, K. Ueyama, *Microporous and Mesoporous Materials*, 66 (2003) 69.

77. G. P. Knowles, S. W. Delaney, A. L. Chaffee, *Ind. Eng. Chem. Res.* 45 (2006) 2626.

78. S. Furukawa, T. Nishiumi, N. Aoyama, T. Nitta, M. Nakano, *J. Chem. Eng. Jpn.*, 38 (2005) 99.

79. T. Aydogmus, D. A. Ford, *J. Membr. Sci.*, 314 (2008) 173.

80. G. Garberoglio, *Chem. Phys. Lett.*, 467 (2009) 270.

81. P. Payra, P. K. Dutta, *Microporous Mesoporous Mater.* 64 (2003) 109.

82. K. Kaneyasu, K. Otsuka, Y. Setoguchi, S. Sonoda, T. Nakahara, I. Aso and N. Nakagaichi, *Sensor Actuator B-Chem.* 66 (2000) 56.

83. C. Chuapradit, L. Ruangchuay Wamatong, D. Chotpattananont, P. Hiamtup, A. Sirivat and J. Schwank, *Polymer* 46 (2005) 947.

84. L. Gonzalez, M. E. Franke, U. Simon, *Stud. Surf. Sci. Catal.* 158 (2003) 2049.

85. A. Dubbe, K. Moss, *Electrochem. Solid State Lett.* 9 (2008) 31.

86. O. Schaf, V. Wernert, H. Ghobarkar, P. Knauth, *J. Electroceram.*, 16 (2006) 93.

87. N. F. Szabo, P. K. Dutta, *Sensor Actuator B-Chem.*, 88 (2003) 168.

88. H. Huang, J. Zhou, S. Chen, L. Zeng, Y. Huang, *Sensor Actuator B-Chem.*, 101 (2004) 316.

89. X. Xu, J. Wang, Y. Long, *Microporous Mesoporous Mater.*, 83 (2005)60.

90. G. Grubert, M. Stockenhuber, O. P. Tkachenko, M. Wark, *Chem. Mater.*, 14 (2002) 2458.

9 Catalytic Reactions

The longstanding goal of computer simulations of materials is the accurate knowledge of structural and electronic properties for realistic condensed matter systems. Because nanomaterials are complex in nature, in the past theoretical investigations were restricted only to simple models. Recent progress in computer technology has made it possible to afford a realistic description of a wide range of materials. The initial chapters of this book have explored the different methodologies involved in explaining the structure–property correlation. In the last few chapters we discussed the synthesis and characterization of nanoporous materials. Now, the designed material, which is synthetically viable and technically applicable in terms of pore architecture, metal loading as established through experimental technique and computer simulation technology needs to be tested. In this chapter we focus on the catalytic activity of nanoporous materials. The discussion will cover the related topics of catalysis starting from shape-selective reactions within nanopores, chemical adsorption reactions, cracking, and, finally, the mechanistic aspect of a reaction using transition state theory, especially the activation barrier, intrinsic reaction coordinate, and the effect of solvent on catalytic reactions. This chapter will provide a detailed explanation of the methodology of simulation to approach reaction mechanism with a special emphasis on catalytic reaction, so that one can follow these cases and be able to approach their own problems for a solution.

9.1 SHAPE SELECTIVITY

The most important aspect in nanoporous catalysts is porosity. The porous structure of nanoporous materials is indeed a complicated, interesting, and most significant topic for future challenges, especially as catalysts. Hence, it will be of interest to discuss the role of voids in catalytic reactions.

Shape selectivity is vital for nanoporous catalysts and it can be divided into three parts: (1) reactant, (2) transition state, and (3) product. Shape selectivity generally arises due to the constraint imposed on the reactant, transition state, or product molecule by various pore architectures. So, for a shape-selective reaction, product selectivity depends on pore structure and the shape of the reactant/transition state/product molecule. Generally, selectivity can be categorized as chemo-, regio-, and stereoselectivity.

Regioselectivity is associated with the isomerization reaction over zeolite catalysts, which depends on multiple overlapping factors. The demanding parameters are relative thermodynamic stabilities of product isomers, relative diffusivities of isomers through the catalyst matrix, and the presence of pore-modifying species within the framework. The crystalline architecture of microporous materials, however, also dictates the way the catalyst material interacts with the reaction products. Within this realm of interactions, the key factor is the entrance of the molecule, which maintains shape and size restrictions. Modeling methods provide valuable insight into the mechanism of shape selectivity through simulation of intermolecular interactions

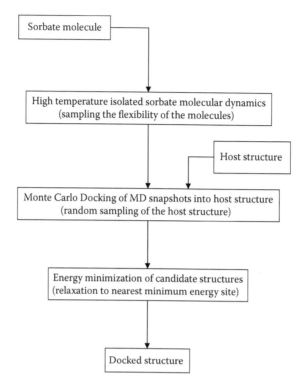

SCHEME 9.1   Workflow for the calculation scheme.

between guest molecules and porous host materials. Tomlinson-Tschaufeser and Freeman [1] have applied molecular modeling technology to investigate butene isomer selectivity over several zeolite structures (fifty-five types) and ranked the predicted performance of potential zeolite catalysts. The procedure for the calculation is shown in Scheme 9.1. The first step is to see a high-temperature dynamics of the sorbate molecule in vacuum to figure out the conformational flexibility in terms of the local minima. Once the local minima of those molecules were obtained, the initial structures were subjected to the docking process by using a Monte Carlo method based on the Metropolis method [2]. The docking process is within the cages of the all possible zeolites chosen for this study. After docking, the structures were minimized by a CVFF force field using the DISCOVER engine of MSI (DISCOVER, Molecular Simulations Inc., San Diego). The sorption energies, as defined above, of each butane isomer were calculated for 100 different configurations for each host structure and from these energies the average sorption energy was evaluated. The ratio of butene to isobutene binding energies are called the *selectivity* and are plotted for different zeolites (Figure 9.1). The ratio of binding energies of butene:isobutene is 0.5:1.8. In TON, isobutene has almost half the binding energy of butene, whereas in MAZ both isomers bind with approximately equal strength. The characteristic that promotes high selectivity is the internal pore geometry, which provides good correlation with the dimension of one adsorbate in comparison with another. For TON,

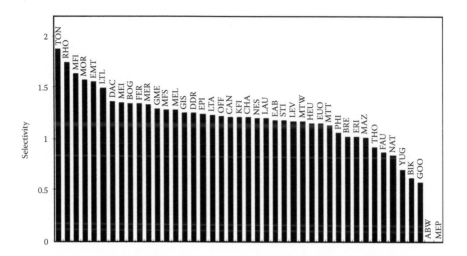

FIGURE 9.1 Zeolites were ranked in terms of the selectivity; only zeolites with channels are considered. [1] With permission.

this is obtained by the cross-channel mode of binding, which is only accessible to the butene isomer. For FAU, no pore features are present in the range of dimensions exhibited by the butene isomers and low degrees of selectivity are obtained by calculations. The simulation studies have shown that the TON framework has the highest selectivity compared to other frameworks evaluated.

The structural role and interaction energy introduced by the zeolite framework in this reaction were studied using spirolactone to enone conversion reaction over zeolite catalysts [3]. The shape-selective behavior of various zeolites is rationalized by comparing the dimension of the molecules and zeolite pore diameter. It is observed that the spirolactone and enone molecules have dynamic freedom to hop among the various sites inside the supercages of zeolite-Y. The abstraction of proton at the Brønsted acid site by ketonic oxygen of the reactant has been indicated as the first step in the reaction mechanism. The interaction energy of the molecules with the framework cluster and the electron redistribution occurring in the reactant molecules due to adsorption are brought out. These results are useful to understand the mechanism of dehydration of spirolactones to enones. The equilibrium geometry of the reactant and product molecules shown in Figure 9.2 was obtained by force field calculations developed by Gelin and Karplus [4]. The total strain energy of the molecule is expressed by the following equation, which is explained in detail in the literature [5].

$$E_{\text{total strain}} = E_{\text{bonded}} + E_{\text{nonbonded}}$$

where

$$E_{\text{bonded}} = E_{\text{bond length}} + E_{\text{bond angle}} + E_{\text{dihedral angle}} + E_{\text{improper torsion}}$$

$$E_{\text{non bonded}} = E_{\text{electrostatic}} + E_{\text{Van der Waals}}$$

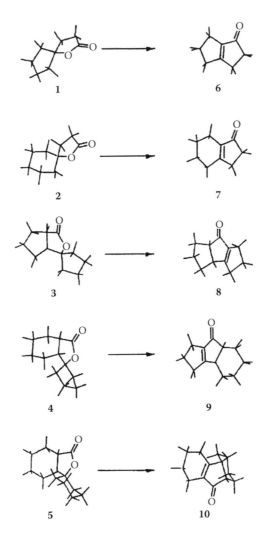

FIGURE 9.2 The three-dimensional structure of spirolactone and enone molecules, as derived from force field calculations.

The respective expressions used to calculate the individual terms were given in earlier studies [6]. The visualization and calculations of energy were performed using Quanta/Charm M software packages distributed by Polygen Corporation. The strain energy of the molecule was minimized as a function of geometry by the steepest descent method to eliminate initial bad contacts and then later by the conjugate gradient and Newton-Raphson methods. The H–Y lattice was modeled from the structure reported by X-ray crystallographic studies [7] followed by semiempirical quantum chemical method using AM1 Hamiltonian [8].

With the advance of computer graphics, a method has been developed that assumes that the molecules are exactly fitting inside a smallest possible rectangular

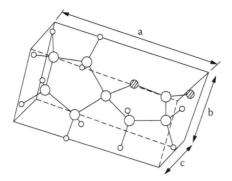

FIGURE 9.3 Three largest dimensions of the molecule as used in the molecular fitting procedure. Spirolactone (1) is shown as a typical example. Shaded circles, large empty circles, and small empty circles are oxygen, carbon, and hydrogen atoms, respectively.

box and the dimensions of the molecules are the dimensions of the box (Figure 9.3) as shown in Table 9.1. The three largest dimensions of the molecule are used in the molecular fitting procedures. Spirolactone (1) is shown as a typical example. Shaded circles, large empty circles, and small empty circles are, respectively, oxygen, carbon, and hydrogen atoms. The product molecules (6–10) are of similar or smaller dimensions than the reactant molecules (1–5). The dimensions of the intermediates

TABLE 9.1

Total Strain Energy and Dimensions of Various Molecules at Their Minimum Energy Configuration

Molecule	No. of Atoms	Dimension (Å)	Strain energy (kcal/mol)		
			Total	Bonded	Nonbonded
1	22	7.00 × 4.25 × 4.00	15.7889	21.5913	−5.8035
2	25	7.00 × 4.25 × 4.25	6.4875	10.2886	−3.8035
3	29	7.50 × 5.50 × 5.50	20.8836	21.8529	−0.9693
4	32	7.50 × 6.00 × 6.00	19.8921	24.8102	−4.9182
5	35	7.75 × 5.65 × 5.65	8.5793	12.3631	−3.7837
11	22	6.25 × 5.25 × 5.00	2.1288	14.2986	−12.1699
12	25	7.25 × 5.25 × 5.25	−10.9613	2.0318	−12.9931
13	29	7.00 × 5.50 × 5.50	17.1363	25.7821	−8.6458
14	32	7.00 × 5.75 × 5.70	5.0534	15.5062	−10.4528
15	35	8.25 × 5.85 × 5.85	−5.5627	3.9035	−9.4662
6	19	6.50 × 4.50 × 4.50	17.8080	19.9514	−2.1435
7	22	6.00 × 4.65 × 4.65	15.6723	15.6289	0.0434
8	26	6.75 × 5.10 × 5.10	27.7583	30.1872	−2.4270
9	29	7.10 × 5.50 × 5.50	21.7622	24.8070	−3.0447
10	32	8.00 × 5.25 × 5.25	21.7622	17.8135	0.7447

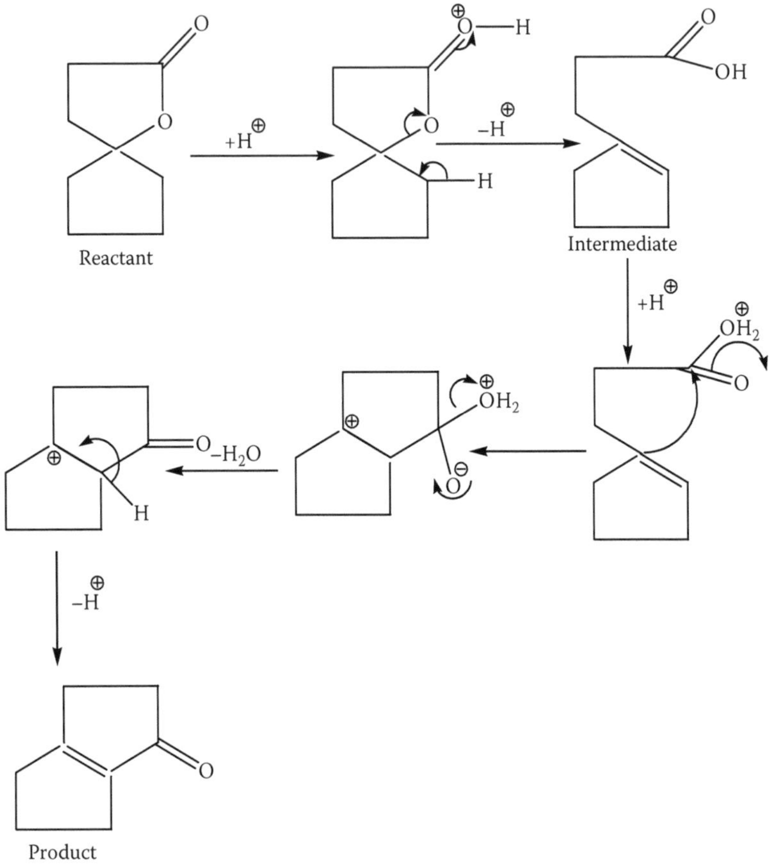

SCHEME 9.2 Proposed mechanism of spirolactone to enone.

(11–15) formed from the reactants (1–5) as shown in Table 9.1 according to the proposed mechanism in Scheme 9.2. It is convention to neglect the largest dimension of the molecule, because the molecules can diffuse into the channels and cages with their largest dimension lying parallel to the channel axis. The intermediate molecules have more conformational flexibility. In general, the intermediates occupy more space and have less strain energy. In contrast, the product molecules occupy less space with more strain energy. In Table 9.1, the strain energy due to bonded and nonbonded interactions is reported. Nonbonded energy arises from electrostatic and van der Waals interactions. Favorable and unfavorable nonbonded term arises due to the attraction between polar groups with unlike charges and repulsion between polar groups with like charges, respectively. The result shows that the bonded energy terms are unfavorable, whereas the nonbonded energy terms are favorable, as indicated by almost zero or negative values. Compared to the reactants, intermediates have favorable nonbonded energy and products have unfavorable nonbonded energy. Thus, relatively unfavorable nonbonded energy for product molecules indicates the

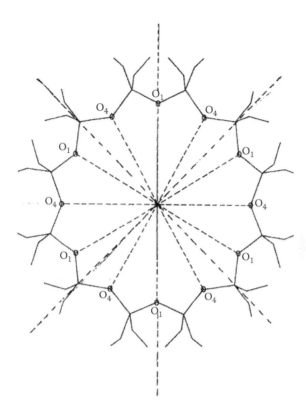

FIGURE 9.4 The molecular graphics view of the twelve-member ring. Pore opening in zeolite H–Y as viewed along the *b* axis.

presence of positively charged groups in the molecule. These results indicate that the positive charges in the molecules have ionic interaction with the oxygen of the zeolite framework, which could lead to their stronger adsorption inside the zeolite void volumes. Moreover, it can be further correlated with the zeolite pore architecture to give a generalized prescription for the selective behavior. To check the electronic interaction between the zeolite matrix and the reactant molecule, the H–Y zeolite (Si/Al = 2.6) was chosen because it gives the best yield and has the best capability in terms of size fitting. The twelve-member pore opening in zeolite H–Y has C6v symmetry along the pore opening as shown in Figure 9.4. Two terameric clusters, each of them centered on O1 and O4, represent two possible interaction sites and the proton is located in the twelve-membered ring. The formula of the representative cluster is $H_{11}Si_3AlO_{13}$. The hydrogen was positioned at the silicon position and kept fixed during the calculation, to keep the environment of zeolite. The interaction energy was calculated as follows and is shown in Table 9.2:

$$\text{Interaction energy} = \text{Total energy of complex} - (\text{Total energy of framework} + \text{Total energy of the interacting molecule})$$

TABLE 9.2

Interaction Energy of Reactants and Products with Framework Cluster Centered on Oxygen Site

Molecule	Total Energy (eV)	Interaction Energy (eV)	Charge on Oxygen Ring	Charge on Oxygen Ketonic	Bond Order Between C and O in the Ring
Cluster centered at O1 (FWC1)	−4588.55	—	—	—	—
Cluster centered at O4 (FWC4)	−4588.39	—	—	—	—
Water molecule	−348.56	—	—	—	—
FWC1 + 1	−6417.67	0.13	−0.28	−0.31	0.933
FWC4 + 1	−6417.62	0.12	−0.25	−0.34	0.922
FWC1 + 2	−6575.15	0.28	−0.26	−0.26	0.922
FWC4 + 2	−6574.86	0.41	−0.23	−0.22	0.919
FWC1 + 3	−6858.86	0.11	−0.21	−0.29	0.920
FWC4 + 3	−6858.73	0.07	−0.24	−0.31	0.922
FWC1 + 4	−7014.41	0.26	−0.25	−0.27	0.920
FWC4 + 4	−7014.12	0.38	−0.24	−0.23	0.922
FWC1 + 5	−7170.35	0.20	−0.26	−0.28	0.921
FWC4 + 5	−7169.72	0.67	—	−0.21	—
FWC1 + 6	−6069.86	−0.03	—	−0.26	—
FWC4 + 6	−6069.55	0.01	—	−0.28	—
FWC1 + 7	−6226.30	−0.04	—	−0.29	—
FWC4 + 7	−6226.05	0.08	—	−0.29	—
FWC1 + 8	−6509.29	−0.01	—	−0.29	—
FWC4 + 8	−6509.05	0.07	—	−0.29	—
FWC1 + 9	−6664.45	0.13	—	−0.25	—
FWC4 + 9	−6664.39	0.09	—	−0.28	—
FWC1 + 10	−6821.36	−0.07	—	−0.29	—
FWC4 + 10	−6821.09	0.04	—	−0.29	—

In the zeolite cluster-organic molecule complex, it was found that the major interaction occurred between the methylene hydrogen of the organic molecule and the oxygen of framework. Consequently, the interaction energy of the enones with the framework are relatively more favorable than the spirolactones. Much importance cannot be laid on the actual numerical values of these results in terms of electronic interaction (as energy value) with the semiemprical method due to use of confined cluster models. However, the qualitative trends predicted for the interaction of various molecules will be reliable. Chemically, the conversion of spirolactones to enones is a dehydration process. The solvation by water molecules and their interaction with the framework of hydrophilic zeolites such as H–Y is poorly understood. The beauty of this work is the simplicity of the model and most importantly the concept. Because

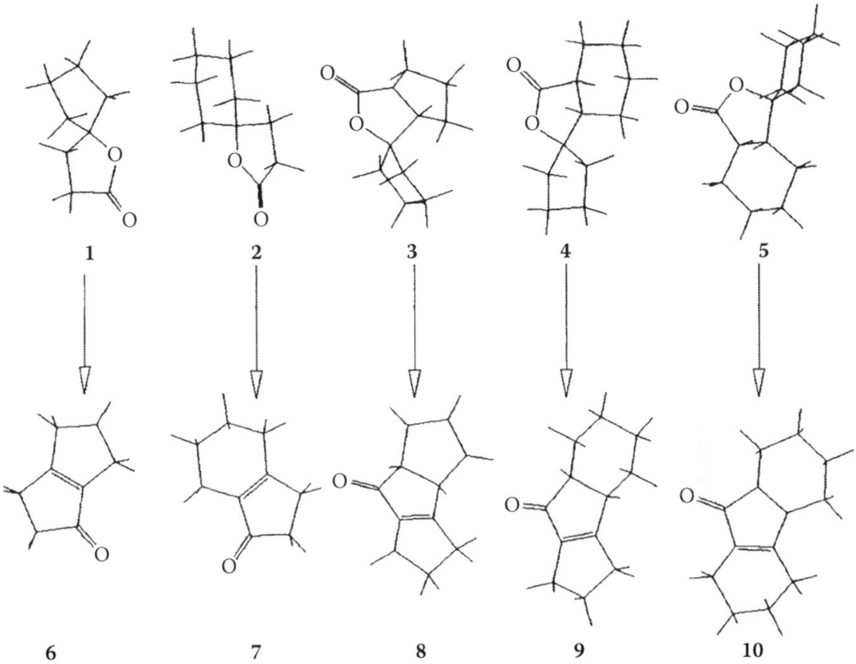

FIGURE 9.5 DFT-optimized structure of the reactants and products for a series of reactions.

the reaction in a nanoporous matrix is pore dependent, this model with interaction energy calculations will be applied to describe the reactivity of nanoporous materials. A follow-up article [9] with an improvement of theory within the helm of DFT to revisit the reaction mechanism with a better accuracy was carried out using the same zeolite H–Y. Because the molecules of spirolactone and enones are less symmetrical, we have chosen to dock the molecules within the H–Y supercage and minimize them within the structure to find the lowest energy configurations of the interacting molecules. The localized intermolecular interactions in terms of the guest and host species were monitored. Initially all the reactants and products labeled in Figure 9.5 were optimized with the DMol³ code of Accelrys [10]. This approximation assumes that the electron density varies slowly in comparison to the exchange and correlation effects. A double numerical with polarization (DNP) basis set with spin-restricted energy calculations was performed with a fine mesh grid and frozen core electrons. Because of the quality of these orbitals, basis set superposition effects [11] are minimized and an excellent description of an even, weak bond is possible. A BLYP-type functional [12] was used for the exchange correlation energy terms in the total energy expression. Then these molecules were docked inside the zeolite H–Y (Si/Al = 2.6) supercage using grand canonical Monte Carlo (GCMC) simulation using the SORPTION module of Accelrys [13]. The created simulation box consists of 2 × 2 × 2 unit cells. The molecule is moved randomly within the confined space of the supercage in all directions with translational and rotational degrees of freedom and

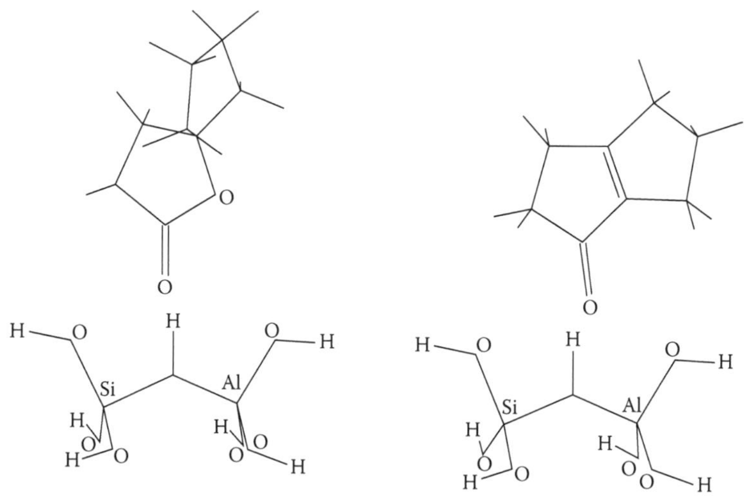

Reactant (1)   Product (6)

FIGURE 9.6   The orientation of typical spirolactone (1) and eneone (6) with respect to the zeolite framework cluster.

the CVFF forcefield parameter was used for the calculation [14]. Once the molecules were docked, we took that geometry out of the periodic box as a cluster with the reactant or product moiety. A dimmer cluster with Si–O–Al with a Brønsted proton was chosen. The clusters remained fixed with only the Brønsted proton and the interacting molecule optimized. The point of the DFT calculation is to look into the relative stability of the reactant and the product while interacting with the zeolite matrix. The models are shown in Figure 9.6a and Figure 9.6b for the reactant and product complex, respectively. The results of the total energy charge and the bond order for the reactant and product are shown in Table 9.3. The result shows that the ketonic oxygen is more electronegative than the ring oxygen and there is a change in the bond order of the C–O within the ring, proposing further reactivity of these centers favoring the product. The interaction energy values revealed two important phenomena: (1) a major interaction between the methylene hydrogen of the organic molecule and the oxygen of the zeolite framework and (2) the product molecule is more polar than the reactant. Therefore, the interaction energy of enones with the framework is more favorable than the spirolactones.

Another interesting reaction to be considered in shape-selective reactions over zeolites is the acylation reaction. It is generally known that acylation reactions are electophilic aromatic substitution reactions. Acylation of toluene and naphthalene was simulated over the acid form of five different zeolites such as ZSM-5, ZSM-12, Beta, mordenite, and zeolite-Y [15]. Selective benzoylation of toluene and naphthalene to 4-methylbenzophenone (4-MBP) and 2-benzoylnaphthalene (2-BON) is of considerable interest due to its commercial importance in the perfumery, dyes, and pesticide industries. Table 9.4 and Table 9.5 show the experimental results of benzoylation of toluene and naphthalene, respectively, over different zeolites [16]. This reaction

TABLE 9.3

Electronic Properties of Reactant and Products (Optimized) Calculated Using DFT

Molecule	Total Energy (kcal/mol)	Charge on O		Bond Order Between C and O	
		Ring	Ketonic	Ring	Ketonic
1	−287,915.46	−0.33	−0.36	0.927	1.831
6	−240,209.06	—	−0.37	—	1.875
2	−312,372.95	−0.33	−0.36	0.914	1.816
7	−264,667.47	—	−0.38	—	1.878
3	−360,103.58	−0.32	−0.35	0.912	1.812
8	−312,414.39	—	−0.37	—	1.872
4	−384,974.12	−0.34	−0.39	0.915	1.814
9	−337,253.45	—	−0.37	—	1.880
5	−409,422.14	−0.34	−0.39	0.913	1.826
10	−361,321.12	—	−0.40	—	1.882

Scheme is shown in Scheme 9.5. Although *para*-substitution predominates in classical acylation, the exclusive formation of *para*-isomers is rare and can only be explained by the shape selectivity of zeolites during product formation. The *ortho*-isomer for the benzoylation of toluene and more bulky 1-BON for the benzoylation of naphthalene would require a greater volume than the space available within the channels or cavities of the catalyst. The results show that (Table 9.4) H–Beta acts as the best zeolite in terms of product selectivity and the product selectivity is on the order 4-MBP > 3-MBP > 2-MBP. Similarly, H-Beta is the best candidate for benzoylation of naphthalene and

TABLE 9.4

Benzoylation of Toluene Over Different Zeolites

Catalyst	Time (h)	Conv. of BOC (wt %)	Activity (mmol/g/h)	Product Selectivity (wt %)		
				2-MBP	3-MBP	4-MBP
H-ZSM-5	18	4.2	0.6	14.6	5.1	80.3
H-ZSM-12	18	41.0	6.7	2.4	2.9	94.7
H-Beta	18	83.4	9.9	3.4	1.3	95.3
H-Mordenite	18	19.5	2.0	14.9	4.4	80.7
H–Y	18	18.8	1.9	22.9	6.1	71.0
AlCl$_3$	1	67.3	14.5	22.0	3.9	74.1
SiO$_2$–Al$_2$O$_3$	18	1.8	0.4	25.4	3.4	71.2

Notes: Reaction conditions: catalyst/C$_6$H$_5$COCl (w/w) = 0.33; temperature = 383 K; toluene/C$_6$H$_5$COCl (mol/mol) = 5; toluene = 0.11 mol.

TABLE 9.5
Benzoylation of Napthalene Over Different Zeolites

Catalyst	Time (h)	Conc. of BOC (%)	Initial Rate	Product Selectivity (mol %)			2-BON/1-BON
				1-BON	2-BON	Others	
H–RE–Y	1	0.2	0.05	62.6	25.4	12.0	0.40
	18	7.2	—	59.8	20.0	20.2	0.33
H-Beta	1	1.5	0.36	17.9	82.1	—	4.58
	18	17.6	—	17.4	74.8	7.8	4.30
AlCl3	1	14.6	3.52	87.3	7.2	5.5	0.08
H-ZSM-5	18	—	—	—	—	—	—

Notes: Reaction conditions: catalyst/BOC = 0.37 (w/w); napthalene/BOC = 2 (mol/mol); napthalene = 0.039 mol; 1-2-dichloroethane = 15 g; temperature = 358 K; others = dibenzoxylnapthalene.

the product selectivity is on the order 2-BON > 1-BON (Table 9.5). Zeolite Beta is generally considered a major catalyst in the organic chemical industry because of its unique acid properties related to local defects. These defects are generated when a tertiary building unit (TBU) is rotated 90 degrees around the c-direction with respect to the neighboring TBUs in the same layer. The rotated TBU connects properly with the adjacent layers and results in T atoms that are not fully coordinated to the framework, thereby creating potential Lewis acid sites [17]. The importance of simulating this type of reaction lies beneath the fact that all the studied zeolites exhibited different selectivity and the reason cannot be explained experimentally. The valuable catalytic and reactive properties of zeolites provide ample reason for establishing a firm theoretical understanding of their structures and behaviors. Computer simulation studies can contribute significantly in achieving an understanding of the structure–property relationship, revealing critical conceptual issues whose resolution demands additional experimentation. There has been a phenomenal growth of interest in theoretical simulations over the past decade and some outstanding reviews were published in this field in the past couple of years [18]. Here, we have used the same procedure of force field-based methodology to find the best zeolite in terms of size fitting. We consider the cluster model of the formula $(HO)_3Si–OH–Al–(OH)_2–O–Si(OH)_3$ for the calculations, which is a replica of the acid site of zeolite and interacts with the reactant molecule. Hydrogen atoms are necessary to maintain the cluster neutrality for a substituted situation that was located at 1 Å along the bond axes connecting the bridging oxygen. Nonlocal density functional (NLDF) calculations were performed using the DMol^3 program of MSI, Inc. [10]. This approximation assumes that the electron density varies slowly in comparison to the exchange and correlation effects. The final geometries were accepted when the norm of the energy gradient was less than 0.002 au. In the optimized structure, nonlocal functions were used to get the energy at the self-consistent level to improrize the accuracy of the energy number. A basis set superposition error was also calculated for the current basis set in nonlocal density approximation using the Boys-Bernardi method [19] and the determined value was 3.45 kcal/mol.

TABLE 9.6
Force Field Calculation Results for the Product Molecules

Molecule	Total Nonbonded Strain Energy (kcal/mol)	Total Electrostatic Strain Energy (kcal/mol)	Total Strain Energy (kcal/mol)	Total Energy (kcal/mol)
2-MBP	1870.65	312.69	507.99	2655.39
3-MBP	1665.04	115.53	481.50	2347.77
4-MBP	1602.23	58.57	481.50	2291.04
2-BON	2325.40	163.96	744.16	3241.36
1-BON	2341.42	250.20	704.26	3478.42

The molecular electrostatic potential (MESP) was calculated by NLDF at given point r in space representing a first-order approximation to the molecular charge distribution with the probe of unit charge at that point; details are described elsewhere [20]. To justify the product selectivity, the product molecules were minimized using a force field procedure, and to explain the mechanism of the reaction, the interaction energy of molecules with a zeolite framework cluster containing an acid site was monitored using DFT methodology. The strain energy for all the components was calculated using the same force field method as mentioned above (Table 9.6). Nonbonded energy arises from electrostatic and van der Waals interactions. Favorable and unfavorable nonbonded terms occur due to the attraction and repulsion between polar groups with unlike charges and like charges, respectively. Here, both bonded and nonbonded strain energy are unfavorable as indicated by the positive value, but the order decreased according to the order of selectivity as observed experimentally. The unfavorable nonbonded energy for product molecules indicates the presence of positively charged groups in the molecule. These results indicate that the positive charges in the molecules have ionic interaction with the basic oxygen of the zeolite framework, which could lead to their change of adsorption mode inside the zeolite void volume. The product yields over various zeolite catalysts correspond well with their structural fitting. It can be generalized that large-pore zeolites with cage structures are efficient catalysts for the acylation reaction and the role of shape selectivity in controlling the yield can be explained from the results of calculations. Figure 9.7a and Figure 9.7b show the two average dimensions of MBP and BON compared with the average diameters of the pore openings of zeolites. Although the void dimensions of the zeolite catalysts control the product yield, the electronic interactions are also expected to play a vital role in the mechanism of this reaction. Corma et al. [21] proposed the general mechanism of the Friedel-Crafts acylation over zeolites as shown in Scheme 9.3. According to this mechanism, the Brønsted acid sites react with the acylating agent to produce an acylium ion or an acylium-like complex. The electrophilic species generated through this interaction attack the aromatic ring in the second step to form a metastable complex and then after the abstraction of hydrogen, transformed into the product. The nature of the product formed depends on the attacking center of the electrophile on the substituted aromatic ring. Because there is not much difference in the dimensions of the product molecules, the selectivity can be accounted for by

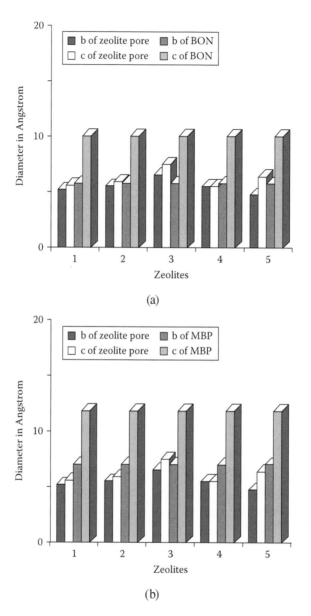

FIGURE 9.7 (a) Schematic representation of pore diameters in various zeolites used as catalyst for the formation of MBP; (b) schematic representation of pore diameters in various zeolites used as catalyst for the formation of BON.

the interaction energy calculations using DFT. The electronic properties of toluene, naphthalene, benzoylchloride (acylating agent), the framework cluster, and all the product molecules were calculated by full geometric optimization using DFT. First we calculated the interaction energy manually by changing the distance between the Brønsted proton of the zeolite framework and the C$=$O of the acylating agent to

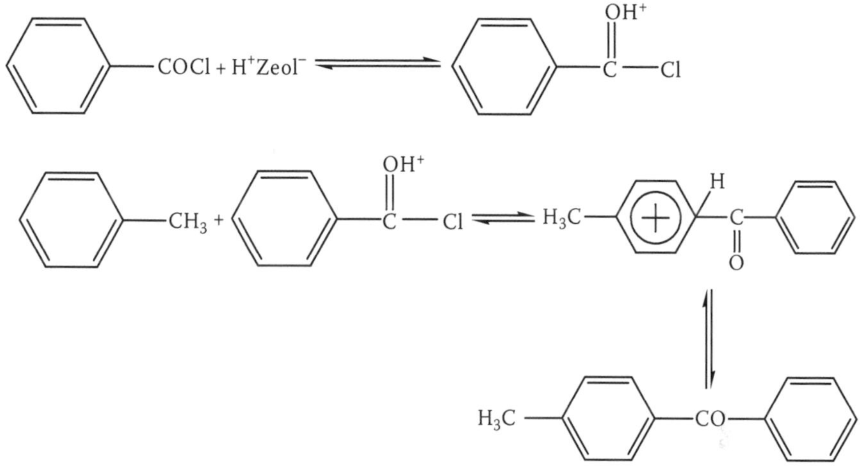

SCHEME 9.3 Friedel–Crafts acylation over zeolites.

locate the hypothetical transition state for proton abstraction. In the next step, the interaction energy of toluene and naphthalene with the zeolite framework was performed followed by the calculation of the interaction energy for individual products with zeolite clusters to compare the product selectivity. Maximum instability resulting from the reaction with toluene will show the activity of the zeolite, and the stability of the product inside the zeolite framework will explain its product selectivity. In the following step, the interaction energy between $[C_6H_5COHCl^+]$, toluene, and naphthalene with a Brønsted proton–abstracted zeolite framework was performed. In this situation the complex is neutral and the zeolite framework having a total charge of -1 can interact with unipositive $[C_6H_5COHCl^+]$. This is a model to replicate the reaction inside the zeolite cage. It is observed that 4-MBP and 2-BON are the most unstable among three isomers (2-MBP, 3-MBP, and 4-MBP) and two isomers (2-BON and 1-BON) for the benzoylation of toluene and naphthalene, respectively. It is difficult to perform the calculation for a metastable intermediate, so interaction of a complex with a zeolite framework (proton abstracted) was performed, which represents the instability in terms of interaction energy and further dictates the possibility of elimination of hydrogen to generate product molecules. The model is shown in Figure 9.8. These energies are significantly higher than those expected (60 kJ/mol) from single hydrogen bonds between carbonyl oxygen atoms and acidic hydroxyl groups on oxide surfaces [22,23]. This can be explained in terms of the fact that the reaction takes place inside the zeolite cage, so the environment of the Brønsted proton really matters. In the case of zeolites, the acidic proton has bridging oxygen linked with silica terminated by hydrogen that results in a steep increase in the energy value, as also observed by other researchers [24]. The order obtained from the interaction of product molecules with a zeolite cage is 4-MBP > 3-MBP > 2-MBP for the acylation reaction of toluene, whereas the product of acylation reaction of naphthalene shows the order of 2-BON > 1-BON. We then compared the trend with experimental observation to assimilate the

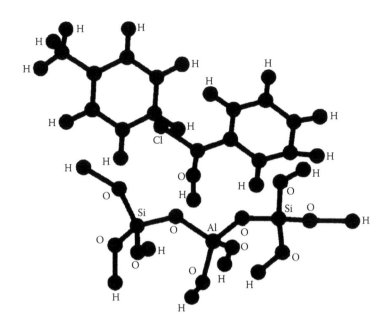

FIGURE 9.8   Interaction of toluene with benzoyl chloride and the proton-abstracted zeolite framework cluster model.

power of computer simulation. The MESP was plotted for a potential range of –0.05 au to +0.05 au. Figure 9.9a and Figure 9.9b show the results of proton abstraction during the interaction of benzoylchloride with the framework and the interaction of product 2-BON with the zeolite framework, respectively. It is observed from Figure 9.9a that the strong negative potential around C=O pulls the Brønsted proton with high positive potential and results in the abstraction of the proton. This validates Scheme 9.3 where the first step is a proton abstraction, which results in an acylium ion. For product interaction, Figure 9.9b shows the elimination of hydrogen back to the basic oxygen site of zeolite framework obtained from the mutual attraction of the positive and negative potentials of the interacting species and zeolite framework. MESP revealed the region of plausible bond breaking and bond formation, which may further help in the future study of locating the transition state. It also shows the proton abstraction form framework as well as hydrogen atom elimination from the intermediate to form the product and supports the reaction mechanism proposed by the experiment.

ZSM-5 type zeolites are known to be an excellent catalysts for aromatization of lower alkanes. The activity of ZSM-5 for aromatization reactions is mainly due to the presence of high acidity and correct pore geometry [25]. Aromatization mostly depends on the amount and strength of the acid site distributed over the catalyst surface. Keeping this in mind, it would be interesting to compare the performance of other medium-pore zeolites like ZSM-22, ZSM-5, and EU-1 as reported experimentally by Bhattacharya and Sivasanker [26]. On the basis of zeolite structure, the respective size of the pores remains same, but interestingly these zeolites performed differently. So it is reasonable to think about the possibility of the difference in pore

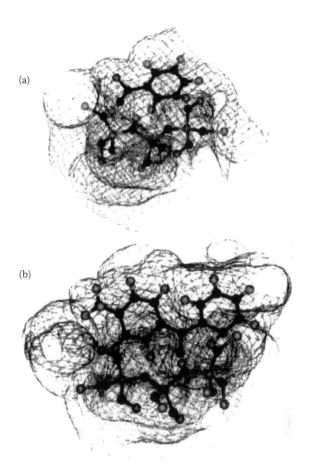

FIGURE 9.9 (For color version see accompanying CD) (a) MESP for benzoylchloride interacting with the zeolite framework. The +ve (+0.05 to 0.00 au) and −ve (−0.05 to 0.00 au) potential contours are shown as green and red shades, respectively; (b) MESP for 4-MBP interacting with zeolite framework. The +ve (+0.05 to 0.00 au) and −ve (−0.05 to 0.00 au) potential contours are shown as green and red shades, respectively.

architecture to explain the difference in catalytic behavior and makes the problem appealing enough to study. We compared the activities of different T sites present in these three zeolites and located the active T sites as T6 (ZSM-5), T4 (ZSM-22), and T8 (EU-1), respectively, by correlating the geometric parameters such as bond length and bond angles with electronic properties [27]. The experimental Brønsted acidity order was compared with calculated deprotonation energy to validate the model. To choose the best zeolite among three zeolites studied in terms of activity toward aromatization, we performed DFT calculations on the organic molecules (both reactant and major products), zeolite clusters, and zeolite organic molecule complexes. The results matched the experimental observations by justifying the suitability of ZSM-5 in aromatization reaction in comparison to other medium-pore zeolites. Calculations

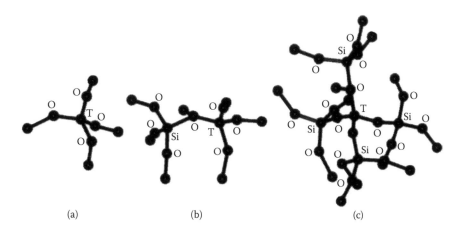

(a) (b) (c)

FIGURE 9.10 (a) Monomer cluster model with formula [T(OH)$_4$], (b) dimer cluster model with formula [(OH)$_3$–Si–T(OH)$_2$–(OH)$_3$], and (c) pentamer cluster model with formula [T(OSiO$_3$H$_3$)$_4$].

were performed with cluster models with one T site (monomer), two T sites (dimer), and five T sites and are presented in Figure 9.10. Hydrogen atoms necessary for cluster neutrality for a substituted situation (T = Al) were located at 1 Å along the bond axes connecting with the bridging oxygen. The hydrogen occupying the position of nearest T site saturates the boundary oxygens. The dimer cluster model consisted of two TO4 groups bridged by commonly shared oxygen. The pentamer cluster model represented a TO4 group, which shares a corner with four adjacent TO4 groups through the bridging oxygen atom. These cluster models were used for calculating the activity of T sites. For the interaction energy calculation the models were chosen to mimic the individual zeolite pore architectures. The cluster formula used for ZSM-5 and ZSM-22 was T(OSiO$_3$H$_3$)4, where T stands for aluminum at the most active T site, whereas for EU-1, which has a side pocket in its typical pore architecture, the cluster chosen has the formula Si$_5$AlO$_{18}$H$_{13}$. The cluster models are shown in Figure 9.11. The calculation uses the same methodology of DFT formalism as described earlier. X-ray crystal data of ZSM-5, ZSM-22, and EU-1 were taken from the literature [28]. The framework dimer clusters were fully optimized for the calculation of deprotonation energy [18] in which the terminal hydrogen was fixed, and for the interaction energy calculation the cluster was also kept fixed. The orientation of n-hexane within individual zeolites was obtained by the Monte Carlo-type of simulation performed using the Solids Docking module of MSI Inc. Three different zeolites have different pore architectures because they represent three different classes of zeolites. To show the variation of geometric parameters present in these three different medium-pore zeolite structures, average Si–MO lengths for each T site of these three zeolites were compared from their respective XRD data. It was observed that aluminum in ZSM-5 favored longer bond lengths and smaller bond angles than silica. So, for all three zeolites the average bond distances were plotted with average bond angles with respect to the individual T sites present in the zeolites, and previous findings were implemented to locate the probable substitution site in all

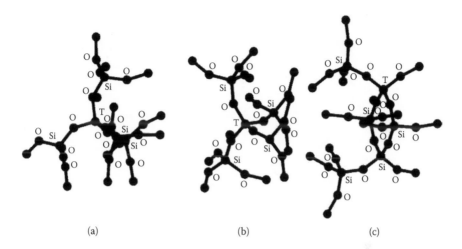

(a)   (b)   (c)

FIGURE 9.11   Cluster model used for interaction energy calculation: pentamer cluster model of (a) ZSM-5, (b) ZSM-22, and (c) Eu-1 with formula $Si_5AlO_{18}H_{13}$.

three zeolites (Figure 9.12a, Figure 9.12b, and Figure 9.12c). To calculate the relative substitution energy for aluminum in place of silica we chose a pentamer cluster for all three zeolites. The substitution energy was calculated by the following process:

$$H_{12}Si_5O_{16} + [H_4AlO_4]^- \rightarrow [H_{12}Si_4AlO_{16}]^- + H_4SiO_4 \qquad (9.1)$$

The results are shown in Figure 9.13. The most preferred sites for aluminum substitution are T6 for ZSM-5, T2 for EU-1, and T3 for ZSM-22. This is undoubtedly a qualitative conclusion, because the activities of these sites vary with the cluster size, the calculation methodology, and other parameters. However, this is a well-established way to locate the active T site present in the zeolites [29]. Once the location of the acid centers within the zeolite is known, Monte Carlo simulation techniques allow us to find the reasonable conformation to locate the position of hexane molecules. The minimum energy conformation of hexane was studied for all three zeolites and the orientation of the hexane molecule was dependent on the different pore architectures of the three zeolites, although the approach of the methyl group of *n*-hexane toward the framework was almost same for all the zeolites. To validate the finding, possibility of hexane molecules inside the zeolite cage docking is a useful technique but a limitation is that the process is not conclusive about global minima. Docking of hexane molecule within zeolite cage is a useful method to adopt to know the local minima of the structure, but that must be followed by geometry optimization to confirm the energetic feasibility. Based on the minimum energy conformation of the hexane molecule inside the zeolite cage, we performed the optimization of the hexane molecules with respect to the different representative framework clusters. Experimentally, it has been observed that the order of acidity among above-mentioned zeolite is as follows: ZSM-5 > EU-1 > ZSM-22. To trust this, we calculated the proton affinity in terms of the deprotonation energy by substituting the energy of

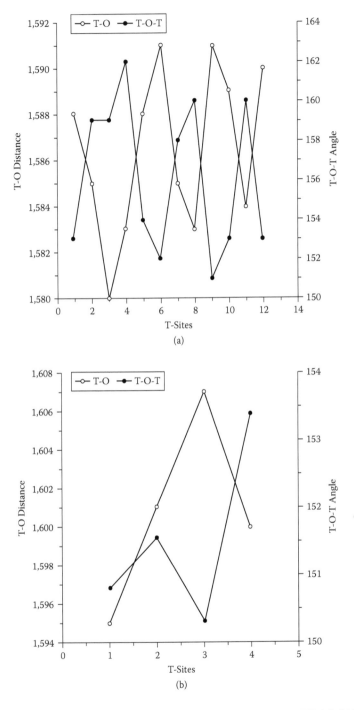

FIGURE 9.12   Average T–O bond lengths and T–O–T bond angles of (a) ZSM-5, (b) ZSM-22, and (c) EU-1 as obtained from crystallographic data.

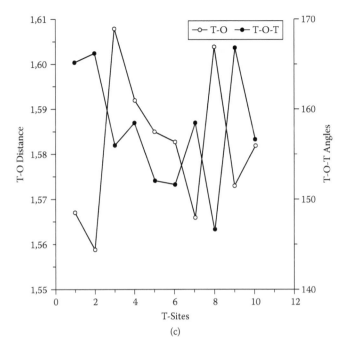

FIGURE 9.12 (Continued).

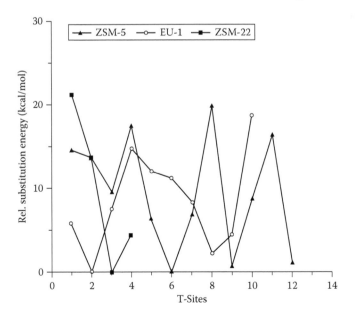

FIGURE 9.13 Relative substitution energy for Al in ZSM-5, ZSM-22, and Eu-1 (pentamer cluster model).

TABLE 9.7

Experimental Acidity Results in Comparison with DFT Result

Catalyst	Acidity Due to Si-OH[a]	Acidity Framework Al[a]	Deprotonation Energy (kcal/mol)
H-ZSM-22	0.198	0.039	−299.2
H-EU-1	0.417	0.099	−302.1
H-ZSM-5	0.546	0.213	−314.6

Note: [a]Brønsted acid site concentrations measured in mmol/g.

the deprotonated cluster from the energy of the protonated cluster (Table 9.7), which represents the individual zeolite structures. The proton affinity was calculated. The obtained result matched the experiment and confirmed the successful application in various types of reactions [30]. So, it is reasonable to predict that not only the pore size but the pore architecture is important and responsible for the acid strength as well as the activity of the nanoporous material. This trend can be validated by optimizing the hexane and cluster complexes to predict the interaction energy or the binding energy of hexane with the specific zeolites. This is an example to show specifically the role of the zeolite architecture and capability of simulation technology to explore and justify the reason for the activity, which is difficult to prove by experiment only.

The next targeted reaction is another type of regioselective reaction, an important class of reactions in organic synthesis, where nanoporpus materials play a crucial role as catalysts. This is again a combination of experiemtal and theoretical studies to explore why one particular zeolite acts better than another for a regioselective reaction. Nitrotoluenes are important intermediates in the chemical industry, produced by liquid-phase nitration of toluene using a mixture of nitric and sulfuric acid as a nitrating agent [31]. A large quantity of dilute sulfuric acid is generated as waste in the conventional process and its disposal or recycle is very expensive; this makes toluene nitration an environmentally harmful process. The replacement of sulfuric acid with solid acid catalyst of high selectivity for *para*-isomer would be an attractive environmentally benign route for the production of nitrotoluenes. Several solid acid catalysts were investigated for the liquid- and vapor-phase nitration process. Among the solid acid catalysts H-mordenite, H-beta, H-ZSM-5, and H–Y, zeolite H-beta are highly effective and offered high conversion along with remarkable selectivity for *para*-isomer. Choudary et al. [32] carried out the nitration of toluene in liquid phase employing nitric acid of 60–90% concentration over solid acid catalyst by azeotropic removal of water. Beta zeolite was proved to be the best candidate in terms of space–time yield (STY) and *para*-selectivity and exhibited consistent activity and selectivity even after five cycles without any dealumination. The activity and selectivity decreased over a period of about 5 h on stream, due to pore filling/blockage by strongly adsorbed products/by-products. Nitration of toluene using H-beta zeolite to replace the sulfuric acid in the conventional process could not be commercialized because of deactivation

TABLE 9.8

Dimensions of Different Molecules in Their Minimum Energy Configuration

Molecule	Dimensions		
	a (Å)	b (Å)	c (Å)
Nitronium ion (1)	4.6	3.0	2.7
Toluene (2)	6.9	5.3	2.8
ONT (3)	8.0	6.9	2.9
MNT (4)	8.3	7.0	2.9
PNT (5)	8.8	5.3	2.8
DNT (6)	9.9	7.1	3.0

of catalyst, leading to low space–time yields. The explanation is demanding and challenging, where experiment might not be sufficient to explain the reason and simulation entered. The first part of this simulation procedure focuses on the effect of size factors by comparing the dimensions of the molecules and zeolites of interest. This is followed by energy minimization of the guest in the host structure, followed by a force diffusion technique to see the diffusion characteristics of the molecules to know the behavior of the molecules. Table 9.8 represents three largest dimensions of nitronium ion (1), toluene (2), *ortho*-nitrotoluene (3), *meta*-nitrotoluene (4), *para*-nitrotoluene (5), and di-nitrotoluene (6) molecules in mutually perpendicular directions. The pore dimension of beta zeolite is 7.6 × 6.4 Å, whereas that of ZSM-5 is 5.1 × 5.6 Å. The energy minimization was performed with the code DISCOVER and the force field CVFF was used as mentioned before. Once the molecular structures were energy minimized, zeolite models were prepared to rationalize the diffusion characteristics. The Beta zeolite with molecular formula $(SiO_2)_{94}$ and cell parameter of 12.66 ($a = b$) and 24.606 (c) was chosen from the XRD data [33]. The ZSM-5 structure was also taken from the XRD [34], with the dimensions of the cells were 20.022, 19.899, and 13.383, representing a, b, and c, respectively, with the cell formula of $(SiO_2)_{96}$. The sizes of the simulation boxes were chosen in such a way that the symmetry along the channel direction was taken care of and the box was just large enough in the other two directions to take care of the nonbonded interactions, whose cutoff distance was taken as 8.5 Å. The diffusion energy profiles symmetrically repeated themselves in each unit cell, indicating the validity of the simulation box size, potential parameters, and energy minimization calculation procedures. The calculations were performed following the well-established forced diffusion procedure. This procedure has been widely used to study the diffusion of methanol [35]. Here the authors investigated the shape-selective production of *para*-nitrotoluene over two zeolites. The sorbate molecule was forced to diffuse in regular steps of 0.2 Å along the diffusion path defined by the initial and final positions within the channel. At each point, a strong harmonic potential constrained the molecule to lie at a fixed distance from the initial position and the energetically favorable conformation and orientation of the molecule were derived by varying the internal degrees of freedom as well as nonbonding interaction of the molecule with the zeolite framework. The interaction energy at each point was calculated using the following equation:

$$\text{Interaction energy} = E_{\text{zeolite+molecule complex}} - (E_{\text{zeolite}} + E_{\text{molecule}})$$

Thus, the diffusion energy profile is a graph showing the variation of interaction energy between a single molecule and the zeolite framework as the molecule diffuses within the channel of the zeolite. The diffusion energy profiles are useful to identify the most favorable (minimum energy) and unfavorable (maximum energy) adsorption sites for the molecules inside the zeolite channels. The difference in energy between the most favorable and most unfavorable sites in the diffusion energy profile gives the diffusion energy barrier for self-diffusivity. A fully siliceous zeolite lattice was considered for the study because the main emphasis for the calculation is to see the effect of pore architecture. For this purpose, two exemplary zeolites with completely different architecture, ZSM-5 and zeolite beta were chosen. The influence of the presence of more molecules on the diffusivity (mutual effect) and the influence of temperature are not considered here. The mean energy is the numerical average of the interaction energy of the molecules at all locations in the diffusion path. The ratio of mean energy to minimum energy (the most favorable site) is a diffusivity index parameter. If the mean energy is close to minimum energy, the situation represents the presence of several minima and the mean/minimum energy ratio will be closer to one. On the contrary, if the mean energy is much higher than the minimum energy, the situation represents the presence of several maxima and the mean/minimum energy ratio will be closer to zero. Thus, the ratio is an indicator of the diffusivity of the molecule; the higher the value, the greater the diffusivity. This method has some approximation, but it is one of the simplest approaches to look into the shape-selective reaction along with the diffusivity calculation to provide some quantification in terms of accuracy in the results. The diffusivity results for the set of molecules compared between ZSM-5 and Beta are shown in Table 9.9. All of the molecules have one maximum when they diffuse through one unit cell of beta; this maximum occurs when the molecule moves from one 12-M intersection to the other 12-M intersection. The mean energy is closer to the minimum energy values, which indicates that the molecule will diffuse selectively. Mean energy values and the diffusivity values are given in Table 9.9. Results indicate that the diffusivity decreases in the order of 5 > 3 > 4. The diffusion energy barrier is more for 4 compared to the barriers for 5 and 3. These results indicate Beta zeolite will exhibit be high selectivity for the diffusion of 5. The diffusion of the 5 showed minimum energy across the unit cell, when the phenyl ring lies at the intersection of straight channels and perpendicular channels. Whenever the phenyl group of the molecule passed through the wall of the zeolite framework, the phenyl ring was found to be parallel to the diffusion plane. The results are consistent with the hypothesis, which states that upon adsorption of the reactant on a rigid surface such as the zeolite framework, steric hindrance would direct the orientation of the aromatic in the adsorbed state in a way that the substituent on the aromatic ring points toward the zeolite cavity where the least repulsion would be expected. During the toluene nitration, aluminum changes its coordination from tetrahedral to octahedral [36], reducing the pore size by ~1 Å. Beta zeolite has

TABLE 9.9

The Deepest Minima, Peak Maxima, Diffusion Energy Barrier, Mean Energy, and Diffusivity for Organic Molecules in Zeolites

Zeolite	Molecule	Deep Minima (kcal/mol)	Peak Maxima (kcal/mol)	Energy Barrier (kcal/mol)	Mean Energy (kcal/mol)	Diffusivity (mean/min)
MFI	1	−10.671	−9.183	1.488	−9.893	0.927
	2	−29.363	−6.550	22.813	−21.474	0.731
	3	−33.175	7.698	41.402	−10.227	0.303
	4	−34.448	19.287	53.735	−11.897	0.345
	5	−34.560	−25.247	9.313	−29.734	0.860
	6	−40.444	96.119	136.563	−0.550	0.014
Beta	1	−8.819	−6.844	1.981	−07.766	0.881
	2	−19.321	−14.887	4.437	−17.010	0.880
	3	−27.948	−17.572	10.176	−23.225	0.831
	4	−28.383	−14.672	14.672	−26.613	0.838
	5	−30.975	−24.023	6.924	−27.204	0.878
	6	−28.626	−27.827	18.443	−31.530	0.807

nitrite ion (1), toluene (2), ortho-nitrotoluene (3), meta-nitrotoluene (4), para-nitrotoluene (5), di-nitrotoluene (6)

a three-dimensional interconnected channel system with 12-M elliptical channels of diameter 7.6 × 6.4 Å. This will further reduce the pore size by ~1 Å and impart the *para*-selectivity in toluene nitration due the diffusional constraints. These results agree with experimental observations. A computer graphics picture is shown in Figure 9.14. It shows a picture of the minimum energy configuration of molecule 5 inside beta. The variation in the interaction energy between 5 and beta is calculated with the linear progression of the molecule. The molecular modeling study indicates that *para*-selectivity is due to faster diffusion of *para*-isomer in the pores of the catalyst.

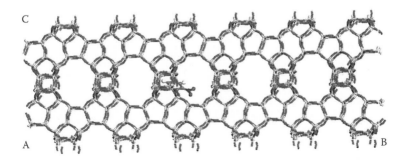

FIGURE 9.14 (For color version see accompanying CD) The minimum energy configuration of molecule 5 inside beta zeolite.

9.2   CHEMICAL ADSORPTION

The aim of this section is to look into the feature of chemisoprtion of different sorbent molecules in nanoporous materials, which are dependent on the reactivity of the substituted or incorporated metals within the nanopores. These metal substitutions in nanoporous materials enhance the adsoption capability of microporous materials and result in some reactivity with a specific mechanism, which we will talk in the next section. This section will focus on those reactions where adsorption behavior with the nanopore is of primary importance. The specific examples are the NO_x type reaction related to the environmentally toxic gas adsorption issues, where nanoporous materials act as good sorption agents through the metal, and hence the metal-to-adsorbent interaction is a major issue.

Environmental pollution caused by chemicals contained in exhaust gases from mobile and stationary sources is a serious problem that needs to be solved. In particular, unwanted pollutants such as nitrogen oxides participate in chemical reactions in the atmosphere and in chemical corrosion such as photochemical smog and acid rain. Historically the NO_x emissions have been controlled by the reduction of nitrogen oxides in the presence of oxygen by ammonia on titania-supported vanadia [37]. Nevertheless, the use of ammonia as a reducing agent is undesirable because large quantities of ammonia must be stored, and the process must be controlled to avoid its toxic and corrosive effects. On the other hand, the selective catalytic reduction (SCR) of NO_x by hydrocarbons using metal-zeolite catalysts, in which ammonia is not used, has been considered as a practical alternative for the elimination of nitrogen oxide pollutants. This is a work at the initial days of NO_x catalysis simulation when modeling was undertaken to understand the chemistry associated with the adsorption of NO_2 on supported metals in zeolites [38]. The analysis of the electronic interaction between the NO_2 molecule and different cluster models (M-zeolite: M = Fe, Co, Ni, Cu, and Zn), using *ab initio* DFT calculations is presented. A tetrahedral model $[H_3SiOAl(OH)_2OSiH_3]$, was chosen as the T3 site model of the ZSM-5 zeolite, with the metallic atom M set on a bridge between two oxygen atoms [39] (Figure 9.15).

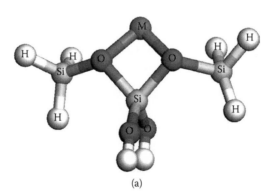

(a)

FIGURE 9.15   (For color version see accompanying CD) General structures obtained (a) for the parallel (NO_2M–T3∥) (b) and perpendicular (NO_2M–T3⊥)adsorption modes (c) of the NO_2 on M–T3 systems (M = Zn, Cu, Ni, Co, and Fe).

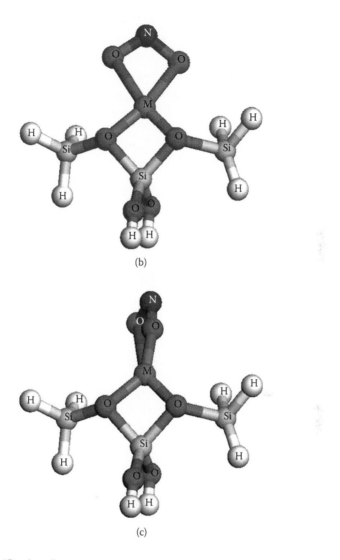

(b)

(c)

FIGURE 9.15 (Continued).

All calculations and geometry optimizations were performed by using the Gaussian 94 program [40], at DFT level using Becke's three-parameter hybrid functional [41] and Lee et al.'s correlation functional [42]. In all calculations the symmetry of the models used was C2, and the total charge of the M-T3 complex was set according to the lowest formal oxidation state of the metallic ion. For all complexes the spin state was given by the number of unpaired electrons on the metallic center.

For the free NO_2 molecule, the calculated N–O bond length and O–N–O angle are 1.19 Å and 134.2 degrees, respectively, in good agreement with the experimental values of 1.19 Å and 133.9 degrees. For the free NO_2 molecule, the calculated values of the N–O bond length and O–N–O angle (1.27 Å, 116.0 degrees) are similar to the

TABLE 9.10

Bond Distances and Angles of the Cu–T3 and NO$_2$ Cu–T3 Structures and Free NO$_2$ and NO$_2$ Molecules[a]

	Cu–O (Å)	Angle O–Cu–O	N–O$_3$ (Å)	Angle O$_3$–N–O$_3$	O$_3$–Cu (Å)	ΔE[b] (kcal/mol)
Cu–T3	2.00	83.2				
	1.99[c]					
NO$_2$–Cu–T3	1.94	80.6	1.26	11.09	2.02	43.3
	1.94[c]		1.26[c]	11.05[c]	2.01[c]	42.7[c]
Free NO$_2$			1.19	134.2		
Free NO$_2^-$			1.27	116.0		

[a] T3 = H$_3$SiOAl(OH)$_2$OSiH$_3$.
[b] Calculated binding energy.
[c] From Li and Hall [108].
Source: [39] With permission.

experimental values of 1.25 Å and 117.5 degrees. The first step is to see the localized effect of NO$_2$ adsoprtion on Cu atoms if there is a change in the localized geometry. We have seen a change in the structural parameter even for the aluminum substitution in the zeolite lattice. Results in Table 9.10 suggest that the interaction of the NO$_2$ molecule with the Cu atom produces an enlargement of the N–O distance (N–O3), from 1.19 to 1.26 Å from 134.2 to 110.9 degrees, of the free NO$_2$ molecule. The geometrical properties of adsorbed NO$_2$ were similar to those of the free ion NO$_2$, which is an indication of a charge transfer from the metal to the NO$_2$. To investigate the general mechanism of the interaction of the NO$_2$ molecule with metallic ions, other metals such as Zn, Ni, Co, and Fe were studied. When the two oxygen atoms interacted directly with the metallic center, the adsorption modes of the NO$_2$ molecule were found to be more stable than the modes where the N atom interacted with the metal. Figure 9.15a, Figure 9.15b, and Figure 9.15c display the original model and general structures obtained for the parallel (NO$_2$M–T3∥) and perpendicular (NO$_2$M–T3 ⊥) adsorption modes of the NO$_2$ on M–T3 systems (M = Zn, Cu, Ni, Co, and Fe). As shown in Table 9.11, the calculated binding energies for the Zn–T3 system were higher than those corresponding to the Cu–T3 system. In contrast to the Cu–T3 system, in the Zn–T3 system the perpendicular adsorption mode was slightly more favored than the parallel mode. The structures, binding energies, and topological properties of the bonds were determined for M–T3 and NO$_2$M–T3 systems (M = Fe, Co, Ni, Cu, and Zn). The coordination mode of NO$_2$ through the two oxygen atoms to the metallic center was the most stable, whereas the other coordination modes were less stable. This agrees with previously published results [43] that suggested that the η$_2$–O, O structure is optimal when NO$_2$ binds to a metallic ion attached to the zeolite framework. In general, a charge transfer from the metal to the NO$_2$ molecule and a weakening of the N–O bond occur. The charge transfer increases the electrostatic interactions of the metal with the negatively charged zeolite, reducing the metal–oxygen (ZSM-5) distance. The bonding in NO$_2$M–T3 can

TABLE 9.11

Bond Distances and Angles of M–T3, NO$_2$M–T3⊥, and NO$_2$M–T3∥ Structures

M–T3, NO$_2$M–T3	M–O (Å)	Angle O–M–O	N–O$_3$ (Å)	Angle O$_3$–N–O$_3$	O$_3$–M (Å)	ΔE[b] (kcal/mol)
Zn–T3	2.04	78.1	—	—	—	—
NO$_2$Zn–T3⊥	1.97	80.9	1.26	113.0	2.09	64.7
NO$_2$Zn–T3∥	1.98	80.0	1.25	113.0	2.12	61.6
Cu–T3	2.00	83.2	—	—	—	—
NO$_2$Cu–T3⊥	1.94	80.6	1.26	110.9	2.02	43.2
NO$_2$Cu–T3∥	2.00	79.5	1.26	109.9	2.00	27.9
Ni–T3	1.96	83.9	—	—	—	—
NO$_2$Ni–T3⊥	1.96	79.3	1.26	111.6	2.07	52.0
NO$_2$Ni–T3∥	1.96	79.1	1.27	111.3	2.05	50.7
Co–T3	1.86	87.1	—	—	—	—
NO$_2$Co–T3⊥	1.81	83.0	1.28	108.1	1.92	20.0
NO$_2$Co–T3∥	2.04	77.3	1.27	111.0	2.04	13.6
Fe–T3	1.90	85.3	—	—	—	—
NO$_2$Fe–T3⊥	1.88	84.8	1.27	109.9	2.07	33.8
NO$_2$Fe–T3∥	1.87	86.0	1.28	106.9	1.93	12.2

Notes: [a] T3 = H$_3$SiOAl(OH)$_2$OSiH$_3$; M = Zn, Cu, Ni, Co, Fe.
[b] Calculated binding energy.
Source: [39] With permission.

be described in terms of the metal 4s orbital and the NO$_2$ HOMO orbital. Therefore, an unpaired electron in the 4s orbital of the metal is necessary to produce the bond. The bonding mechanism of Sauer and coworkers [43] explains the interaction of NO$_2$ with Cu$^+$ and Ni$^+$ but not with Zn$^+$, Co^{2+}, or Fe^{2+}. The topological properties of the N–O3 bond were similar in all the NO$_2$M–T3 complexes studied here and were closer to the properties of the NO$_2^-$ ion. Therefore, the weakening of the N–O bond in the NO$_2$ molecule (N–O3) could be explained in terms of a charge transfer from the complex M–T3 to the HOMO orbital of NO$_2$.

In continuation with the issue of adsorption, metal sitting in zeolites is another important topic. The characterization of Cu-zeolites and their complexes with probe molecules (e.g., CO, N$_2$, or NO) by microcalorimetry and by EPR, FTIR, UV-vis-NIR, and EXAFS spectroscopies brought indirect proofs that the copper ions exist in the high-silica zeolites with several site types [44]. In zeolites, the sites are important because they represent each individual environment with different neighbors; as we mentioned before, the internal distance between Si–O and Si–O–Si varies with this environment. Some experimental techniques can prove the presence of the different types of environments within zeolite matrixes. These sites differ in localization, inner volume of zeolite, structure, and/or coordination to framework oxygen atoms. The structure and localization of the active sites are not fully understood at the atomic-scale level. Nichtigall et al. [45] have worked on this sitting and a comparison

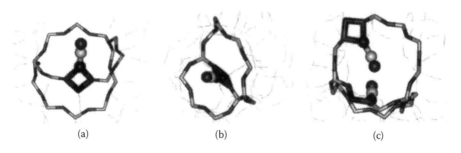

(a) (b) (c)

FIGURE 9.16 (For color version see accompanying CD) CO molecule adsorbed on Cu⁺ in MFI. The Cu, C, and O atoms are depicted as balls. Zeolite framework atoms are depicted in black, light grey, and dark grey (Al, Si, and O atoms, respectively). Ten-membered rings on the channel intersection are shown in "tube" mode. Monocarbonyl adsorbed on the intersection site (a), on the channel wall (b), six-membered ring on the channel wall shown in tube mode), and two monocarbonyls on the channel intersection (c) are shown. [45] With permission.

with CO adsorption was checked by TPD and then validated by DFT calculation. A combination of methodologies was also used by Cvetanovic et al. The TPD results were simulated by a formal kinetic model [46]. From the results some assumptions were made: (1) CO can diffuse through the zeolite channel and there is no influence of the shape or the architecture of the channel, (2) the readsoprtion of CO can be neglected because the desorption is strongly affected by the readsoprtion of CO on the unoccupied Cu^{+3} site, (3) a preexponential factor $A = 10^{13}$ was assumed as a initial guess [47], (4) Langmuir-type adsorption isotherm is observed for Cu^+ sites in zeolite for CO adsorption, and (5) a complete coverage of all Cu^+ sites was assumed. The adsorption energies of the CO with Cu^+ sites were performed with the QM pot method of Sauer and Sierka [48] by considering two typical models. In modeling, the choice of the model based on the problem is a primary aspect because it dictates the approach toward the solution. Here the authors have shown an interesting modeling method for the Cu sitting. They have chosen Cu^+ sites in the vicinity of single aluminum atoms, say, the channel intersection, as shown in Figure 9.16(c), and sites on the channel wall (two aluminum atoms at least in the vicinity) as shown in Figure 9.16b. Because these results will be compared with TPD of CO it is better to consider a lateral interaction of two CO molecules with a pair of Cu^+ at channel intersection (Figure 9.16c). Three types of Cu^+ sites were distinguished from the analysis of TPD spectra of CO in high-silica zeolites and the details of the structure and coordination of Cu^+ at these sites were obtained from calculations. QM pot is a combination of QM and MM technology. One can address or incorporate the outer effect of the zeolite structure, where the main focus can be on the metal sites. Within the QM pot scheme, the B3LYP exchange-correlation functional [41] with combined valence-double(triple)-z basis set with polarization functions [49] was used for the inner part description together with the core-shell model potentials optimized previously [50] was used to describe the outer part of the domain. The GULP methodology was used [51] for description of the outer part of the system. Inner parts with sizes ranging from the 3-T cluster ($COCuAlSi_2O_{10}H_8$) to the 8-T cluster

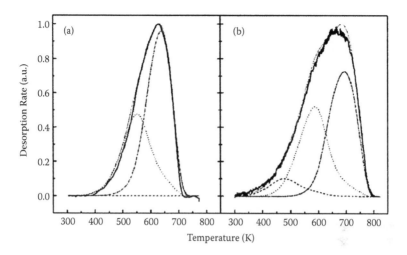

FIGURE 9.17 Experimental and simulated CO–TPD spectra of Cu–MFI zeolite with Si/Al = 14.1; (a) Cu–MFI-0.2, (b) Cu–MFI-0.44; full line — simulated TPD curve; - - - desorption peak with $E_D = 77$ kJ/mol; ... desorption peak with $E_D = 96.7$ kJ/mol; - desorption peak with $E_D = 121.8$ kJ/mol. [45] With permission.

($COCuNaAl_2Si_6O_{23}H_{14}$) were used. The Na cation was used to neutralize the cluster resulting due to the substitution of silica by aluminum and the resulting TPD curves along with the simulation curve for MFI are shown in Figure 9.17. It was observed that the amount of desorbed CO was proportional to the copper content and the results are shown in Table 9.12 along with simulated TPD. The three site models

TABLE 9.12

Experimental and Simulated TPD Characteristics and Distribution of Cu Ions Between the Different Sites in Cu–MFI Zeolites with Si/Al 14.1

Sample[a]	w_{Cu} (wt %)	CO–TPD Experiment		CO–TPD Simulation		
		n_{CO}/n_{Cu}	T_{max}/K	x_A	x_B	x_C
Cu-MFI-0.02	0.12	0.58	527	0.0	100.0	0.0
Cu-MFI-0.19	1.13	0.95	620	0.0	39.6	60.4
Cu-MFI-0.20	1.26	0.93	625	0.0	35.4	64.6
Cu-MFI-0.37	2.23	0.98	665	4.4	41.9	53.7
Cu-MFI-0.44	2.75	1.00	664	9.0	40.4	50.6
Cu-MFI-0.52	3.05	1.00	687	20.0	28.8	51.2
Cu-MFI-0.56	3.45	0.98	689	17.2	29.3	53.5
Cu-MFI-0.59	3.63	0.68	680	13.4	37.8	48.9
Cu-MFI-0.63	3.93	0.88	683	19.9	30.7	49.4

Note: [a]Cu/Al ratio.
Source: [45] With permission.

TABLE 9.13

Arrhenius parameters for desorption and readsorption of CO on Cu-MFI zeolites

TPD Site Type	CO–TPD Simulation			QM Pot Calculations	
	$k_A{}^a$	$A_D{}^b$	$E_D{}^c$	$E_{ads}{}^c$	Site Details
A	2.7×10^5	1.1×10^9	77.0	−83.8	Al pair
B	2.2×10^6	6.4×10^{10}	96.7	−108.0	Channel wall site
C	4.8×10^6	7.5×10^{11}	121.8	−134.4	Intersection wall site

Notes: [a] Rate constant of adsorption (in dm^3 mol^{-1} s^{-1}).

[b] Preexponential factor (in s^{-1}).

[c] Desorption/adsorption energies (in kJ/mol).

reproduce the experimental TPD, attributed to the fact that the described model can reproduce desorption and adsorption successfully to account for the experimental behavior. This can surely then be verified by the site-specific adsorption energy calculated by QM pot methodology. The CO adsorption energies are given in Table 9.13. Calculated $E_{ads}(CO)$ and CO desorption energies obtained from TPD spectra simulations are in very good agreement with experimental results. Furthermore, the individual CO adsorption sites determined from the TPD experiment can be undoubtedly linked to the particular Cu^+ site type found at the quantum mechanical level. The calculations of lateral interaction between CO molecules adsorbed on the pair of Cu^+ atoms show that there is no lateral repulsion. For the samples with high Si/Al ratio investigated in this work, there is enough space in the MFI channel system for adsorbed CO molecules to avoid lateral repulsion. Analysis of CO TPD spectra based on the quantum chemical calculations brings information about the character and population of Cu^+ sites in zeolites. Because the character of the Cu^+ sites depends on the position of a framework aluminum atom, this analysis also brings information about the localization of the framework aluminum. The Cu^+ intersection sites are populated when framework aluminum atom is located in T1, T2, T3, T5, T6, T7, T9, or T12 positions. Again, the Cu^+ sites on the channel wall were populated when framework aluminum atom wes in T4, T8, T11, and T10 positions. This methodology probably works for other zeolitic systems with the Cu^+ ions, and a combination of TPD and QM calculations can probably help resolve the issues of quantitative conclusion with the sitting, which then further helps prescribing the reaction mechanism. Nachtigall et al. have followed up an exhaustive study to go deeper and check the effect of Cu^+–NO binding and compared the results with EPR signals. They studied the spin orbit coupling scenario with DFT and looked into the spin population and suggested that in the gas phase Cu^+–NO complexes result from the interactions of the singly occupied orbital of NO with the unoccupied 4s and 3d orbitals of Cu^+. However, in the Cu/zeolite system, electrostatic attraction increases the Pauli repulsion and pushes up the occupied 3d orbitals and hence weakens the NO bond, which was otherwise strong at the gas phase. The different character of ON...Cu^+ interaction in the Cu^+–NO gas-phase complex and the ON–CuZ adsorption complex

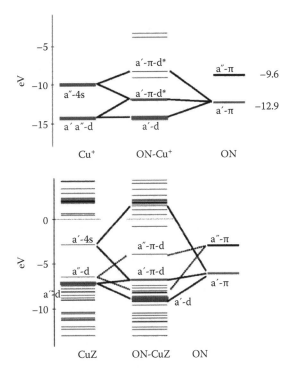

FIGURE 9.18   (For color version see accompanying CD) MO correlation diagrams for the ON–Cu^+ gas-phase complex (top) and the ON–CuZ adsorption complex (bottom, Z-shell-1.5 model). In the top diagram the NO orbital energies are given for NO in the potential of a point charge at the position of Cu^+ ion in the ON–Cu^+ gas-phase complex. For systems with open-shell electronic structures the R-spin orbital energies are shown. [48] With permission.

can also be discussed in terms of the MO diagram as shown in Figure 9.18. This is a very interesting picture to look into the intrinsic bonding scenario and that is the one of main reasons to use DFT to rationalize these interactions. Now, one thing we know here is that there is a charge transfer to occur between Cu^+ and NO when they form a bond. Note that opening of the d-shell on Cu^+ (3d-4s hybridization) is a prerequisite to this charge transfer because the 3d orbitals are fully occupied in Cu^+. This is illustrated by the occupations of the Cu 3d (9.85) and Cu 4s (0.21) orbital in Cu^+–NO. Because the singly occupied orbital of NO is antibonding, depopulation of this MO by partial electron transfer to Cu^+ makes the NO bond in the Cu^+–NO gas-phase complex stronger (Wiberg bond index 2.17) than that in the NO radical (Wiberg bond index 2.11). This is the reason for a shorter NO bond (by 0.014 Å) and a higher vibrational wave number (by 18 cm^{-1}) in the Cu^+–NO complex. The same system was also used for the channel wall (coordination to 3–4 framework oxygen atoms) and on the channel intersection (coordination of two oxygen atoms). Upon the interaction with NO, the Cu^+ ion stays coordinated with only two framework oxygen atoms, which belong to the AlO_4 tetrahedron and the structures of all ON–Cu/zeolite adsorption complexes become very similar. EPR hyperfine coupling constants were calculated and compared with experiment. The isotropic component of the hyperfine coupling constant strongly depends on CuNO angle, but due to small variations of this angle, it is a challenge to resolve EPR signals (or other spectroscopic signals) for different sites. Hence, a qualitative comparison was tried. The interaction of NO with the Cu$^+$ sites on the channel intersection is significantly stronger (29.5–27.1 kcal/mol) than the interaction with the Cu$^+$ sites on the channel wall (22.6–15.0 kcal/mol) because the bonding results a framework deformation and can be interpreted by the subtraction of the energies to calculate the cost of the deformation.

It has to be mentioned that CuZSM-5 is the most effective catalyst for NO$_x$ decomposition. The mechanism is challenging and despite of lots of theoretical work that has already been done, it is still significant to compare with experiments and reconfirm the process. One recent work by Pietrzyk et al. [52] investigated the interaction of NO with CuIZSM-5 catalyst in static (IR and EPR) and flow (IR) regimes, complemented by DFT quantum chemical calculations. DFT calculations were carried out by means of DMol^3 (Insight II, Release 2000.1, Accelrys Inc., San Diego) software using BPW [53] exchange-correlation functional and DNP basis set. Various cluster models of increasing size used for the quantum chemical investigations were cut off from ZSM-5 structure [54] and terminated with OH groups. The authors modeled the CuIZSM-5 sites with the following clusters: CuI[Si$_2$AlO$_2$(OH)$_8$](CuI–I2), CuI[Si$_4$AlO$_5$(OH)$_{10}$](CuI–M5), CuI[Si$_5$AlO$_6$(OH)$_{12}$] (CuI–Z6), and CuI[Si$_6$AlO$_8$ (OH)$_{12}$](CuI–M7). Vibrational analysis was performed within the double-harmonic approximation. The Hessian matrices were evaluated by numerical differentiation of the analytic energy gradients. To be directly compared with experimental IR results, the calculated harmonic frequencies were corrected by a scaling factor (NO$_{exptl}$/NO$_{DFT}$) to account for the anharmonicity. For the stretching mode, this factor was derived from the frequency of the free NO molecule, because it fits the experimental data quite well [55]. The ratio of the experimental to the computed harmonic frequencies for the NO molecule was equal to 1,876/1,843 = 1.018. Calculations of the hyperfine coupling constants were performed with the Gaussian 98 program (Revision A.6, Gaussian, Inc., Pittsburgh, PA) for the previously optimized cluster structures. From IR one will be able to determine the coordination of the NO$_x$ and zeoite coordination through Cu and the peak positions proposes two possibilities: (i) {CuI(NO)$_2$}ZSM-5 complex and (ii) {CuI(NO)$_2$}ZSM-5 complex. It is further observed that these two complexes exist as a mixture and equilibrium can be shifted with a larger dose of NO. DFT optimization confirms the fact in terms of the adsorption energy the mono NO complex is −17 kcal/mol more stable than that of the (NO)$_2$ complex, validating the IR observation in terms of the co-ordination. An *in situ* IR study describes that mono NO complex forms readily and disappears fast with the increase in temperature and there is no existence of di-nitro complex. This provides a justification of the earlier observation that di-nitro is unstable and hence the formation of either nitro is favored by the presence of more NO molecules. It was then followed by the EPR measurement and the hyper-coupling constants were calculated from the spin interactions. It was observed that the spin density is centered on the NO ligand in the {CuINO}M5 complex, as only 10% of the total spin density was delocalized over the metal-based orbitals. Therefore, the spin density repartition suggests that the attack of the paramagnetic NO molecule should be directed to the ligand. This was further studied with di-nitro complex {CuI(NO)$_2$}M7, and two binding scenarios of (1) repulso and (2) attracto were observed. In the case of the repulso conformation, the NO ligands were bent outwards, giving rise to the proximal-nitrogen and distorted oxygen atomic arrangement, whereas in the attracto form the NO ligands were bent inwards, bringing both oxygen close to (d_{O-O} = 2.40 Å) and the nitrogen atoms away from each other. The attracto conformer was favored by 7.7 kcal/mol over the repulso one. This result is valid for all the copper hosting sites considered here (I2, M5, Z6, M7). The relative intensities of the symmetric and antisymmetric stretchings were markedly altered, whereas the band positions did not vary much with the conformation change for all the investigated sites. Reliable prediction of IR intensities is much more demanding than locating the frequencies of the vibrational transitions, within the so-called double-harmonic approximation [56]. With that information the authors then explored the mechanistic part and trying to figure out which could be the best scenario. The spatial proximity of both oxygens in the attractive conformation may suggest that the O–O bond formation could initiate the NO decomposition. However, the Mulliken population results for the DFT showed that once the oxygen–oxygen distance decreased, the positive charge on both terminal oxygen atoms gave rise to steadily increasing energy due to the growing Coulombic repulsion. This may help one to conclude that the attractive conformation is an essentially inert form, and the NO decomposition cannot be initiated by the oxygen–oxygen bond formation step. The alternative N–N bond making route entails a prior transformation of the attractive to repulsive conformation. From the above results the reaction between CuIZSM-5 and NO can be described as follows:

$$\{CuI\}\ Z + NO(g) \rightarrow \{\ CuINO\}Z \tag{9.2}$$

$$\{CuINO\}\ Z + NO(g) \rightarrow cis\text{-}[CuIN_2O_2]Z \rightarrow trans\text{-}[CuIN_2O_2]\ Z \tag{9.3}$$

$$\{CuI\ N_2O_2\}Z \rightarrow \{(CuO)I\}Z + N_2O(g) \tag{9.4}$$

where Z is the zeolite. The total energy change for the overall reaction of {CuI} M5 + 2NO → {CuO}M5 + N$_2$O was equal to ~47.3 kcal/mol. The fascinating thing here is that one can now identify the N–N bond formation and can decisively discuss the bond knitting process, which includes exclusive formation of the mononitrosyl complex (1), its transformation into {CuIN$_2$O$_2$}Z transient via an outer-sphere NO coupling (2), and finally decomposition of copper bound N$_2$O$_2$ species into N$_2$O and the copper-oxo site (3). The latter process is kinetically constrained, because the spin singlet {1Cu N$_2$O$_2$}/Z intermediate is transformed into the {3 CuO}Z center, exhibiting a triplet ground state. One can therefore look into the metal sitting in nanoporous materials through their binding with the adsorbate to compare with different experimental measurements and validate the mechanism of a reaction, which is dependent not only the architecture of the nanopore but the position of the active species within it.

9.3 TRANSITION STATE

Let us now concentrate on transition-state theory to elucidate chemical reactions occurring in nanoporous materials, particularly the microporous material acting as a catalyst. A catalyst is defined as a substrate capable of changing the kinetics of a chemical reaction, increasing its rate by lowering the energy barrier to activate the reactant molecules. Catalysts are not consumed during a chemical reaction, and their activity is defined as the reaction rate for conversion of reactants into products. Understanding at a molecular level the relation between the catalyst surface composition and its activity assists in the catalyst design in industrial chemical processes. A schematic diagram is shown in Figure 9.19 to show the importance of transition-state theory through the reaction coordinate. R_G and P_G describe the reactant and product at the ground state. Experimentalists need to know the barrier of the reaction in the presence and in absence of catalysts so that they can determine whether the desired reaction is a single-step or a multi-step reaction. One can therefore get the numbers in terms of the energy, which the computational chemist calls *barrier height* and is labeled as Z1 or Z2.

When a molecular or crystal structure is built, it usually needs to be refined to bring it to a stable geometry. The refinement process is known as optimization (or minimization) and is an iterative procedure in which the coordinates of the atoms are adjusted so that the energy of the structure is brought to a stationary point; i.e., one in which the forces on the atoms are zero. A transition state is a stationary point that is an energy maximum in one direction (the direction of the reaction coordinate) and an energy minimum in all other directions. During the course of a chemical reaction, the total energy naturally changes. Starting from the reactants, the energy increases to a maximum and then decreases to the energy of the products. The maximum energy along the reaction pathway is known as the *activation energy*; the structure corresponding to this energy is called the *transition state*. Synchronous

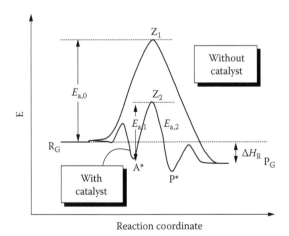

FIGURE 9.19 (For color version see accompanying CD) Schematic diagram to show the importance of transition-state theory through the reaction coordinate.

transit methods are used to find a transition state (TS) when reasonable structures for the reactants and products exist, but the location of the TS is unknown. Starting from reactants and products, the synchronous transit methods interpolate a reaction pathway to find a transition state. Linear synchronous transit [57] is a common method of interpolating geometrically between a reactant and a product to generate a reaction pathway. It can be used in conjunction with single-point energy calculations to perform transition-state searches. In addition, the LST method implemented in Materials Studio of Accelrys has been generalized to handle periodic systems, so that reactions on surfaces and within crystal lattices may be studied.

In the original LST method, an idealized set of structures connecting reactants and products is obtained by linearly interpolating the distances between pairs of atoms in the reactant and product according to:

$$r_{ab}^i(f) = (1-f)r_{ab}^R + fr_{ab}^P$$

where a and b are the reactant and product, respectively, and f is an interpolation parameter that varies between 0 and 1.

The number of distinct internuclear separations in a molecule with N atoms is $N(N-1)/2$. Because this is usually greater than the $3N$ Cartesian degrees of freedom of the system (as shown in the above equation) over specifies the geometry. Halgren and Lipscomb defined the LST path by finding the molecular geometry with internuclear distances as close as possible to the idealized values. This is obtained by minimizing the function S:

$$S(f) = \frac{1}{2}\sum_{a \neq b}\frac{\left[r_{ab} - r_{ab}^i(f)\right]^2}{\left[r_{ab}^i(f)\right]^4} + 10^{-6}\sum_a \left[X_a - X_a^i(f)\right]^2$$

where x_a^i is the interpolated Cartesian position of an atom and x_a is the actual coordinate. The function S is necessarily always greater than or equal to zero and the reactant and product geometries therefore minimize S when f is 0 and 1, respectively. Such an interpolation is purely geometrical; it involves no calculation of energy. The pathway calculated by a TS confirmation corresponds to an intrinsic reaction path (IRP). This is the path that would be produced by a molecular dynamics (MD) simulation starting at the TS using completely damped velocities and infinitesimal steps. Each point on the IRP is an energy minimum in all directions except one, which defines the direction of the IRP. This pathway also corresponds to the intuitive minimum energy pathway (MEP) connecting two structures. The TS confirmation begins by approximating the IRP with the QST and then performs subsequent refinements. Each such refinement is called a *macroiteration*. During each macroiteration, the method performs a number of constrained geometry optimizations, called *microiterations*. The technology underlying the TS confirmation algorithm is based on the nudged-elastic band (NEB) algorithm of Henkelman and Jonsson [58]. However, the NEB locates a TS and reaction pathway based on two endpoints, and, in contrast, the TS confirmation calculation begins with a TS obtained by using another algorithm (e.g., QST) and returns the IRP through three points (R, TS, P).

TABLE 9.14

Simulation Results of ZrCl$_4$ Dissociating at the Si(1 0 0) Surface: Energetics

	E_{reac} (kcal/mol)	ΔE_b (kcal/mol)	v_{TS} (cm^{-1})
$\Theta_{OH} = 0.00$	33.3	42.5	—
$\Theta_{OH} = 0.25$	−0.7	3.7	−330
$\Theta_{OH} = 0.50$	−1.1	3.7	−311
$\Theta_{OH} = 0.75$	−2.0	3.8	−387
Si$_9$	−0.2	4.3	−175
Si$_{15}$	0.6	3.1	−905
Si$_{21}$	−1.4	5.0	−49

The methodology has been explored further by Govind et al. [59]. This paper describes a generalized synchronous transit method for locating transition-state structures or first-order saddle points. The algorithm is based on the established scheme of combining the linear or quadratic synchronous transit method with conjugate gradient refinements but generalized to deal with molecular and periodic systems in a seamless manner. This method is further applied to atomic layer deposition growth of ZrO$_2$ [60]. Table 9.14 summarizes the values for reaction energies (E_{reac}), activation energies ΔE_b, and the frequency of the bond breaking mode. The upper section of Table 9.14 discusses the slab models. The reaction of ZrCl$_4$ at the nonhydroxylated Si (1 0 0) surface (ΘOH = 0:0) was found to be associated with a cost of energy of 33.3 kcal/mol. For the partially hydroxylated surfaces, the reactions are characterized by reaction energies in the order of 0.7–2.0 kcal/mol, indicating a weak binding of ZrCl$_3$ to the surface that increases slightly with increasing OH coverage. ΔE_b was found to be independent of the OH coverage at 3.7 kcal/mol. The atomistic structure of the transition states is essentially independent of the OH coverage, which can be seen from Table 9.15.

TABLE 9.15

Simulation Results of ZrCl$_4$ Dissociating at the Si(1 0 0) Surface: Structural Parameters

	Si–O (Å)	O–Zr (Å)	O–H (Å)	∠Si–O–Zr (°)
$\Theta_{OH} = 0.25$	1.73	2.16	1.33	149.5
$\Theta_{OH} = 0.50$	1.76	2.16	1.31	147.6
$\Theta_{OH} = 0.75$	1.72	2.07	1.44	149.2
Si$_9$	1.78	2.08	1.14	163.3
Si$_{15}$	1.71	2.06	1.43	145.9
Si$_{21}$	1.80	2.27	1.03	156.6

With this background we want to explore the reaction mechanisms in nanoporous materials with different catalytic reactions. Let us start with a very interesting reaction of alkene epoxidation over the TS-1 catalyst [61]. The isomorphous substitution of silica by titanium into the microporous matrix, especially ZSM-5 with an MFI structure, results in a TS-1-type structure. These titanium-containing porous silicates exhibited excellent catalytic activities in the oxidation of various organic compounds in the presence of hydrogen peroxide under mild conditions. One of the main catalytic reactions by TS-1 is epoxidation of alkenes [62]. A number of reaction mechanisms and active oxidizing intermediates in the ethene epoxidation process over Ti-silicate catalysts have been presented and are still in debate. Another example is the synthesis of cyclohexenoneoxime (an industrially important ingredient of Nylon-6) over TS-1 catalyst in the presence of ammonia and hydrogen peroxide, which involves a ketone. There are two plausible mechanism proposed: (1) the reaction can proceed through an unstable intermediate ketoimine [63] or (2) through the hydroxylamine formation followed by the formation of oxime [64]. Some experimental results showed the formation of the ketoimine over TS-1 and Ti-mordenite catalysts, but the unstable ketoimine decomposed in the presence of a large amount of water to reproduce the starting material of the ketone. Kinetic studies were carried out on the hydroxylamine formation over the Ti-mordenite catalyst in the presence of ammonia, hydrogen peroxide, and ketones. Different reaction rates were observed over successive formations of the ketooximes from various ketones. On the basis of those results, the reaction pathway via the hydroxylamine formation was proposed. This is very difficult process to conclusively predict the reaction mechanism, so DFT calculations to simulate and quantify the mechanism and DMol^3 of Accelrys Inc. were used. Not a TS algorithm but a reaction energy calculation methodology based on the following equation was used:

$$E_{react} = \Delta E_{product} - \Delta_{Ereactant},$$

where E_{react}, $E_{product}$, and $E_{reactant}$ represent the binding energies of the corresponding reaction, a product, and a reactant, respectively. The negative value of E_{react} indicates that the formation of product is energetically favorable. The cluster model was constructed by extracting the titanium atom from the titanium-substituted MFI-type silicalite crystal together with the nearest-neighbor of four SiO$_4$ groups. Hydrogen was added to the dangling bond of the terminal oxygen along the bonding direction to the adjacent silica atom, to give the cluster model of Ti [−O−Si(OH)$_3$]$_4$. The cluster models were constructed for the titanium-substituted T1, T6, T8, and T12 sites of the MFI-type silicalite crystal of the TS-1 catalyst. These T sites were selected based on published reports [65]. The positions of the T sites in TS-1 are shown in Figure 9.20. The energetic profile for activation of hydrogen peroxide over the TS-1 catalyst cluster model is shown in Figure 9.21. The energetic profile for ethene epoxidation with the hydrated peroxo-TS-1 complex is shown in Figure 9.22. The energetic profile for the formation of hydroxylamine from ammonia, using the hydrated peroxo-TS-1 as an oxidizing agent, is shown in Figure 9.23. In the activation of hydrogen peroxide over the TS-1 cluster model, the hydrogen peroxide-adsorbed TS-1 complex proceeded to form the intermediate complex, TS-1[(Ti)−OOH][HO−(Si)], with an activation

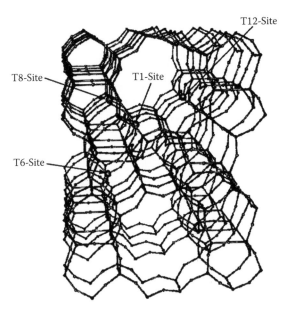

FIGURE 9.20 Locations of T site (tetrahedrally coordinated Ti site) models in the MFI-type silicalite crystal. [61(a)] With permission.

barrier of 69 kJ/mol relative to the energy of TS–1(HOOH) complex. The intermediate complex, TS-1[(Ti)–OOH][HO–(Si)], was then transformed into the hydrated peroxo-TS-1 complex, containing a (Ti)–O–O–(Si) peroxo-moiety, with an activation barrier of 58 kJ/mol relative to the energy of the intermediate complex. Using the hydrated peroxo TS-1 complex, containing a (Ti)–O–O–(Si) peroxo moiety, as an oxidizing agent, the epoxidation of ethene in the postulated reaction pathway

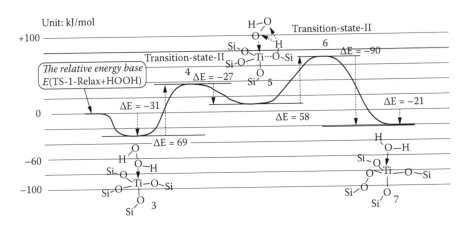

FIGURE 9.21 Energy profile for the pathway for the activation of hydrogen peroxide over TS-1-relax to form the hydrated-peroxo-TS-1 complex. [61(a)] With permission.

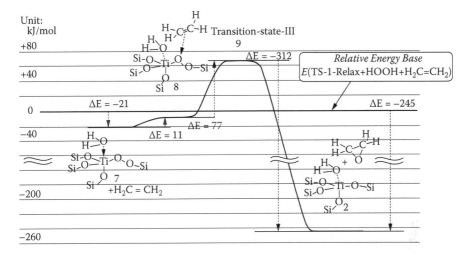

FIGURE 9.22 Energy profile for the pathway for the ethane epoxidation using hydrated-peroxo-TS-1 complex. [61(a)] With permission.

proceeded to produce ethylene epoxide, with activation barrier of 77 kJ/mol relative to the energy of the ethane, weakly bound hydrated peroxo-TS-1 complex. The water molecule of the hydrated peroxo-TS-1 complex was substituted with ammonia to give the ammonia adsorbed peroxo-TS-1 complex. The ammonia-adsorbed peroxo-TS-1 complex was transformed into the ammonia-N-oxide TS-1 complex, with an activation barrier of 56 kJ/mol relative to the energy of the ammonia-adsorbed peroxo-TS-1 complex. The (ammonia-N-oxide)-TS-1 complex was transformed into

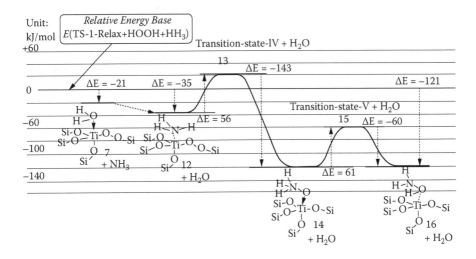

FIGURE 9.23 Energy profile for the pathway for ammonia oxidation using peroxo-TS-1 complex. [61(a)] With permission.

the (hydroxylamine)-TS-1 complex, with the hydrogen transfer from the nitrogen to the oxygen in the ammonia-*N*-oxide moiety, with an activation barrier of 61 kJ/mol, relative to the energy of the (ammonia-N-oxide)-TS-1 complex. The DFT calculation results exhibited the possibility to form the peroxo-TS-1 complex, containing the (Ti)–O–O–(Si) moiety, as an active oxidizing agent, in the activation of hydrogen peroxide over the TS-1 catalyst. Using the peroxo-TS-1 complex as an active oxidizing agent, catalytic cycles for the epoxidation of ethene and for the formation of hydroxylamine from ammonia over the TS-1 catalyst in the presence of hydrogen peroxide are proposed. So the bottom line of this work is to prove the reaction mechanism with the energy scenario. Now there are some more demands of the trade with frequency or thermodynamics relation, but the stability and the barrier height of a reaction can thus predict the feasibility of the reaction.

Let us move to another reaction scenario: the conversion of methanol to gasoline known as the MTG process, where again microporous zeolites play an important role. The main interest is in the mechanism of the first C–C bond formation and cleavage of the C–O bond of methanol. This mechanism is revisited and explained by Andzelam et al. [66]. Payne et al. have published a series of papers to explain the methodology by exploring the reaction mechanism through periodic model and the dynamics with CASTEP [67]. *Ab initio* MD simulations [68] have been performed for the formation of di-methyl ether (DME). All technical details of the simulations are as described in Sïtich et al. [69]. It suffices to say that simulations were run in the (N,V,T) ensemble by using DFT in its plane wave pseudo-potential formulation. Hence, periodic boundary conditions allow us to consider the full zeolite topology. Thermodynamic integration was performed by using the so-called Blue Moon ensemble [70] to overcome the large reaction barrier and to evaluate the entropic contribution along the reaction coordinate. This is a well-known approach; however, it has not been applied very often to complex chemical systems. This new approach adopted here combines the well-known technique of thermodynamic integration, required to extract the entropy contribution beyond the harmonic approximation, with *ab initio* MD, needed to sufficiently describe the breaking/forming of chemical bonds in the chemical reaction. The main complication with this approach is the high computational cost. However, the class of systems and processes with entropically controlled behavior and/or with complicated multidimensional, difficult-to-locate transition states is large, and the techniques of thermodynamic integration [70] and transition-path ensemble [71] will play an increasingly important role in a realistic study of chemical reactions.

Going back to the discussion of reaction of Andzelam et al. [66], there are many proposed mechanisms both by experiment and simulation. There is a belief that the initial physisorption occurs at the Brønsted acid site of the aluminum-substituted zeolite. As the concentration of methanol increases, clusters of hydrogen-bonded methanol molecules are formed in the zeolite cage [72]. At that stage, DME can be formed [73], as confirmed by experiments [74]. However, it is unclear whether DME is a necessary intermediate for the first C–C bond formation. As for the cleavage of the C–O bond of methanol, it can occur through the formation of surface methoxyl species [75]. It is very hard to monitor the mechanism of the reaction by cluster model, because this is important to see the effect of the architecture of the zeolite matrices. So this calculation was performed using the plane wave code CASTEP of Accelrys Inc.; for example, the physisorption and clustering of methanol molecules leading to the formation of DME [73]. Andzelm et al. have performed the first periodic calculation was to investigate the formation of a surface methoxyl species through the formation of an initial COC bond within a zeolite cage, which ultimately leads to ethanol [78]. The possibility of -ylide formation close to a Brønsted acid site was investigated as well. Figure 9.24a displays the most stable structure, with adsorption energy of 18.5 kcal/mol, and compares the experimental estimates of the heat of adsorption in acidic zeolites, ranging from 15 to 27 kcal/mol [76]. No protonation of methanol by the Brønsted acid site was found, which is in agreement with the recent study by Haase and Sauer [77]. The calculated transition state for the methylation of surface oxygen at the aluminosilicate Brønsted acid site of FER zeolite is presented in Figure 9.24b. Clearly this is a concerted reaction involving breaking of the COO bond in methanol and bond formation between C and surface oxygen. Simultaneously the proton from the Brønsted acid site is transferred to the hydroxyl group, thereby forming a water molecule (Figure 9.24c). This is a strained SN2-type reaction with an activation barrier of 54 kcal/mol and, interestingly, the presence of a second methanol molecule lowers the above activation barrier to 44 kcal/mol, as the TS now corresponds to an unstrained SN2 pathway (see Figure 9.25a). At the TS the surface oxygen, methyl carbon, and oxygen of the leaving the water molecule are roughly collinear (Figure 9.25b). The water molecule is formed as a result of proton transfer to the methanol hydroxyl group from the methoxonium ion. Finally, the methoxyl goes closer to the surface in the presence of water and methanol as shown in Figure 9.25c. These figures are a representative model of the reaction path, and the original figures are available in the paper. The advantages of current approach [78] are summarized below:

- It allows us to calculate hydrogen bonding with a reasonable accuracy of ~0.5 kcal/mol for interaction energies.
- Periodic DFT calculations can describe well the neighborhood of a Brønsted acid site in aluminosilicate faujasite zeolite. The experimentally known positions of the bridging hydroxyl groups and their relative stabilities are correctly reproduced in our calculations.
- A single methanol molecule is adsorbed in zeolite cages via hydrogen bonds, whereas the presence of the two methanol molecules allows for spontaneous formation of the methoxonium ion.
- Formation of surface methoxyl species occurs in an SN2-type concerted reaction with a barrier of 44 kcal/mol if two methanol molecules are present.
- Surface methoxyl species can undergo a reaction to form an ethanol molecule. The barrier for such a reaction is 25 kcal/mol, if the water molecule is present.
- Formation of the -ylide that is built into the zeolite cage surface has to be ruled out as the barrier because such a reaction exceeds 78 kcal/mol.
- Synchronous transit coupled with conjugated gradient refinement to optimize transition states and delocalized internals to optimize structures of reaction intermediates [78], under periodic boundary conditions and internal constraints, allowed fast and accurate exploration of reaction pathways and associated heats and barriers.

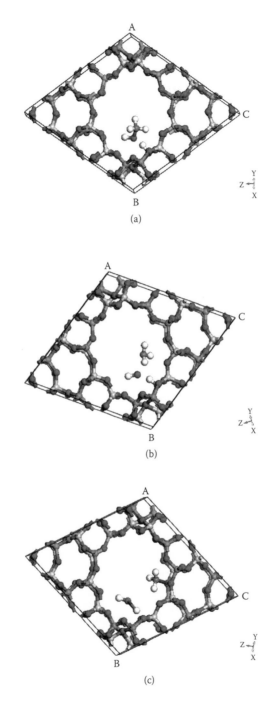

FIGURE 9.24 (For color version see accompanying CD) The reaction pathway model for surface methylation with single methanol. (a) Methanol–hydrogen complex at Brønsted acid site; (b) transition state; (c) surface methoxyl and water.

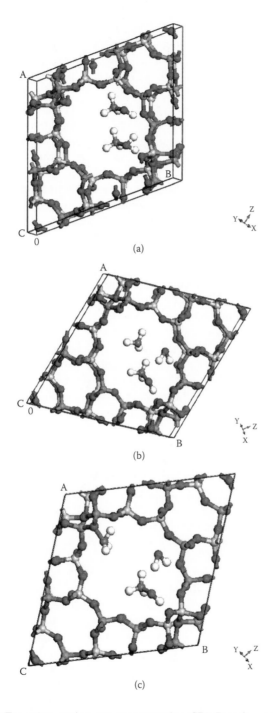

FIGURE 9.25 (For color version see accompanying CD) Reaction pathway model for surface methylation with two methanol molecules. (a) Methoxonium–methanol complex; (b) transition state; (c) surface methoxyl, methanol, and water complex.

The next topic we wish to cover where Lewis acid is detrimental, as we have just see the effect of Brønsted acid sites in the reaction and one can therefore monitor the effect of metal substitution or loading, which changes the ionic equilibrium of the system. In the case of Lewis acid, the metal substitutents may act as a center of activity. In terms of Lewis acidity, alkane conversion is one of the significant and most studied reactions. Pidko and Van Santen [79] have explored the mechanism of alkane dehydrogenation over Zn/ZSM-5 zeolite with DFT using the cluster models. Modification of zeolite with zinc ions can in principle result in formation of various cationic species depending on the type of zeolite, its Si/Al ratio, the method of preparation of the catalyst, and the conditions of the subsequent thermochemical activation. Zn/ZSM-5 prepared using conventional techniques (incipient wetness impregnation, ion exchange in aqueous solution) can, therefore, contain (1) isolated Zn^{2+} ions stabilized in cation sites of the zeolite, (2) binuclear $[ZnOZn]^{2+}$ species, and (3) intrazeolite clusters of zinc oxide. Comparison of the activity of these species for dehydrogenation of alkanes is therefore important for understanding the catalytic properties of zinc-exchanged high-silica zeolites. Most researchers have concluded that isolated Zn^{2+} species are present in the cationic positions of zeolites [80]. Using *in situ* zinc K-edge X-ray absorption techniques, Biscardi et al. [80] have shown that zinc in the Zn/ZSM-5 zeolite (Si/Al = 14.5) prepared via the conventional ion-exchange technique is present as isolated zinc–oxygen species located at the zeolitic exchange sites coordinated to four oxygen nearest neighbors and without next nearest zinc neighbors. Computational studies [81] on location and stabilization of Zn^{2+} ions in ZSM-5 report a preferential fourfold coordination of Zn^{2+} to the zeolite oxygen. Despite all the data available, the nature of the active site and the mechanism of the catalytic dehydrogenation of light alkanes are not very well justified. It is necessary to visualize and acquire an understanding to determine the mechanism of the catalytic dehydrogenation reaction over nanoporous catalysts like zeolites. To clarify the different questions regarding the mechanism, DFT was used. Two different reaction paths have been considered. At the initial step for both paths, molecular ethane adsorbed over the Zn cation and dissociated following the alkyl mechanism. The resulting intermediate Zn-alkene and O–H can be decomposed via elimination of ethylene with the subsequent hydrogen recombination from intermediate. It is also predictable that the reaction could proceed via an alternative route of one-step process of desorption of H_2 and the formation of a molecular complex of ethylene with the active site. Both mechanisms have been studied for three different representations of the active sites in Zn/ZSM-5: (1) binuclear $[ZnOZn]^{2+}$ ion, (2) isolated Zn^{2+} stabilized in the convention ion-exchange site of the zeolite and (3) in the cation site with distantly separated framework aluminum. The models are very important for simulation, and a success to reproduce reaction mechanism depends on the quality of the model. The starting geometry of the clusters to represent the lattice of the ZSM-5 zeolite is normally considered according to the X-ray diffraction data [82]. The next is to figure out where the probable site for the Zn addition is, because it will be a Lewis acid site one needs to locate an interconnected secondary building unit. The cluster model shown in Figure 9.26a is labeled with a T site and its connecting oxygen to show the zeolite connectivity. The authors have chosen the wall of the straight channel and the first model is with a zinc atom bonded with one of the

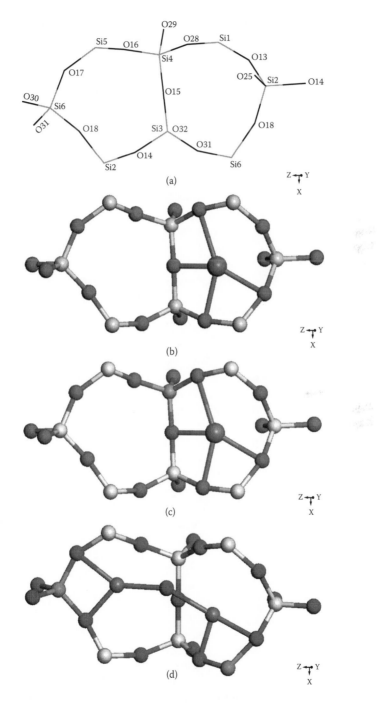

FIGURE 9.26 (For color version see accompanying CD) (a) Cluster model of Zn-ZSM-5, (b) Zn atom bonded with one of the five-membered rings to model silicalite, (c) location of Al as nearest neighbor, and (d) location of Al atom at the other and adjacent five-membered rings.

five-membered rings to model silicalite (Figure 9.24b). This is followed by the model with one silicon replaced by aluminum and the location of Zn^{2+} ion remains the same as with silicalite. There is one more issue in this Si/Al manipulation scheme, which is the location of aluminum it can be very close substitution (Figure 9.26c), meaning nearest neighbors like Si–O–Al, or can be a distant aluminum to compare chemical stability (Figure 9.26d). The calculations were performed with DFT using the regular procedure by Gaussian03 as described elsewhere (revision B.05, Gaussian, Inc., Pittsburgh, PA) The point by simulation is to see the adsorption followed by the reaction of the molecules concerned. As we know that zeolites can be formed in different ways, hence it is better to check the effect of activity of Zn in the isolated form or in the bonded state. The authors have shown by adsorption energy that there is no difference in activity between zinc dust and zinc loaded in ZSM-5. Thus, it needs to determine the factor responsible for the stability of Zn in zeolite matrix. Zn^{2+} might be stabilized in a situation when two Si^{4+} are replaced by Al^{3+} and how far the Al^{3+} cations are. Two different locations were chosen such as one at the charge alternating site and the one with the conventional substitution site. The model chosen mimicked the zeolite lattice with specific T sites to have a proper understanding of the structure. For this purpose, IR frequency of the related sites with respect to models were compared and it was concluded that the shape and geometry of cation sites influence the ligand field stabilization of the exchanged cation, which was zinc. This, in turn, would affect the strength of donor–acceptor interactions between the adsorbate and the exchanged cations. Moreover, it was recently shown [83] that depending on the shape of the cation site and the size of the exchanged cations, the effective electrostatic field of the cations can be significantly varied. This would result in different polarizations and redistribution of charges within the adsorbed molecule. When the adsorption is confirmed, the next step is the elimination of hydrogen to get molecular ethylene and hydrogen. The catalytic cycle is closed by consecutive desorption of C_2H_4 and H_2. However, both present calculations and recent experimental results [84] show that H_2 dissociative adsorption on Zn/ZSM-5 is a very easy process. The dimeric Zn–O–Zn side has been explored and a very low stability of the dimeric sites in comparison to situations when Zn^{2+} cations are independent was reported [85]. So, one can conclude that due to significantly lower Lewis acidity, the initial heterolytic dissociation of ethane is less favorable. Moreover, subsequent steps of the catalytic cycle exhibit higher activation barriers and reaction energies.

High-silica zeolites modified with Ga also act as an effective catalyst for promoting the dehydrogenation of light alkanes and their subsequent aromatization [86]. Interaction of methane with the GaO^+ cations in zeolite was theoretically first studied by Himei et al. [87] and Broclawik et al. [88]. The gallyl ions were shown to be highly active for the initial C–H bond cleavage. However, the subsequent regeneration of the active site was found to be strongly disrupted by high stability surface species. This does not match well with the experimentally suggested high initial catalytic activity of the selectively oxidized Ga/ZSM-5 zeolite. This is the reason to discuss this example along with Lewis acidity, which is very important for catalytic activity of nanoporous materials. In a recent work by Pidko and van Santen [79], a detailed comparative analysis of the various reaction paths over the oxidized form of gallium, the gallyl GaO^+ ion stabilized in ZSM-5 zeolite, was studied.

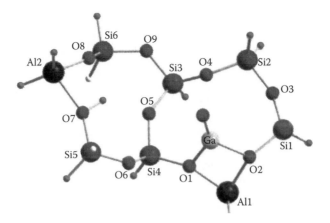

FIGURE 9.27 (For color version see accompanying CD) Optimized structure of the GaO$^+$ ion stabilized in the cluster model representing ZSM-5 zeolite. [79] With permission.

The importance in this modeling is the model building; a periodic model with all the atoms is undoubtedly the best choice but the drawback is the high computational cost. We have seen so far that the best results could be obtained by a cluster model where the boundary remains constrained to mimic and control the microporous structure, to keep a balance between cost and efficiency to reproduce the pore architecture. The GaOHAl$_2$Si$_6$O$_9$H$_{14}$ cluster model shown in Figure 9.27 represents two adjacent five-membered rings from the wall of the straight channel of ZSM-5 zeolite. The starting geometry of the cluster corresponds to the lattice of ZSM-5 zeolite derived from crystal structure data [82]. Aluminum atoms were positioned in the T12 and T8 lattice positions (Al1 and Al2, respectively, as shown in Figure 9.27) [89]. The T12 site is located at the cross section of the straight and sinusoidal channels of ZSM-5 because that is the best position to model, which gives the scenario very close to reality. We have seen that even for the zinc case, the extraframework species stabilized in the vicinity of this site are considered to have the highest accessibility. Therefore, the charge-compensating gallyl ion was located in the five-membered ring containing aluminum at the T12 position (Al1 in Figure 9.27). Hydrogen atoms were used to saturate the dangling Si–O bonds at the boundary of the cluster. The hydrogen atoms also saturate the negative charge due to the aluminum atom at the T8 site. Similar to the case of dehydrogenation over Zn-ZSM-5, DFT with the B3LYP hybrid exchange-correlation functional was used for the calculations. Geometry optimization and saddle-point searches were all performed using the Gaussian 03 program. The gallium and oxygen atoms of the gallyl ion GaO$^+$ represent a Lewis acid base pair, which is able to polarize and cleave the C–H bond of ethane due to the presence of partial charges (C$^{\delta-}$H$^{\delta+}$). This leads to dissociative adsorption of ethane resulting in the formation of an alkyl fragment (C$_2$H$_5$$^{\delta-}$) and a hydroxyl group attached to the Ga cation (IIa as shown in Figure 9.28). This competing path (I + C$_2$H$_6$ → TS$_a$1*) is suggestive of the mechanism proposed for reduced gallium species in ZSM-5 (i.e., Ga$^+$ and GaH^{2+} cations) [90]. The computed activation energy for this reaction path

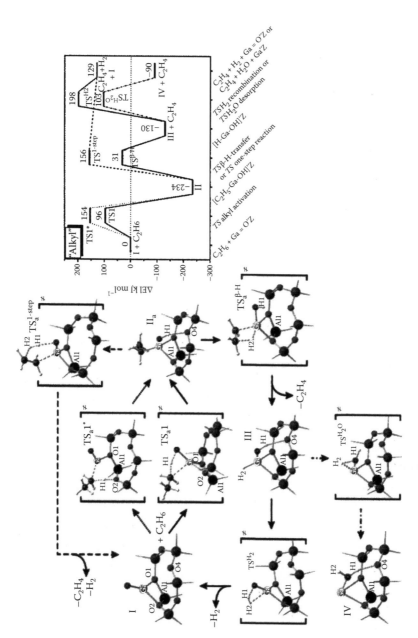

FIGURE 9.28 (For color version see accompanying CD) Reaction paths for the alkyl activation mechanism of ethane dehydrogenation over gallyl ion in ZSM-5 zeolite. [79] With permission.

TABLE 9.16

Calculated Changes in ZPE-Corrected Energy and Gibbs Free Energy Changes (298 K, 1 atm; and 823 K, 1 atm) for the Elementary Reaction Steps Involved in the Alkyl Path of C_2H_6 Dehydrogenation

	ΔE	ΔG_{298}	ΔG_{823}
$I + C_2H_6 \rightarrow TS1_a \rightarrow II_a$	−234	−189	−114
$II_a + TS_a^{\beta\text{-}H} \rightarrow III + C_2H_4$	104	62	10
$III \rightarrow TS^{H2} \rightarrow I + H_2 + C_2H_4$	259	226	151
$III \rightarrow TS^{H2O} \rightarrow IV + C_2H_4$	40	35	20
$II_a \rightarrow TS_a^{1\text{-step}} \rightarrow I + H_2 + C_2H_4$	363	288	141

Note: All values are in kJ/mol.

$(I + C_2H_6 \rightarrow TS_a1^*$ (as shown in Figure 9.28) is significantly higher (by 58 kJ mol^{-1}) than that of the reaction $I + C_2H_6 \rightarrow TS_a1$. The higher basicity of the extraframework oxygen atom (O) is the reason for this difference. Moreover, this mechanism results in the formation of a highly unstable neutral C_2H_5–Ga=O species in the vicinity of an acidic proton. This species easily rearranges to intermediate IIa. The catalytic cycle is closed via recombination of molecular hydrogen from [HO–Ga–H]$^+$ ions III (III → TS2 → I + H2). This reaction can be considered as a Brønsted acid–base reaction, where the proton from the HO–(Ga) group attacks the hydride ion resulting in hydrogen desorption and regeneration of the GaO$^+$ site. This reaction is calculated to be highly endothermic ($\Delta E = 259$ kJ/mol). Moreover, the very high activation energy of 328 kJ/mol has to be overcome. One can therefore easily conclude that because of the very high basicity of the extraframework oxygen atom, the dissociation of the H1–O bond in III is very unfavorable (Table 9.16). The decomposition of IIc-2 has been proposed to be a one-step process (IIc-2 → TSc 1-step → I + C$_2$H$_4$ + H$_2$); in this case, the O–C1 bond breaks and one hydrogen atom (H2) of the ethyl group coordinates to the H1 atom of the Ga–H group. As a result in a one-step, desorption of molecular hydrogen and ethylene with simultaneous regeneration of the initial GaO$^+$ active site takes place. Similar to the one-step decomposition of IIa, this reaction IIc is improbable (Table 9.17). The calculated activation energy is equal to 354 kJ/mol. One can see that the overall activation energy for the dehydrogenation of ethane via "carbenium" mechanism until the formation of III is less than 109 kJ/mol. Therefore, one expect the generation of these species via this route under reaction conditions (Table 9.17). Although the initial alkyl activation of ethane is slightly more favorable compared to the carbenium mechanism, subsequent chemical transformations of the reaction intermediates are easier in the latter case because of their lower stability. The calculated activation barrier for the initial oxidation step of the carbenium activation is only slightly higher than that for the heterolytic C–H bond cleavage over the gallyl ion. On the other hand, the former reaction is less exothermic. Taking into account that catalytic ethane dehydrogenation takes place at rather high temperatures, one can expect that the oxidative carbenium route is the most likely reaction path.

TABLE 9.17

Calculated Changes in ZPE-Corrected Energy and Gibbs Free Energy Changes (298 K, 1 atm; and 823 K, 1 atm) for the Elementary Reaction Steps Involved in the "Carbenium" Path of C_2H_6 Dehydrogenation

	ΔE	ΔG_{298}	ΔG_{823}
$I + C_2H_6 \rightarrow TS1_c\text{-}1 \rightarrow II_c\text{-}1$	−141	−100	−30
$II_c\text{-}1 + TS_c\text{-}2 \rightarrow II_c\text{-}2$	−21	−17	−8
$II_c\text{-}2 \rightarrow TS_c^{\beta\text{-}H} \rightarrow III + C_2H_4$	32	−9	−86
$II_c\text{-}2 \rightarrow TS_c^{1\text{-step}} \rightarrow I + H_2 + C_2H_4$	291	216	65

Note: All values are in kJ/mol.

However, because of a small difference in energetics of the processes, the alkyl activation cannot be excluded. Molecular adsorption of ethanol to the gallyl ion (I-1) is rather strong ($\Delta E = -83$ kJ/mol). During this process, a hydrogen bond is formed between the H1 atom of C_2H_5OH and the extraframework oxygen (O) of the gallyl ion. In addition, the negatively charged O atoms coordinate to the positively charged gallium. Both these interactions result in significant weakening of the $O^{\delta-}$-H1 bond and facilitate its subsequent heterolytic cleavage. The $O^{\delta-}$-H1 bond of the ethanol heterolytically dissociates (I-1 → TSe1 → IIe, Figure 9.29) on the oxogallyl species, resulting in very stable [HO–Ga–OC$_2$H$_5$]$^+$ species (IIe) without a notable activation barrier. Ethanol dissociation is strongly thermodynamically favored (Table 9.18). Further possible chemical transformations are very similar to those discussed above for the dehydrogenation of ethane. The calculated reaction paths for the concerted ethane dehydrogenation pathway over intermediates IIc-2 and III as well as their energetics in comparison with the competing primary dehydrogenation or reduction routes are shown in Figure 9.30. Different reaction paths for ethane dehydrogenation over the gallium species in oxidized Ga/ZSM-5 (i.e., GaO$^+$ ions stabilized at cationic sites of the zeolite) have been discussed. This example shows how one can approach the problem, what are the main issues in the problem, and how one can resolve the issue. Two different reaction paths were found for the initial C–H activation. The first one is the heterolytic dissociation of ethane on the gallyl ion following the alkyl ($C^{\delta-}$-$H^{\delta+}$) polarization mechanism. This reaction path exhibits a rather low activation barrier and results in formation of a very stable intermediate [C_2H_5–Ga–OH]$^+$ that can be decomposed via ethylene desorption to form a [H–Ga–OH]$^+$ species. Further regeneration of the active site via H_2 desorption is strongly disfavored both thermodynamically and kinetically. Another possible reaction path for ethane dehydrogenation is initiated by partial oxidation of an ethane molecule by the extra lattice oxygen atom. The resulting ethanol molecule then interacts with the adsorption site (Ga$^+$), leading to formation of a product supporting the carbenium mechanism [H–Ga–O–C$_2$H$_5$]$^+$.

The direct dissociative adsorption of ethane on GaO$^+$ ions as present in Ga/ZSM-5 following such a polarization mechanism ($C^{\delta+}$-$H^{\delta-}$) does not happen. Similar to the

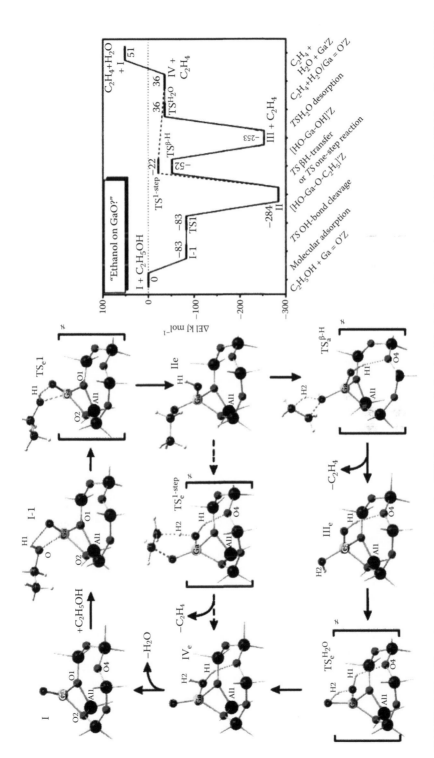

FIGURE 9.29 (For color version see accompanying CD) Reaction path for the dehydration of ethanol over gallyl ion in ZSM-5 zeolite. [79] With permission.

TABLE 9.18

Calculated Changes in ZPE-Corrected Energy and Gibbs Free Energy Changes (298 K, 1 atm; and 823 K, 1 atm) for the Elementary Reaction Steps Involved in Ethanol Activation

	ΔE	ΔG_{298}	ΔG_{823}
$I + C_2H_5OH \rightarrow I_e\text{-}1$	-83	-38	35
$I_e\text{-}1 + TS1_e \rightarrow II_e$	-201	-198	-191
$II_e \rightarrow TS_e^{\beta\text{-}H} \rightarrow III_e + C_2H_4$	31	-9	-83
$III_e \rightarrow TS^{H2O} \rightarrow IV_e$	217	216	214
$IV_e \rightarrow I + H_2O$	87	45	-38
$II_e \rightarrow TS_e^{1\text{-}step} \rightarrow IV_e + C_2H_4$	248	207	131

Note: All values are in kJ/mol.

Scheme 1:

(c)

case of the alkyl mechanism, subsequent ethylene desorption leads to highly stable intermediate [H–Ga–OH]$^+$. Although catalytic cycles of these primary dehydrogenation routes cannot be closed because of formation of very stable intermediates, several catalytic cycles may occur via a concerted dehydrogenation mechanism. Involvement of these secondary reactions over the intermediates formed via the primary route leads to estimation of the overall activation energy for the catalytic ethane dehydrogenation over GaO$^+$ ions below 150 kJ/mol. This value is significantly lower than the overall activation energy calculated for the reduced gallium-containing active sites. This fits well with the experimentally observed high initial activity of the oxidized Ga/ZSM-5 zeolite. However, despite this simplistic pathway, gallyl ions reduce very rapidly via water desorption to produce univalent gallium ions. This latter reaction faces activation energy lower by 95 kJ/mol, compared to the regeneration of the oxidized active site, and is significantly more favorable thermodynamically. From the results presented, one cannot expect the formation of more than three ethylene molecules per GaO$^+$ site before being reduced to Ga$^+$. Therefore, gallyl ions cannot be considered as catalytically active sites for dehydrogenation of light alkanes.

The Meerwein-Ponndorf-Verley (MPV) reduction of aldehydes and ketones and the complementary Oppenauer oxidation (gentle oxidation method of alcohols to ketones) of alcohols (MPVO) are examples of chemoselective reactions and possess a significant status in organic synthesis. Generally, both reactions can be performed under mild conditions. Carbonyl groups are selectively reduced using secondary alcohols as hydrogen donors in the MPV reduction in the presence of other reducible groups such as unsaturated C=C bonds, C-halogen bonds, or nitro groups. Thus, for example, α, -β unsaturated alcohols, which are important starting materials for the production of fine chemicals, can be easily synthesized. The catalyst used so far was a homogeneous catalyst containing zirconium and hafnium, but some heterogeneous catalyst was also applied especially for MPVO reactions. The best heterogeneous

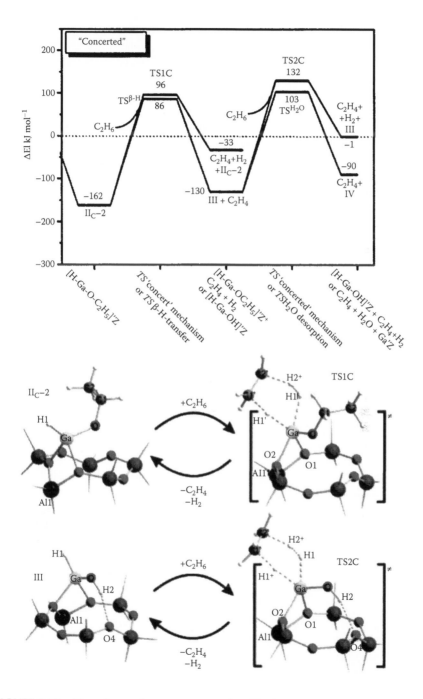

FIGURE 9.30 (For color version see accompanying CD) Concerted reaction mechanism of ethane dehydrogenation over [H–Ga–OR]+ ions, where R = H, C_2H_5. [79] With permission.

catalytic materials found are tin or zirconium-substituted zeolites with well-defined and isolated Lewis acid centers [91]. The nature of the active sites in tin-beta has been experimentally and theoretically investigated [92] and it has been found that the activity of this catalyst for the Baeyer-Villiger (BV) oxidation of ketones with H_2O_2 is related to the presence of tetrahedrally coordinated framework tin atoms having one Sn–O–Si bond. These sites are bifunctional and consist of the Lewis acidic Sn atom that activates the carbonyl group and the adjacent basic oxygen atom of the Sn–OH group, which is able to form a hydrogen bond with H_2O_2. A recent work of Corma et al. [93] has shown the MPV reduction of cyclohexanone with 2-butanol catalyzed by Sn-beta and Zr-beta zeolites. This reaction, as we have discussed with the zinc case, will be similarly poised with the activity of the metal center, its location, the amount present, and the architecture to dictate the reaction process. Several possibilities for adsorption of cyclohexanone and 2-butanol on the Lewis acid active site have been considered, and the different reaction paths starting from these adsorption complexes have been investigated. Moreover, an experimental catalytic study has been performed in order to clarify (1) whether the active sites for the Sn-beta catalyzed MPV reaction are the same as partially hydrolyzed Sn–OH groups that were found active for the BV reaction, (2) whether the mechanism involves an alcoholate intermediate formed by deprotonation of the alcohol, and (3) the catalytic differences between Sn-beta and Zr-beta for the MPV reaction. The T9 position was chosen for the metal incorporation based on preliminary molecular mechanics study [92] of the location of Sn in BEA zeolite. It was indicated that this was the most favorable position of Si for Sn substitution. More recently, two different periodic DFT studies found either no preferential location for tin [94] or preferential occupation of the T2 position [95] in BEA zeolite. It is therefore a general conclusion, based on the exhaustive study of Pal et al. [95], that this is not a local effect of the tin in the unit cell. Now, metal location or sitting is a very important aspect for the zeolite, which is hard to rationalize and even harder to conclude from the simulation. However, one thing that has been proved by simulation is that these sites are very much dependent on the local neighbors and hence we can see the effect when the gas loading changes. The calculations in [92] was optimized by means of the ONIOM scheme [96] as implemented in the Gaussian 03 computer program. As mentioned before, the success of any simulation lies in the model generation and transition state is no exception. In this case there are two issues: first adsorption and second the reaction, and in both the cases the first step is to optimize the centers and the complexes that result. There can be more options to guide the best possible scenario, which then can further be validated by experimentation. The simulation was started with five possible structures. In structure A1, both the O_a atom of the ketone and the O_b atom of the alcohol are coordinated to the metal center. In A2 and A3, cyclohexanone is also directly interacting with Sn, but the alcohol is coordinated by hydrogen bonding either to the basic O_c atom (A2) or to the H_c proton (A3) of the Sn–OH group. In the two other adsorption complexes, the O_b atom of the alcohol is directly coordinated to the Sn atom, whereas the O_a atom of the carbonyl group forms a hydrogen bond either with the H_c of the Sn–OH group as in A4 or with the alcohol hydroxy H_a atom as in A5. The activation of the carbonyl group by the active site is also different in the five complexes considered. The models are shown in Scheme 9.4. It is known that the Lewis acid–base interaction

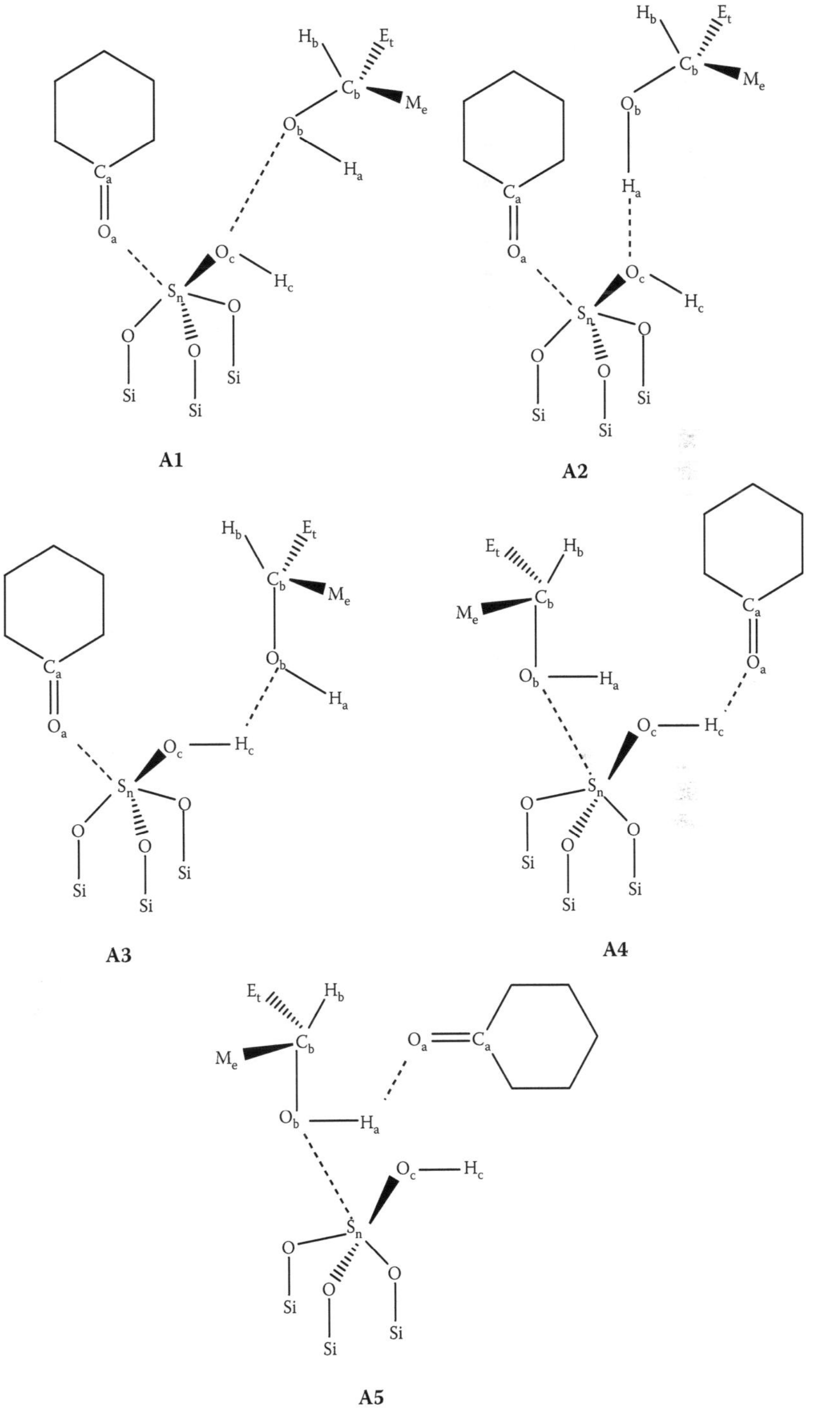

SCHEME 9.4 Interaction of cyclohexanone with the Sn-beta active site.

of cyclohexanone with the Sn-beta active site involves not only the electron density transfer from the organic molecule to the catalyst but also a back-donation from the catalyst to the antibonding $\eth^*(CO)$ orbital of cyclohexanone. This back-donation causes a lengthening of the $C_a–O_a$ bond reflected in an experimentally measured shift of the $\hat{\imath}(CO)$ vibration frequency to lower values [97] and an increase in the positive atomic charge on the C_a atom. The computational study shows that, among the seven different ways in which cyclohexanone and 2-butanol can adsorb on the bifunctional Sn-beta active site, only those in which the carbonyl group is directly coordinated to the Lewis acid metal center are able to activate the CO bond for the MPV reduction. The reaction pathway with the lowest activation energy, and therefore the most probable, is similar to that reported for the Al (III) and La(III) catalyzed reactions, and consists of three steps: (1) deprotonation of 2-butanol, (2) hydride shift through a six-membered cyclic transition state in which both the ketone and the deprotonated alcohol are directly bonded to the Sn atom, and (3) proton transfer from the catalyst to form cyclohexanol. The role of the hydrolyzed Sn–OH group in the MPV reaction is not to bind the alcohol molecule through hydrogen bonding as was the case for H_2O_2 in the BV reaction but to allow the deprotonation of the alcohol to give an alcoholate intermediate bonded to the Sn center and a water molecule. The mechanism for the Zr-beta-catalyzed reaction is equivalent to that found for Sn-beta and involves the same initial and intermediate species. The activation energy of the rate-determining step — i.e., the hydride transfer — is about 15 kcal/mol in both cases, and therefore it can be concluded that both catalysts are active for the MPV reaction and in agreement with the experimental results.

Finally, we wish to mention the role of solvent in the reaction. This is a relatively confined branch due the cost of the calculation and the postulated procedure is complicated. The issue needs to be looked at by incorporating the solvent molecules themselves in the reaction, like water molecules are added in the MTG process, or Ti-silicate were looked upon by hydrolysis like phenomenon when the active Ti centers are surrounded by few water molecules. The popular way of addressing the solvent situation is by a salvation mechanism, when the solvent model is considered as a continuum. This is called the *conductor-like screening model* (COSMO) [98], which is a continuum solvation model (CSM) [99], in which the solute molecule forms a cavity within the dielectric continuum of permittivity (ε), represents the solvent. The charge distribution of the solute polarizes the dielectric medium. The response of the dielectric medium is described by the generation of screening (or polarization) charges on the cavity surface. In contrast to other implementations of CSMs, COSMO does not require solution of the rather complicated boundary conditions for a dielectric in order to obtain screening charges but instead calculates the screening charges using a much simpler boundary condition for a conductor. These charges are then scaled by a factor, $f(\varepsilon) = (\varepsilon - 1)/(\varepsilon + 1/2)$, to obtain a good approximation for the screening charges in a dielectric medium.

The deviations of this COSMO approximation from the exact solution are small. For strong dielectrics like water, they are less than 1%, whereas for nonpolar solvents with $\varepsilon \sim 2$, they may reach 10% of the total screening effects. However, for weak dielectrics, screening effects are small, and the absolute error therefore amounts to less than 1 kcal/mol. Altogether, COSMO is a considerable simplification of the

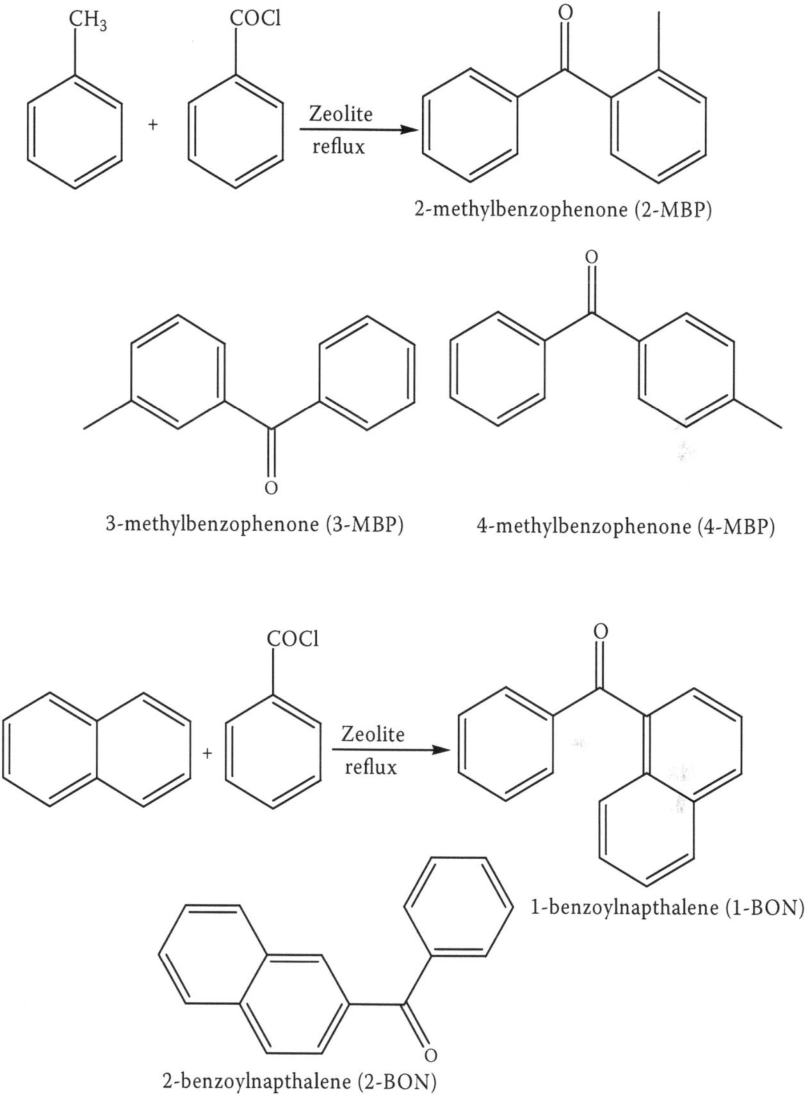

SCHEME 9.5 Selective benzoylation of toluene and naphthalene.

CSM approach without significant loss of accuracy. Because of this simplification, COSMO allows for a more efficient implementation of the CSM into quantum chemical programs and for accurate calculation of gradients, which allows geometry optimization of the solute within the dielectric continuum.

The screening charges are determined from the boundary condition of vanishing potential on the surface of a conductor. If we define q as a vector of the screening charges on the surface of the cavity, and $q = \rho + Z$ for the total solute charges such as electron density, ρ, and nuclear charges, Z, then the vector of potentials, V_{tot}, on the surface is $V_{tot} = BQ + Aq = V_{sol} + V_{pol}$, where BQ is the potential arising from the solute charges, Q, and Aq is the potential arising from surface charges, q. B and A are Coulomb matrices. For a conductor, the relation $V_{tot} = 0$ must hold, which defines the screening charges as:

$$q = -A^{-1}BQ$$

COSMO provides the electrostatic contribution to the free energy of solvation. In addition, there are nonelectrostatic contributions to the total free energy of solvation that describe the dispersion interactions and cavity formation effects. The results depend mainly on the choice of the van der Waals radii used to evaluate the cavity surface.

Supercritical carbon dioxide ($scCO_2$) is currently considered a promising "green" reaction medium for rapid and selective organic synthesis and several other industrial processes. Over the past few years, a number of heterogeneously catalyzed reactions have been successfully carried out in $scCO_2$, often with higher reaction rates and different product distributions, as well as high selectivity, in comparison with those in conventional organic solvents [100]. The origin of the observed selectivity is of importance in view of the current developments in the understanding of the solvent attributes of $scCO_2$ [101] and the interesting supramolecular interactions were also reported for CO_2. Most often, $scCO_2$ is described as a nonpolar solvent with a low dielectric constant (and zero dipole moment), comparable to that for hydrocarbons. However, recent papers [101] have indicated that the charge polarized CO_2 molecule can act as a Lewis acid or a Lewis base. In fact, it has been reported that the oxygen atoms of CO_2 can participate in conventional and unconventional hydrogen bonds with various proton donor systems. The supramolecular interactions of CO_2 give rise to interesting electronic and optical properties in various systems and govern the solvation of several molecular systems carrying relatively polar functional groups, such as acetates [102]. In addition, some previous reports show that CO_2 also plays the role of a protecting group in chemical reactions. Thus, it is pertinent to explore whether CO_2 is capable of mediating any selectivity in homogeneously or heterogeneously catalyzed organic reactions, as opposed to its conventional description as an inert reaction medium. The unique physical properties of $scCO_2$ are also relevant in this context. We described the highly selective Ni(II) catalyzed hydrogenation of α, β-unsaturated aldehyde (citral) into the corresponding unsaturated alcohol in $scCO_2$.

This demands focusing into the reaction in the supercritical solvent medium, which is still beyond the scope of simulation, but we wish to explore the reaction in a solvent continuum model to look into the $scCO_2$ medium. The dipole moment of $scCO_2$ varies with pressure and temperature but for simplicity is taken as 1.6, close to hexane, a common organic solvent. The environment can guide us to predict the role of solvent during the reaction. Selective hydrogenation of α, β-unsaturated to form unsaturated alcohols is an important issue in the manufacturing of fine chemicals and pharmaceutically important molecular systems. For example, unsaturated alcohols such as geraniol have been widely used for flavor and fragrance, as well as in the production of biologically active systems such as phomactin (a platelet-activating factor antagonist). Although, there are several homogeneous and heterogeneous catalytic approaches reported for this transformation, in most cases selective hydrogenation

of the C=O bond in preference to the C=C bond is still a challenge. In search of
suitable catalysts, extensive studies have been carried out in conventional organic
solvents. However, those methods require extreme conditions like high temperature
and raise environmental concerns as described by Chatterjee et al. [103]. In order to
follow the "green" footsteps, $scCO_2$, ionic liquid, and a combination of these were
used for the selective hydrogenation of α, β-unsaturated aldehydes using noble metal
catalyst on various supports. Nickel-catalyzed selective hydrogenation in $scCO_2$ is
not so well documented. Here, we used Ni(II) and Ni(0) supported on mesoporous
silica (MCM-41) as catalysts for the selective hydrogenation of citral and extended
the method to other α, β-unsaturated aldehydes in $scCO_2$ [103]. Although experimental evidence can be correlated to intuitive models to some extent, it is important
to identify the transition states, which could have a decisive influence on the selectivity of a particular reaction. In this context, we carried out calculations using DFT
to give an accurate description of transition states and the reaction mechanism. In
view of the reported geometries for the transition metal complexes of CO_2, we have
considered two different types of geometries in the present case: (A) coordination
via O only and (B) C=O coordination. However, based on energy considerations, the
B-type geometries were excluded. Figure 9.31 presents the probable transition states
obtained from the calculations for the interaction of the Ni(II)–CO_2 complexes of
type A with the C=O (TSAC=O) and C=C (TSAC=C) bonds of citral, respectively.

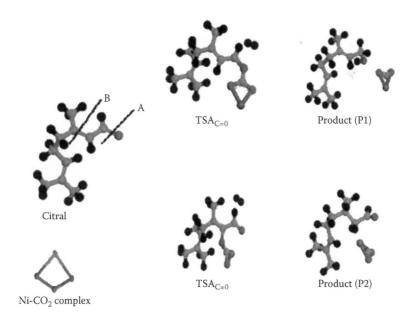

FIGURE 9.31 (For color version see accompanying CD) Interaction of citral with Ni–CO_2
complex: (A) through CLO, proposed reaction intermediates TSACLO and the corresponding product (P1); (B) through CLC, proposed reaction intermediates TSACLC and the corresponding product (P2) in $scCO_2$. The color schemes for various atoms are red (O); black (H);
blue (Ni); and green.

The calculations showed that the interactions involving the C=O bond were more favorable than those with the C=C bond. A comparison between the two transition state energies (TSAC=O and TSAC=C) indicates that the $TSA_{C=O}$ state is energetically favored by ~25.63 kcal/mol over the $TSA_{C=C}$ state, suggesting a preferential hydrogenation of the carbonyl group. This makes the C=O of citral more susceptible to Ni(II)-catalyzed hydrogenation. The energy differences for the $TSA_{C=O}$ and $TSA_{C=C}$ with the corresponding products are 61.175 and 99.150 kcal/mol, respectively. This suggests that the low barrier height of $TSA_{C=O}$ dictates the preferential C=O hydrogenation of citral. Thus, in the Ni(II)-catalyzed reaction there could be a possibility of the chemical participation of the medium. In organic solvents, Ni(0) selectively hydrogenates the C=C bond of citral, producing citronellal. It also catalyzes reduction of cinnamaldehyde producing hydrocinnamyl alcohol (in which both the C=C and C=O bonds underwent reduction). On the other hand, crotonaldehyde and acrolein were converted to the corresponding saturated aldehydes (reduction of C=C) selectively in $scCO_2$. For Ni(0) it is tempting to assume that it favors C=C (146.3 kcal/mol) compared to C=O (177.6 kcal/mol) to hydrogenate as supported by previous *ab initio* MO studies [104].

In continuation with our exploration with $scCO_2$, we applied self-consistent reaction field study to rationalize the selective formation of 3, 7-dimethyloctanal by hydrogenation of citral (3, 7-dimethyl 2, 6-octadienal) exclusively in supercritical carbon dioxide medium [105]. We determined that the dielectric constant of $scCO_2$, which varies with the change in density resulting from the CO_2 gas pressure variation as observed by Arai et al. [106a]. In the literature, it has been reported that for lauric acid at low densities, the solvent–solute clustering is more remarkable than that at higher densities of $scCO_2$, which results in the decrease in activity coefficients of monomer and dimmer [106b]. Gupta et al. [107] performed exhaustive investigation using IR to monitor the solvent effect on H-bonding of methanol and triethylamine throughout the gas, supercritical, and liquid states in the relatively inert solvent SF6. We wish to rationalize the product selectivity of citral hydrogenation in $scCO_2$ in terms of solvent effect. Our explanation was based on the reactivity index calculations performed within the self-consistent reaction field methodology. Condensed Fukui functions for citral molecules at gas phase and in solvents such as hexane, amyl alcohol, and methanol have been calculated in the framework of DFT within the helm of the HSAB principle. The geometry of the citral molecule in all cases was optimized with DFT-based methodology, explained elsewhere [27]. Experimentally it has been observed that the product distribution in $scCO_2$ is different from that in organic solvents. Better selectivity to citronellal is obtained in the solvent-less or organic liquid-phase conditions, but hardly any dihydrocitronellal was formed, which is the major product in $scCO_2$ media. Hence, it is obvious from the results that $scCO_2$ media have a major role in the high selectivity to dihydrocitronellal. To rationalize the fact we have performed DFT calculation both in gas phase and as well in the presence of solvent. We have used a wide range of solvents starting from a nonpolar solvent with dielectric constant of 1.6 to a polar solvent of with dielectric constant 32.6 representing methanol. The optimized geometry of citral in gas phase, in the presence of a solvent with dielectric constants 1.6, 15.0, and 32.6, respectively, are shown in Figure 9.32. We have taken these three solvated structures to compare them and with the gas-phase conformer.

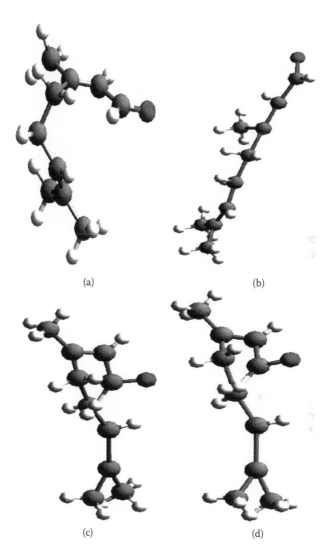

FIGURE 9.32 (For color version see accompanying CD) The geometries of three conformers of citral using DFT with B3LYP functional and 6-311G** basis set: (a) gas phase, (b) liquid phase with $\varepsilon = 1.6$, (c) liquid phase with $\varepsilon = 15.0$, and (d) = liquid phase with $\varepsilon = 32.6$. The color schemes for various atoms are red (O); white (H); gray (C).

The structures of the liquid phases are different form the gas phase. The conformer obtained at very low dielectric constant close to scCO$_2$ is very linear. Citral has two carbon–carbon double bonds. One is between C2 and C3 close to the terminal C=O and the other is between C6 and C7, which is at the other side close to the terminal methyl group. As mentioned before, it is important to compare the relative nucleophilicity and electrophilicity of each center of a molecule to find the active site, which will propose that the molecule may preferably interact with another molecule with comparable activity. Comparing the results in the presence of solvents with different dielectric constants, we observed that a solvent whose dielectric constant is very close to $scCO_2$ has the highest s_x^+/s_x^- value for C7 (2.36) and the second highest is C2 (1.827). This shows that in $scCO_2$ it is possible to hydrogenate both of the C1 and C2, whereas for the other two solvents the activity remains either on C3 (dielectric constant 15.0) or C2 (dielectric constant 32.6). Hence, formation of dihydrocitronellal is improbable in the organic solvent. Therefore, the solvents contribute toward the activity of a reactant molecule in terms of its configuration and reactivity.

In this chapter the focus was to explain the reaction mechanism and the role of nanoporous materials to actively participate in a reaction with an understanding throughout the other chapters describing the synthesis, characterization, and applications of nanoporous materials. One can successfully use computer simulation methodologies to explore the feasibility of the reaction and look into the reaction mechanism in terms of adsorption, shape selectivity, cracking, and transition-state situations. We also explored the effects of solvents and focused through supercritical carbon dioxide to rationalize the polarization and the relation between product selectivity to the dielectric constant of a solvent. Reaction mechanisms and its understanding are a challenge, simulation can guide experiment to design successful reactions. The design by computer simulation will certainly reduce the number of trial and errors in experiment and thereby reduce the usage of chemicals, which in other ways will provide a better environment for future generations.

REFERENCES

1. P. Tomlinson-Tschaufeser, C. M. Freeman, *Catal. Lett.*, 60 (1999) 77.
2. N. Metropolis, A. W. Metropolis, M. N. Rosenbluth, A. H. Teller, E. J. Teller, *J. Chem. Phys.*, 21 (1953) 1087.
3. A. Chatterjee, R. Vetrivel, R. Sreekumar, Y. V. S. N. Murthy, C. N. Pillai, *J. Mol. Catal. Chem.*, 127 (1997) 153.
4. B. R. Gelin, M. Karplus, *Biochemistry*, 18 (1979) 1256.
5. A. Chatterjee, R. Vetrivel, R. Sreekumar, Y. V. S. N. Murthy, C. N. Pillai, in *Catalysis: Modem Trends*, N. M. Gupta, D. K. Chakraborty (eds.), Narosa Publishing House, New Delhi (1996).
6. B. M. Bhawal, R. Vetrivel, T. I. Reddy, A. R. A. S. Deshmukh, S. Rajappa, *J. Phys. Org. Chem.*, 7 (1994) 377.
7. D. H. Olson, *J. Phys. Chem.*, 74 (1970) 2758.
8. M. J. S. Dewar, E. G. Zoebisch, E. F. Healy, J. J. P. Stewart, *J. Am. Chem. Soc.*, 107 (1985) 3902.
9. A. Chatterjee, T. Iwasaki, T. Ebina, R. Vetrivel, *J. Mol. Graph. Model.*, 15 (1997) 216.
10. W. Kohn, L. J. Sham, *Phys. Rev. A*, 140 (1995) 1133.
11. B. J. Delly, *J. Chem. Phys.*, 92 (1990) 508.
12. C. Lee, W. Yang, R. G. Parr, *Phys. Rev. B*, 37 (1988) 786.
13. R. L. June, A. T. Bell, D. N. Theodorou, *J. Phys. Chem.*, 94 (1990) 8232.
14. P. Dauber-Osguthorpe, V. A. Roberts, D. J. Osguthorpe, J. Wolff, M. Genest, A. T. Hagler, *Protein. Struct. Funct. Genet.*, 4 (1988) 31.
15. A. Chatterjee, D. Bhattacharya, T. Iwasaki, T. Ebina, *J. Catal.*, 185 (1999) 23.
16. A. P. Singh, D. Bhattacharya, S. Sharma, *J. Mol. Catal.*, 102 (1995) 139; D. Bhattacharya, S. Sharma, A. P. Singh, *Appl. Catal.*, 150 (1997) 53.

17. J. C. Jansen, E. J. Creyghton, S. L. Njo, H. V. Koningsveld, H. V. Bekkum, *Catal. Today*, 38 (1997) 205.
18. (a) S. P. Bates, R. A. van Santen, *Adv. Catal.*, 42 (1998) 1; (b) J. F. Haw, T. Xu, *Adv. Catal.*, 42 (1998) 115; (c) J. Sauer, P. Ugliengo, E. Garrone, V. R. Saunders, *Chem. Rev.*, 95 (1995) 637; (d) J. Sauer, *Chem. Rev.*, 89 (1989) 199.
19. S. F. Boys, F. Bernardi, *Mol. Phys.*, 19 (1970) 553.
20. R. Vetrivel, R. C. Deka, A. Chatterjee, M. Kubo, E. Broclawik, A. Miyamoto, *Theor. Comput. Chem.*, 3 (1996) 509.
21. A. Corma, M. J. Climent, H. Garcia, P. Primo, *Appl. Catal.*, 49 (1989) 109.
22. W. Hertl, M. L. Hair, *J. Phys. Chem.*, 72 (1968) 4676.
23. M. Allian, E. Borello, P. Ugliengo, G. Spano, E. Garrone, *Langmuir*, 11(1995) 4811.
24. R. A. Van Santen, G. J. Kramer, *Chem. Rev.* 95 (1995) 637; G. J. Kramer, R. A. van Santen, *J. Am. Chem. Soc.*, 115 (1993) 4811.
25. M. Guisnet, N. S. Gnep, *Appl. Catal. Gen.*, 89 (1992) 1.
26. D. Bhattacharya, S. Sivasanker, *J. Catal.*, 153 (1995) 353.
27. A. Chatterjee, D. Bhattacharya, M. Chatterjee, T. Iwasaki, *Microporous and Mesoporous Materials*, 32 (1999) 189.
28. W. M. Meier, D. H. Olson, C. H. Baerlocher, *Atlas of Zeolite*; *Zeolites*, 17 (1996) A1.
29. A. S. Alverado-Swasigood, M. K. Barr, J. F. Hay, A. Redondo, *J. Phys. Chem.*, 95 (1991) 10031.
30. A. Chatterjee, T. Iwasaki, T. Ebina, A. Miyamoto, *Microporous and Mesoporous Materials*, 21 (1998) 421.
31. (a) S. P. Dagade, S. B. Waghmode, V. S. Kadam, M. K. Dongare, *Appl. Catal. Gen.*, 226 (2002) 49; (b) K. Winnacker, L. Kuchler, in *Chemische Technologie (Band 6): Organische Technologie II* (4th ed), H. Harnisch, R. Steiner, K. Winnacker (eds.), Carl Hanser Verlag, Munchen (1982).
32. B. M. Choudary, M. Sateesh, M. Lakshmi Kantan, K. Koteswara Rao, K. V. Ram Prasad, K. V. Raghavan, J. A. R. P. Sharma, *Chem. Comm.*, 1 (2000) 25.
33. J. M. Newsam, M. M. J. Treacy, W. T. Koetsier, C. B. De Gruyter, *Proc. R. Soc. London, Ser. A*, 420 (1988) 375.
34. H. van Koningsveld, H. van Bekkum, J. C. Jansen, *Acta Crystallogr. B*, 43 (1987) 127.
35. R. Vetrivel, C. R. A. Catlow, E. A. Colbourn, *Stud. Surf. Sci. Catal.*, 49 (1989) 231.
36. M. Haouas, A. Kogelbauer, R. Prins, *Catal. Today*, 70 (2000) 61.
37. (a) A. T. Bell, *Catal. Today*, 38 (1997) 151; (b) Z. Zhu, Z. Liu, H. Niu, S. Liu, T. Hu, T. Liu, Y. Xie, *J. Catal.*, 197 (2001) 6; (c) T. Maunula, J. Ahola, H. Hamada, *Appl. Catal. B Environ.*, 26 (2000) 173.
38. A. Sierraalta, R. Anez, M. Brussin, *J. Catal.*, 205 (2002) 107.
39. P. Nachtigall, D. Nachtigallova, J. Sauer, *J. Phys. Chem. B*, 104 (2000) 1738.
40. M. J. Frisch, G. W. Trucks, H. B. Schlegel, G. E. Scuseria, M. A. Robb, J. R. Cheeseman, G. Scalmani, et al. Gaussian, Inc., Wallingford CT, 1995.
41. A. D. Becke, *J. Chem. Phys.*, 98 (1993) 5648.
42. C. Lee, W. Yang, R. G. Parr, *Phys. Rev. B*, 37 (1988) 785.
43. L. Rodriguez-Santiago, M. Sierka, V. Branchadell, M. Sodupe, J. Sauer, *J. Am. Chem. Soc.*, 120 (1998) 1545.
44. (a) J. Dedecek, B. Wichterlova, *J. Phys. Chem.*, 98 (1994) 5721; (b) M. Iwamoto, Y. Hoshino, *Inorg. Chem.*, 35 (1996) 6918; (c) Y. Kuroda, Y. Yoshikawa, R. Kumashiro, M. Nagao, *J. Phys. Chem. B*, 101 (1997) 6497; (d) P. Kumashiro, Y. Kuroda, M. Nagao, *J. Phys. Chem. B*, 103 (1999) 89; (e) V. Bolis, S. Maggiorini, L. Meda, F. D'Acapito, G. T. Palomino, S. Bordiga, C. Lamberti, *J. Chem. Phys.*, 113 (2000) 9248; (f) A. Gervasini, C. Picciau, A. Auroux, *Microporous and Mesoporous Materials*, 35–36 (2000) 457; (g) C. Lamberti, G. T. Palomino, S. Bordiga, G. Berlier, F. D'Acapito, A. Zecchina, *Angew. Chem. Int. Ed.*, 39 (2000) 2138; (h) J. Datka, P. Kozyra, *Stud. Surf. Sci. Catal.*, 142 (2002) 445.

45. R. Bulanek, P. Cicmanec, P. Knotek, D. Nachtigallova, P. Nachtigall, *Phys. Chem. Chem. Phys.*, 6 (2004) 2003.
46. R. J. Cvetanovic, Y. Amenomiya, *Adv. Catal.*, 17 (1967) 103.
47. V. P. Zhdanov, J. Pavlicek, Z. Knor, *Catal. Rev. Sci. Eng.*, 30 (1988) 501.
48. J. Sauer, M. Sierka, *J. Comput. Chem.*, 21 (2000) 1470.
49. A. Schafer, H. Horn, R. Ahlrichs, *J. Chem. Phys.*, 97 (1992) 2571.
50. M. Sierka, J. Sauer, *Faraday Discuss.*, (1997) 41.
51. J. D. Gale, *J. Chem. Soc., Faraday Trans.*, 93 (1997) 629.
52. P. Pietrzyk, B. Gil, Z. Sojka, *Catal. Today*, 126 (2007) 103.
53. (a) A. D. Becke, *J. Chem. Phys.*, 88 (1988) 2547; (b) J. P. Perdew, Y. Wang, *Phys. Rev. B*, 45 (1992) 13244.
54. D. Nachtigallová, P. Nachtigall, M. Sierka, J. Sauer, *Phys. Chem. Chem. Phys.*, 1 (1999) 2019.
55. A. P. Scott, L. Radom, *J. Phys. Chem.*, 100 (1996) 16502.
56. R. D. Amos, Molecular property derivatives, in *Ab Initio Methods in Quantum Chemistry—Part I*, K. P. Lawley (ed.), Wiley, Chichester (1987).
57. T. A. Halgren, W. N. Lipscomb, *Chem. Phys. Lett.*, 49 (1977) 225.
58. G. Henkelman, H. Jonsson, *J. Chem. Phys.*, 113 (2000) 9978.
59. N. Govind, M. Petersen, G. Fitzgerald, D. King-Smith, J. Andzelm, *Comput. Mater. Sci.*, 28 (2003) 250.
60. (a) N. Govind, J. Andzelm, K. Reindel, G. Fitzgerald, *Int. J. Mol. Sci.*, 3 (2002) 423; (b) J. Andzelm, N. Govind, G. Fitzgerald, A. Maiti, *Int. J. Quant. Chem.*, 91 (2003) 467; (c) A. Maiti, N. Govind, P. Kung, D. King-Smith, J. Miller, C. Zhang, G. Whitwell, *J. Chem. Phys.*, 117 (2002) 8080.
61. (a) H. Munakata, Y. Oumi, A. Miyamoto, *J. Phys. Chem. B*, 105 (2001) 3493; (b) H. Munakata, A. Miyamoto, *J. Appl. Phys. Jpn.*, 39 (2000) 4323.
62. (a) M. G. Clerici, G. Bellussi, U. Romano, *J. Catal.*, 129 (1991) 1; (b) M. G. Clerici, G. Bellussi, U. Romano, *J. Catal.*, 129 (1991) 159; (c) G. Bellussi, A. Carati, M. G. Clelici, G. Madinelli, R. Millini, *J. Catal.*, 133(1992) 220; (d) M. G. Clelici, P. Ingallina, *J. Catal.*, 140 (1993) 71; (e) C. B. Khouw, C. B. Dartt, J. A. Labinger, M. E. Davis, *J. Catal.*, 149 (1994) 195.
63. A. Thangaraj, S. Sivasanker, P. Ratnasamy, *J. Catal.*, 131 (1991) 394.
64. P. Wu, T. Komatsu, T. Yashima, *J. Catal.*, 168 (1997) 400.
65. (a) G. N. Vayssilov, R. A. van Santen, *J. Catal.*, 175 (1998) 170; (b) T. Oumi, K. Matsuba, M. Kubo, T. Inui, A. Miyamoto, *Microporous Matter*, 4 (1995) 5.
66. J. Andzelam, N. Govind, G. Fitzgerald, A. Maiti, *Int. J. Quant. Chem.*, 91 (2003) 467.
67. (a) M. Hytha, I. Sïtich, J. D. Gale, K. Terakura, M. C. Payne, *Microporous and Mesoporous Materials*, 48 (2001) 375; (b) *Chem. Eur. J.*, 7 (2001) 2521.
68. M. C. Payne, M. P. Teter, D. C. Alan, T. A. Arias, J. D. Joannopoulos, *Rev. Mod. Phys.*, 64 (1992) 1045.
69. I. Sïtich, J. D. Gale, K. Terakura, M. C. Payne, *J. Am. Chem. Soc.*, 121 (1999) 3292.
70. (a) E. A. Carter, G. Ciccotti, J. T. Hynes, *Chem. Phys. Lett.*, 156 (1989) 472; (b) G. Ciccotti, M. Ferrario, Monte Carlo and molecular dynamics of condensed matter systems, in *Proceedings of Euroconference on Computer Simulation in Condensed Matter Physics and Chemistry*, K. Binder, G. Ciccotti (eds.), (1995); (c) M. Sprik, G. Ciccotti, *J. Chem. Phys.*, 109 (1998) 7737.
71. (a) C. Dellago, P. G. Bolhuis, D. Chandler, *J. Chem. Phys.*, 108 (1998) 9236; (b) *J. Chem. Phys.*, 110 (1999) 6617; (c) C. Dellago, P. G. Bolhuis, D. Chandler, *Faraday Discuss.*, 110 (1998) 421.
72. (a) P. E. Sinclair, C. R. A. Catlow, *J. Chem. Soc. Faraday Trans.*, 92 (1996) 2099; (b) R. Shah, J. D. Gale, M. C. Payne, *J. Phys. Chem.*, 100 (1996) 11688.
73. E. Sandre, M. C. Payne, J. D. Gale, *Chem. Comm.*, (1998) 2445.

74. C. D. Chang, A. J. Silvestri, *J. Catal.*, 47 (1977) 249.
75. (a) S. R. Blaszkowski, R. A. van Santen, *J. Phys. Chem.*, 99 (1995) 11728; (b) C. M. Zicovich-Wilson, P. Viruela, A. Corma, *J. Phys. Chem.*, 99 (1995) 13224; (c) P. E. Sinclair, C. R. A. Catlow, *J. Chem. Soc. Faraday Trans.*, 93 (1997) 333.
76. F. Haase, J. Sauer, *J. Am. Chem. Soc.*, 117 (1995) 3780.
77. F. Haase, J. Sauer, *Microporous and Mesoporous Materials*, 35 (2000) 379.
78. J. Andzelm, D. King-Smith, G. Fitzgerald, *Chem. Phys. Lett.*, 335 (2001) 321.
79. E. A. Pidko, R. A. van Santen, *J. Phys. Chem. C*, 111 (2007) 2643.
80. (a) J. A. Biscardi, E. Iglesia, *Catal. Today*, 31 (1996) 207; (b) J. A. Biscardi, G. D. Meitzner, E. Iglesia, *J. Catal.*, 179 (1998) 192; (c) J. A. Biscardi, E. Iglesia, *J. Catal.*, 182 (1999) 117; (d) J. A. Biscardi, E. Iglesia, *Phys. Chem. Chem. Phys.*, 1 (1999) 5753; (e) A. L. Lapidus, A. A. Dergachev, V. A. Kostina, I. V. Mishin, *Russ. Chem. Bull. Int. Ed.*, 52 (2003) 1094; (f) V. B. Kazansky, I. R. Subbotina, N. Rane, R. A. van Santen, E. J. M. Hensen, *Phys. Chem. Chem. Phys.*, 7 (2005) 3088; (g) R. L. De Cola, R. Gläser, J. Weitkamp, *Appl. Catal. Gen.*, 306 (2006) 85; (h) S. Y. Yu, J. A. Biscardi, E. Iglesia, *J. Phys. Chem. B*, 106 (2002) 9642; (i) Y. Sun, T. C. Brown, *Int. J. Chem. Kinet.*, 34 (2002) 467.
81. (a) M. V. Frash, R. A. van Santen, *Phys. Chem. Chem. Phys.*, 2 (2000) 1085; (b) A. A. Shubin, G. M. Zhidomirov, A. L. Yakovlev, R. A. van Santen, *J. Phys. Chem. B*, 105 (2001) 4928; (c) N. A. Kachurovskaya, G. M. Zhidomirov, R. A. van Santen, *Res. Chem. Intermed.*, 30 (2004) 99; (d) G. M. Zhidomirov, A. A. Shubin, V. B. Kazansky, R. A. van Santen, *Int. J. Quant. Chem.*, 100 (2004) 489; (e) G. M. Zhidomirov, A. A. Shubin, V. B. Kazansky, R. A. van Santen, *Theor. Chem. Accounts*, 114 (2005) 90; (f) L. A. M. M. Barbosa, G. M. Zhidomirov, R. A. van Santen, *Catal. Lett.*, 77 (2001) 55; (g) L. A. M. M. Barbosa, R. A. van Santen, *J. Phys. Chem. B*, 107 (2003) 4532; (h) L. A. M. M. Barbosa, R. A. van Santen, *J. Phys. Chem. B*, 107 (2003) 14342; (i) L. Benco, T. Bucko, J. Hafner, H. Toulhoat, *J. Phys. Chem. B*, 109 (2005) 20361.
82. D. H. Olson, G. T. Kokotailo, S. L. Lawton, W. M. Meier, *J. Chem. Phys.*, 85 (1981) 2238.
83. E. A. Pidko, R. A. van Santen, *Chem. Phys. Chem.*, 7 (2006) 1657.
84. (a) V. B. Kazansky, V. Yu. Borovkov, A. I. Serykh, R. A. van Santen, B. G. Anderson, *Catal. Lett.*, 66 (2000) 39; (b) V. B. Kazansky, V. Yu. Borovkov, A. I. Serykh, R. A. van Santen, B. G. Anderson, *Catal. Lett.*, 74 (2001) 55; (c) V. B. Kazansky, A. I. Serykh, B. G. Anderson, R. A. van Santen, *Catal. Lett.*, 88 (2003) 211.
85. H. A. Aleksandrov, G. N. Vayssilov, N. Rösch, *Stud. Surf. Sci. Catal.*, 158 (2005) 593.
86. (a) A. Hagen, F. Roessner, *Catal. Rev.*, 42 (2000) 403; (b) G. Caeiro, R. H. Carvalho, X. Wang, M. A. N. D. A. Lemos, F. Lemos, M. Guisnet, F. Ramôa Ribeiro, *J. Mol. Catal. Chem.*, 255 (2006) 131.
87. H. Himei, M. Yamadaya, M. Kubo, R. Vetrivel, E. Broclawik, A. Miyamoto, *J. Phys. Chem.*, 99 (1995) 12461.
88. E. Broclawik, H. Himei, M. Yamadaya, M. Kubo, A. Miyamoto, R. Vetrivel, *J. Chem. Phys.*, 103 (1995) 2102.
89. H. Lermer, M. Draeger, J. Steffen, K. K. Unger, *Zeolites*, 5 (1985) 131.
90. E. A. Pidko, V. B. Kazansky, E. J. M. Hensen, R. A. van Santen, *J. Catal.*, 240 (2006) 73.
91. (a) A. Corma, M. E. Domine, L. Nemeth, S. Valencia, *J. Am. Chem. Soc.*, 124 (2002) 3194; (b) Y. Zhu, S. Jaenicke, G. K. Chuah, *Chem. Comm.*, (2003) 2734.
92. (a) M. Boronat, P. Concepción, A. Corma, M. Renz, S. Valencia, *J. Catal.*, 234 (2005) 111; (b) M. Boronat, A. Corma, M. Renz, G. Sastre, P. M. Viruela, *Chem. Eur. J.*, 11 (2005) 6905.
93. M. Boronat, A. Corma, M. Renz, *J. Phys. Chem. B*, 110 (2006) 21168.
94. S. R. Bare, S. D. Kelly, W. Sinkler, J. J. Low, F. S. Modica, S. Valencia, A. Corma, L. Nemeth, *J. Am. Chem. Soc.*, 127 (2005) 12924.
95. S. Shetty, S. Pal, D. G. Kanhere, A. Goursot, *Chem. Eur. J.*, 12 (2006) 518.

96. (a) M. Svensson, S. Humbel, R. D. J. Froese, T. Matsubara, S. Sieber, K. Morokuma, *J. Phys. Chem.*, 100 (1996) 19357; (b) S. Humbel, S. Sieber, K. Morokuma, *J. Chem. Phys.*, 105 (1996) 1959.

97. M. Renz, T. Blasco, A. Corma, V. Fornés, R. Jensen, L. T. Nemeth, *Chem. Eur. J.*, 8 (2002) 4708.

98. (a) A. Klamt, G. Schüürmann, *J. Chem. Soc., Perkin Trans.*, 2 (1993) 799; (b) B. Delley, *Mol. Simul.*, 32 (2006) 117.

99. J. Tomasi, M. Persico, *Chem. Rev.*, 94 (1994) 2027.

100. (a) A. Baiker, *Chem. Rev.*, 99 (1999) 453; (b) A. Fürstner, L. Ackermann, K. Beck, H. Hori, D. Koch, K. Langeman, M. Liebl, C. Six, W. Leitner, *J. Am. Chem. Soc.*, 123 (2001) 9000.

101. (a) P. Raveendran, Y. Ikushima, S. L. Wallen, *Accounts Chem. Res.*, 38 (2005) 478; (b) L. Reynolds, J. A. Gardecki, S. J. V. Frankland, M. L. Horng, M. Maroncelli, *J. Phys. Chem.*, 100 (1996) 10337; (c) S. Saharay, S. Balasubramanian, *Chem. Phys. Chem.*, 5 (2004) 1442; (d) D. M. Rudkevitch, *Angew. Chem. Int. Ed.*, 43 (2004) 558; (e) P. Raveendran, S. L. Wallen, *J. Am. Chem. Soc.*, 124 (2002) 12590.

102. (a) P. Raveendran, S. L. Wallen, *J. Am. Chem. Soc.*, 124 (2002) 7274; (b) M. A. Blatchford, P. Raveendran, S. L. Wallen, *J. Am. Chem. Soc.*, 124 (2002) 14818.

103. M. Chatterjee, A. Chatterjee, P. Raveendran, Y. Ikushima, *Green Chem.*, 8 (2006) 445.

104. S. Sasaki, K. Mine, D. Taguchi, T. Arai, *Bull. Chem. Soc. Jpn.*, 66 (1993) 3289.

105. A. Chatterjee, M. Chatterjee, Y. Ikushima, F. Mizukami, *Chem. Phys. Lett.*, 395 (2004) 143.

106. (a) M. Arai, Y. Nishiyama, Y. Ikushima, *Journal of Supercritical Fluids*, 13 (1998) 149; (b) J. Lu, B. Han, H. Yan, *Phys. Chem. Chem. Phys.*, 1 (1999) 449.

107. R. B. Gupta, J. R. Combes, K. P. Johnston, *J. Phys. Chem.*, 97 (1993) 707.

108. Y. Li, W. K. Hall, *J. Catal.*, 129 (1991) 202.

Index